Springer Theses

Recognizing Outstanding Ph.D. Research

Aims and Scope

The series "Springer Theses" brings together a selection of the very best Ph.D. theses from around the world and across the physical sciences. Nominated and endorsed by two recognized specialists, each published volume has been selected for its scientific excellence and the high impact of its contents for the pertinent field of research. For greater accessibility to non-specialists, the published versions include an extended introduction, as well as a foreword by the student's supervisor explaining the special relevance of the work for the field. As a whole, the series will provide a valuable resource both for newcomers to the research fields described, and for other scientists seeking detailed background information on special questions. Finally, it provides an accredited documentation of the valuable contributions made by today's younger generation of scientists.

Theses may be nominated for publication in this series by heads of department at internationally leading universities or institutes and should fulfill all of the following criteria

- They must be written in good English.
- The topic should fall within the confines of Chemistry, Physics, Earth Sciences, Engineering and related interdisciplinary fields such as Materials, Nanoscience, Chemical Engineering, Complex Systems and Biophysics.
- The work reported in the thesis must represent a significant scientific advance.
- If the thesis includes previously published material, permission to reproduce this must be gained from the respective copyright holder (a maximum 30% of the thesis should be a verbatim reproduction from the author's previous publications).
- They must have been examined and passed during the 12 months prior to nomination.
- Each thesis should include a foreword by the supervisor outlining the significance of its content.
- The theses should have a clearly defined structure including an introduction accessible to new PhD students and scientists not expert in the relevant field.

Indexed by zbMATH.

More information about this series at https://link.springer.com/bookseries/8790

Maria Elena Castiello

Computational and Machine Learning Tools for Archaeological Site Modeling

Doctoral Thesis accepted by University of Bern, Switzerland

Springer

Author
Dr. Maria Elena Castiello
Institute of Archaeological Sciences
University of Bern
Bern, Switzerland

Supervisor
Prof. Albert Hafner
Institute of Archaeological Sciences
and Oeschger Centre for Climate Change
Research (OCCR)
University of Bern
Bern, Switzerland

ISSN 2190-5053 ISSN 2190-5061 (electronic)
Springer Theses
ISBN 978-3-030-88569-4 ISBN 978-3-030-88567-0 (eBook)
https://doi.org/10.1007/978-3-030-88567-0

This Springer imprint is published by the registered company Springer Nature Switzerland AG
The registered company address is: Gewerbestrasse 11, 6330 Cham, Switzerland

In the loving memory of my best friend, Anto.

Supervisor's Foreword

I have had the pleasure to supervise Dr. Maria Elena Castiello's Ph.D. thesis, accepted for publication within the Springer Theses series and awarded with a prize for outstanding original work. She joined our Department of Archaeological Sciences, granting an outgoing research fellowship awarded by the Italian Ministry of Education, University and Research, to start her Ph.D. on a very innovative topic never approached before by Swiss academy. The Ph.D. thesis was successfully defended in September 2020.

In this book, Dr. Castiello focuses on the application of Quantitative and Machine Learning tools for modeling archaeological information and environmental proxies. In particular, the book reports on several case studies represented by different Swiss regions and Switzerland as a whole and generates archaeological predictive maps to support cultural heritage management and archaeological research. The study gathers, analyzes and cross-references diverse and separated institutional archaeological data collections with their regional and supra regional environmental settings. An in-depth review of past and recent works on quantitative applications in the context of archaeological heritage management and the most recent advancements in Artificial Intelligence is followed by a range of statistical and model-driven engineering approaches, including exploratory exercises, to address the most traditional archaeological questions, such as the analysis of ancient settlement patterns. Finally, an innovative and original application of Random Forest algorithm is presented, resulting in archaeological site location predictions and in the ranking of environmental features importance for the site modeling procedure.

The author offers a novel conceptual framework and a methodological protocol that can formalize and reveal several levels of knowledge and uncertainty in archaeology. This technological apparatus is meant to assist archaeologists and cultural heritage managers in complying with inventory requirements, forecast effects, and make proactive streamlined management decisions.

The dissertation of Dr. Castiello reaches further than classical approaches in the disciplines of Archaeology and Humanities and is strongly connected with the research domain of Natural Sciences. Computational and Machine Learning tools for

Archaeological Site Modeling includes significant and original scientific contributions and represents the first comprehensive and critical study of technological developments and advancements in archaeological research. Moreover, the book provides interesting insights of what digital archaeology should be, and how it does relate to the field of Natural Sciences and Computational Social Sciences. The manuscript is indeed a clear testimony of the rising importance of Artificial Intelligence and Machine Learning techniques in Archaeology. Science and technology came in this book to be embedded in Archaeology, deeply renewing its methods of investigation and transforming its structure within a quantitative dynamic movement.

I would like to emphasize the originality, innovation, and comprehensiveness of the work accomplished by Dr. Castiello. Her thesis is set within a rarely attained degree of interdisciplinary. At the same time, it opens a door to the Digital Humanities. In light of these results, this book helps to handle knowledge, covering theory as well as practice of modeling archaeological sites over different landscapes and epochs and makes it easier to bridge the gap between natural science and humanities. It provides a structure for communicating comprehensive knowledge inside and between disciplines.

Bern, Switzerland
September 2021

Prof. Albert Hafner

Parts of this thesis have been published in the following articles

Castiello M. E. and Tonini M. 2021. *An Explorative Application of Random Forest for Archaeological Predictive Modeling. A Swiss Case Study.* Journal of Computer Applications in Archaeology 4(1), 110–125. DOI: https://doi.org/10.5334/jcaa.71.

Acknowledgements

I would like to express my sincere gratitude to my supervisor Prof. Dr. **Albert Hafner**, who gave me the possibility to pursue this Ph.D. on such an interesting topic. He unconditionally believed in me and in this research from the beginning to the end, regardless the many constraints and difficulties that this research had to deal with. He always gave me extreme freedom and confidence during these years and I would not have achieved this goal without his guidance, exceptional support, and availability.

My utmost gratitude goes to Dr. **Marj Tonini** from the Faculty of Geoscience, at the University of Lausanne, who introduced me to the Random Forest world. She has been the main source of inspiration toward the construction of a solid methodology for this research. This manuscript would not have been possible without her contribution, insightful inputs, and constructive feedback. Our fruitful discussions have been vital for this research and for my intellectual development. I feel that my knowledge and skills have reached a level that I could have never accomplished alone. I cannot thank her enough for the advices and help.

My gratitude goes to Dr. **Cesar Gonzalez-Perez** from the Instituto de Ciencias del Patrimonio (Incipit) for his brilliant advice, for reviewing my papers, and helping in several matters. Especially, I would like to thank him for teaching me the fundamentals of conceptual modeling and for having introduced me to the study and processing of uncertainty and vagueness in archaeological data.

I gratefully acknowledge Prof. Dr. **Felipe Criado Boado** for his warm hospitality when I joined the Incipit and for delightedly accepting to read and evaluate my thesis.

A particular thought to Dr. **Fabio Veronesi** and **Carma Satchwell** for their availability and advice on GIS analyses.

My heartfelt thanks to **Maryam Lotfian** from HEIG-VD the School of Management and Engineering of Vaud and **Josquin Rosset** from ZHAW Zurich University of Applied Sciences. Without them and their precious contribution, this thesis would have never seen the end. I am forever indebted.

Last but not least, I thank **Etienne** for his unconditional encouragement, exceptional support, and love. I could not have done it without him.

Acknowledgement to the Cantonal Archaeological Departments of Zurich, Aargau, Grisons, Geneva, Vaud, and Fribourg for allowing me to use their data, and to the University of Bern and Dr. Josephine de Karman Foundation for the financial support to this research.

Contents

Abbreviations

AG	(Canton of) Aargau
AI	Artificial Intelligence
APM	Archaeological Predictive Modeling
AUC	Area Under the Curve
BPN	Basic Probability Number
BRT	Boosted Regression Tree
CART	Classification and Regression Tree
CH	Switzerland (all sites)
CHM	Cultural Heritage Management
CNRS	Centre National pour la Recherche Scientifique
ConML	Conceptual Modeling
DB	(Geographically referenced) Database
DBs	Databases
DEM	Digital Elevation Model
DST	Dempster Shafer Theory
DT	Decision Tree
DTM	Digital Terrain Model
FN	False Negative
FP	False Positive
FPR	False Positive Rate
FR	(Canton of) Fribourg
FR	Frequency ratio
GE	(Canton of) Geneva
GIS	Geographic Information Systems
GR	(Canton of) Grisons / Graubünden
IDE	Integrated Development Environment
IKAW	Indicatieve Kaart van Archeologische Waarden
IQR	Interquartile Range
JbSGUF	Jahrbuch der Schweizerischen Gesellschaft für Urund Frühgeschichte
kNN	K-Neural Network
LR	Logistic Regression

ML	Machine Learning
NN	Neural Networks
OOB	Out of Bag
Q1	1st quartile (25th percentile)
Q3	3rd quartile (75th percentile)
RF	Random Forest
ROC	Receiver Operating Curve
SDLC	Software Development Life Cycle
SNSF	Swiss National Science Foundation
SQL	Structured Query Language
SVM	Support Vector Machine
TIF	Tagged Image File
TN	True Negative
TP	True Positive
TPR	True Positive Rate
VD	(Canton of) Vaud
WofE	Weight of Evidence
ZH	(Canton of) Zurich

Section I

Chapter 1
Introduction

1.1 Research Context

In our era of global turmoil, characterized by several crisis and destructions, our cultural and archaeological heritage has received its share of damages, either because of human aggression or intervention or as a result of natural disasters [1]. Even without this negative prospect, the preservation and conservation of archaeological evidences should be a natural concern, since archaeological records represent our most complete compendium on human history and are a nonrenewable resource [2].

In this regard, archaeology as a scientific discipline should play a primary role in addressing some of the most relevant contemporary issues, such as the role of human impact and the effects of climate change on the destruction of cultural heritage. These issues are consistently determining contemporaneous academic research lines [2–7].

Many international studies deal with the assessment of natural hazards and their effects on cultural heritage, covering all geographical areas (to cite only a few: [8–21]. However, despite the extensive results obtained thanks to modern technologies, which have moreover allowed an overall increase in scale and impact as well as a more accurate quantification of these destructions, archaeological risk assessment seems to have had little impact on the actual public policy discussion [2]. Nevertheless, archaeological risk assessment has become an urgent matter for modern cultural heritage managers and for our Governments in general.

Switzerland is experiencing a constant growth of infrastructures and modern settlements since the last decades [22]. Many areas are critically threatened by modern development and human impact, resulting in a permanent destruction of any possible archaeological remains not yet unearthed. Due to Switzerland's political structure characterized by a decentralized organization, each Canton applies its own specific procedures regarding archaeological heritage management [23–26]. Hence, a multiplicity of approaches exists to fulfill the task of protection and conservation of archaeological evidences. This situation further strengthens the need to explore solutions and develop a tool that could effectively help in identifying and protecting archaeological sites.

M. E. Castiello, *Computational and Machine Learning Tools for Archaeological Site Modeling*, Springer Theses, https://doi.org/10.1007/978-3-030-88567-0_1

As analytical tools, predictive models and archaeological site distribution maps can help to make informed decisions regarding current issues and are well suited to applications in urban planning. Among other things, they have successfully demonstrated to provide new and considerable knowledge concerning the location of past human activities and the potential relationship with the natural or social environment by further identifying hidden patterns between the sites and their spatial and physical location. Finally, they can be used to highlight sensitivities and risks to which archaeological sites can be exposed [27–32]. Similar to susceptibility maps in the field of natural hazards management, predictive maps as visual outputs of quantitative and computational modeling involving archaeological information, statistical analyses, mathematical and geospatial processing can represent a relevant automated decision-making and probabilistic reasoning tool [33–40].

Predictive modeling is based on the principle of a causal relationship between the landscape and human activities. The landscape plays a fundamental role in the development of ancient societies and the environmental factors influence where human activities occur. Space, spatial relationships and causalities have such strong impact on human behavior that we take almost them for granted and are unaware of their importance in our everyday life. The increasing presence of locational media and the developments in digital geospatial technologies have strengthened this familiarity even more [41, 42].

Though archaeological predictive modeling is still a highly debated field, "it has been identified as one means of streamlining inventory, evaluation, and project design, that could result in considerable cost and time savings for cultural heritage management", as Heilen and Altschul [10] have rightly pointed out.

Over the last decades, the academic research has made some headway in this direction, borrowing methods and tools from other disciplines [43, 44]. The complexity of the archaeological datasets we deal with today, as well as the complexity of the questions they rise, call for collaboration and exchange with experts from different disciplines: geography, geo-informatics, mathematics, statistics, computer sciences [45–48]. As a consequence, we can today rely on many innovative examples of spatial and statistical computing for site distribution modeling and detection in archaeology [49–51], as well as on the power and effectiveness of Artificial Intelligence (AI) and Machine Learning (ML) techniques.

With this in mind, various Swiss administrative units have been selected as case studies: the Cantons of Aargau, Zurich, Grisons, Fribourg, Vaud, and Geneva. The choice of these regions was primarily justified by their diversified urban and rural landscapes.

The data modeled in this Thesis originates from unique cantonal inventories, which cover large regions and contain information on thousands of archaeological sites. Owing to the large number of sites considered, a global approach prone to investigate the full databases bound to a single epoch was preferred. The historical epoch chosen to implement the analysis is the Roman period. The types of archaeological evidence to be analyzed are the *settlements* and *single finds*, as derived from the conceptual framework elaborated in this Thesis and outlined in Chaps. 4 and 5, because of the ease to detect them in the modern landscape at regional and supra regional scale thanks to their clearly defined structures.

One of the main goals of the study was to find out which factors influenced the location of Roman settlements in the study areas and then to identify other sites with the same set of features, based on the environmental characteristics.

A Locational Preference Analysis and a predictive modeling procedure were then performed to quantify aspects of the natural environment (independent variables) and process the archaeological data (dependent variables) provided by the cantonal departments. Computations were carried out within Geographic Information System (GIS) and R (professional software for graphical and statistical analysis) environments, coupling Information modeling, Machine Learning, Statistics and Geospatial analyses.

The procedural methodology increases the rigor of the results obtained while making the underlying concepts more transparent, allowing the reuse of such protocol for subsequent implementations. The overall structure of the protocol defined in this Thesis can be pertinent to many areas, and the solutions found in the selected case studies could serve as a general model and strengthen the implementation of large-scale archaeological research and sustainable preventive archaeology in other regions equally confronted with such challenges.

This Thesis represents a first exploration of the data gathered from a variety of sources (open access environmental data and institutional archaeological collections) with a view to gauging their use within quantitative analyses and predictive modeling. In addition to its exploratory nature, this research finally aims at developing criteria for the creation of a computational tool—Archaeological Predictive Maps—valuable for Cultural Heritage Management and for academic research. Furthermore, it intends to construct a solid information model and define a clear methodology for the encompassment of uncertainty and vagueness associated with the data used, as well as to provide transparent and objective guidelines and hence to produce a digital instrument for archaeological site protection.

1.2 Swiss Archaeological Heritage Management

Archaeological heritage management is an extremely complex matter in Switzerland. The administrative fragmentation derived from the Swiss federalism has clear repercussions on the organization of archaeological activities and its results. Often, different sectors are involved in the decisional process concerning archaeological heritage management: land-use planning, culture, scientific research, public high education system, natural sciences and humanities, and finally the governmental administration [52].

According to the Swiss Constitution (Art. 78 Cst.), the Swiss Confederation has a subsidiary role with regard to the protection of nature and cultural heritage, which is a matter of Cantonal laws. Although the Confederation does provide subsidies to the Cantons, these are often limited to long-term projects of major importance, priority projects and certain archaeological activities.[1] In other words, Swiss archaeology is entrusted to 26 different authorities (20 Cantons and 6 half-Cantons) sovereignly responsible for the management of their archaeological heritage [61]. As a consequence, several archaeological departments has developed at its own rhythm and according to its own financial resources.

Many of these archaeological departments were created during the major '*rescue*' excavation period that coincides, as in almost every European country, with important construction works. These refer to major development projects and to highway and railway works dating back to 1970–1980, that have radically modified the functioning of Swiss archaeology [53, 54]. In this context, the *Academia*, apart from a few exceptions (Geneva, Lausanne), remained on the sidelines of the "highway archaeology" [55], what may have contributed to a regionalization of Swiss archaeology.

The project "Rail 2000", a large-scale project of the Swiss Federal Railways (SFR) still in progress, employed thousands of archaeologists in almost all Swiss Cantons concerned by those works for approximately 20 years [25, 56]. This program was partly financed by the federal government through the Swiss National Science Foundation (SNSF). Although the Confederation provided part of the financial resources necessary for the fieldworks, the Cantons were responsible for setting up the administrative and logistical infrastructures as well as for the implementation of the research. Finally, this situation has somehow contributed to accentuate the diversity in departmental archaeological management. In fact, the resources invested varied considerably from one Canton to another, consequently *enforcing the power of some* [26, 52]. In such context, the management and conservation of the archaeological heritage can be seriously threatened whenever financial difficulties arise.

Already in 1976, a report from the SNSF resumes the situation of Swiss archaeology at a national scale:

> Il y a autant de régimes légaux que de Cantons. Quelques Cantons ont des services archéologiques à la hauteur de leur tâche, voire excellents. D'autres en sont encore à l'âge des amateurs dévoués et des compétences relatives. Les grands travaux publics de ces trente dernières années et l'extension des agglomérations ont multiplié les fouilles d'urgence; elles se sont succédé; ceux qui auraient normalement eu mandat d'en publier les résultats n'en ont généralement pas eu le temps. [...] On le voit, l'infrastructure est décevante. Il n'y a pas encore de conception d'ensemble sur ce qui devrait être entrepris. Les priorités ne sont pas établies.[2]

According to Delley [52], the absence of coordination between the Cantons is still an issue today. Archaeologists have so far not felt the need to develop a common research plan reaching beyond the "cantonal" operations (with adequate resources) or even the need for consultation between "neighbors". The cantonal borders have rather been a barrier towards an integrated research with scientific objectives [61].

[1] Swiss Federal Council: Message on the promotion of culture for the period 2012–2015 (Message concernant l'encouragement de la culture pour la période 2012 à 2015 (Message culture) (23.02.2011)).

[2] Fonds national suisse de la recherche scientifique, 25e rapport, 1er janvier-31 décembre 1976, 59. In: Delley 2014.

Fragmented and decentralized territorial organization induces a certain compartmentalization in archaeology. Kaenel [57] has clearly outlined the associated risks for archaeological research that can occur while examining a small regional unit without subordinating the research program to a supra-regional scientific coordination: to repeat the same investigations, overlap the same data and finally accumulate the same results and conclusions.

At the date, there are still remarkable differences between cantonal laws. In March 2017, Archäologie Schweiz/Archéologie Suisse/Archeologia Svizzera [58][3] published an overall evaluation of the state of the cantonal archaeology[4] in regard to the *European Convention for the Protection of the Archaeological Heritage* measures.[5] Conforming to these measures, all Cantons are active in the protection and conservation tasks regarding the archaeological remains discovered within their administrative units, but some issues still need to be fixed: the unauthorized transfer of archaeological objects; the obligation to document any archaeological site that cannot be preserved in situ before the destruction. This obligation is only embedded in the legislation of one third of all Cantons at the date of this report. In conclusion, the results of this report show that the *European Convention for the Protection of the Archaeological Heritage* is only partially implemented by the cantonal legislations, despite its ratification by the Confederation and that several measures have still not been taken into account by a majority of the Cantons (Archaeologie Schweiz 2017).

One notable exception to the decentralization of archaeological heritage management was the project of *Corrections des Eaux du Jura* between the 19th and the twentieth centuries. This project involved the Confederation and five Cantons (Vaud, Fribourg, Neuchatel, Solothurn, Bern) and consisted of extensive hydrological works carried out in the region of the three lakes: Morat/Mürten, Neuchatel and Bienne/Biel [59, 60]. For the first time, the archaeological activities were led by a single archaeologist, a supra-cantonal supervisor [61]. Even though this centralized system disappeared with the end of the confederate works, as the federal structure of the Confederation does not allow for the implementation of a centralized preventive archaeology, the country could at least experience the establishment of a preventive archaeology system for a limited time. Excavations could be planned in advance of the destruction, resulting in a wide range of innovations. The focus of archaeological research lay not only on monumental sites, but on all kinds of settlements, simple and complex, probably producing a more balanced picture of the human occupation of the landscape and its development in Switzerland [52].

The Swiss federalism, described above, challenges the accessibility and availability of archaeological data. Archaeological records in general exist in a variety of

[3] Archäologie Schweiz is an association rooted in the Swiss Civil Code that promotes the awareness of archaeological research among the people and authorities of Switzerland; seeks to maintain and protect the archaeological monuments; promotes the collaboration between the universities and colleges, the federal and cantonal authorities in charge and other expert institutions, circles and associations that are active in the field of archaeology in Switzerland and abroad.

[4] Archäologie Schweiz: Die Stellung der Archäologie in den kantonalen Gesetzen. Aktualisierung März 2017, Basel, 2017.

[5] Ratified under the dome of the European Commission in 1992 and signed by Switzerland in 1996.

forms collected across several years and by different actors. This is even more true in Switzerland, where site records are stored and dispersed among several cantonal institutions. The Swiss archives are as varied as the institutions that store them. The cantonal units keep their own records in local archives, which are not duplicated at the level of a federal State repository. Extant archaeological data varies considerably in quality and quantity. A multiplicity of definition methods to classify objects exists, as well as a variety of classification procedures to store the information [61].

For the Archaeological Site Modeling procedure, developed here, aiming at being transferable and interoperable in order to be considered effective, this variety in data quality and structure represents a major challenge, for supra regional research and heritage management.

Nevertheless, the multiplicity of regional approaches also provides a very high level of local archaeological knowledge and produces highly specialized studies. The cantonal organization of Swiss archaeology shows evident advantages: a local archaeology embodied in a very tight network of actors in the field, unique example in Europe, resulting in an impressive number of scientific publications,[6] exhibitions and cultural activities addressed to the wider public [61].

1.3 Motivation—Research Questions

Archaeological record has come under increasing threat from different factors. Climate change-induced storms and sea level rise cause erosion, the spread of infrastructures, urbanization and looting (illegal collecting/excavation), especially in areas fraught with conflict [62–65], destroy the nonrenewable archaeological heritage.

Rick and Sandweiss [66] have recently well highlighted how modern times are in general characterized by a socio-political, socio-economic and environmental uncertainty. Climate change, the quest for food security, the decline of biodiversity and the extinction of species [67], alterations of the ecosystem in general as well as the progressive urban sprawl, threaten the irreplaceable and irreproducible cultural property represented by archaeological sites to a great extent. This is even more problematic, as the archaeological record is the richest and most extensive source of information on human social experience we have [2]. Continued site destruction often means that legacy collections are the only remaining source of information about past human societies in localities where sites once existed. For this reason, the preservation and protection of archaeological sites against any kind of damage and destruction is of vital importance in order to ensure a long-term availability of invaluable resources for science and society. All over the world, these are challenges that cannot be ignored.

[6] In 1974, "Cahiers d'archéologie romande" was first published, followed by the cantonal series: "Archéologie fribourgeoise" which became "Cahiers d'archéologie fribourgeoise", "Archéologie neuchâteloise", "Cahiers d'archéologie genevoise", "Archäologie und Denkmalpflege im Kanton Solothurn", "Archäologie im Kanton Bern/Archéologie dans le Canton de Berne", etc.

Archaeological and historical sciences have effectively demonstrated to have an important role in this context. While they can offer a certain perspective in the long-term environmental evolution of human distribution, they can also help in contextualizing present-day environmental and social issues [68]. Ortman [2] pointed out that: "The process has just barely begun, but if we believe the archaeological record is at least partly systematic, that human behavior is at least partly predictable, and that scientific reasoning can be employed to improve the human condition overall, this [the process of modeling human behavior] seems like a very good thing to incorporate into an expanding scope of archaeological practice".

In Switzerland, the most tangible threat to archaeological heritage lies in urban development. As reported by the Federal Office for Spatial Development,[7] the construction activities have constantly risen in Switzerland and the urbanization destroys one square meter of open space every second. It has been ascertained that settlement areas have grown by 24% or 600 km^2 since the mid-1980s, an area that corresponds to the surface of Lake Geneva. The public authorities are responsible for the management of this archaeological heritage. To fulfill this task, they should be able to provide a balance between the inevitable growth of settlements and industrial areas and the protection and preservation of archaeological elements. Cultural heritage management is consequently strongly affected by these threats and struggles to ensure the protection and conservation of archaeological record.

The choice of this Thesis topic was mainly driven by the need to explore solutions and develop a tool that could effectively help in identifying and protecting sites in areas constantly jeopardized by modern development and human impacts in particular. At the same time, this Thesis attempts to conciliate the objectives, methods and results of two different entities—academic research and archaeological heritage management.

With this perspective, this Study aims at compiling an effective Archaeological Site Modeling procedure, by looking at multidisciplinary collaborations in order to find the most adequate methodological approach. Such kind of research is challenging, but at the point where we are today, it is not possible to face complex problems and challenges individually, which is also why the research proposed in this context takes advantage of the most cutting-edge techniques in the field of mathematics, computer science, statistics and geography.

Moreover, in the wider context of predictive modeling, other emerging challenges are the construction of solid information models and the incorporation of uncertainty and vagueness associated with the data. In this study, particular attention has been paid to the development of a clear procedure, the definition of an explanatory methodology and the compilation of a reusable guideline and information.

While researching and exploring suitable methodologies to achieve the goals proposed, several more specific questions have risen:

1. Can modern data and institutional survey data can be used to model the location of ancient sites?

[7] Bundesamt für Raumentwicklung (ARE): Trends und Herausforderungen. Zahlen und Hintergründe zum Raumkonzept Schweiz, Bern, Mai 2018.

2. Can any pattern be determined in the archaeological site locations?
3. Are there any geo-environmental features showing a strong impact on site distribution?
4. Can predictive maps provide further understanding of settlement patterns at a regional and supra-regional scale? If so, are these supra-regional processes evident in the model outputs?
5. Are modeled outputs corresponding to the current knowledge or do they provide new information?
6. Can data-driven modeling contribute to enhance current knowledge of site distribution and to cultural heritage management practices?

Unlike many international examples (which will be described in the following Chapters), research dealing with these problems and their focal issues is highly disadvantaged in Switzerland. Although many projects have been funded over the past years regarding the introduction and the use of highly advanced technologies in archaeology (dendrochronological analysis, geophysical prospections, archive digitization, musealization and virtual heritage reconstructions), a common and systematic multi-disciplinary approach has not yet been developed. Such an approach aims at "managing" buried archaeological heritage in advance and in terms of planning and safeguarding, which implies a priori acknowledgement of the location of the sites to protect.

This study aims at filling this gap and moreover at promoting interactions between sciences and humanities. It is furthermore an attempt to propose an innovative approach by combining archaeological knowledge and resources derived from sister disciplines. To cite Rick and Sandweiss [66]: "A key direction is a continued collaboration across disciplines, fostering open dialogue and recognition that the human past provides a roadmap for how we got to the present and signposts for where we would like to go in the future."

1.4 Challenges and Objectives

Supra regional analyses, incorporating data from different Cantonal units are rare in Switzerland, even more in the context of Archaeological Site modeling. Although predictive modeling studies are very scarce in Switzerland, in other Countries they have proven to be robust and are backed by at least thirty years of research (to cite a few: [41, 69–76]. These studies try to take advantage of a range of cutting-edge techniques and advancements, both established and emerging in sister disciplines, to develop archaeological predictive maps, with the aim to preserve and protect archaeological evidence and to deepen our knowledge about ancient societies.

As human systems are generally believed to be too complex to be cast in simple rules and their behavior regulated by "independent" and conscious (thus not always rational) decision-making [77], Archaeological Site Modeling studies have tried to address this challenge by following examples and approaches developed in other disciplines [78–80]. Indeed, modeling phenomena determined by human behavior is one of the numerous challenges that digital archaeologists and this Thesis in particular try to meet.

Just like for natural scientists, the archaeologist's work consist in answering research questions using evidence, evidence or corpuses, whether cuneiform tablets or remains of Roman villas, to construct arguments about people's lives, today or in the past [81–83].

Without neglecting the independent and conscious aspects of human behavior mentioned above, Site Modeling has proved its effectiveness for landscape analysis and has been successfully applied by many authors in search for hidden rules guiding the most probable placement of settlements (to cite only a few: [14, 17, 20, 84–99]).

Recently, Machine Learning (ML) methods have been applied to produce highly accurate predictive models in many research fields and disciplines. Ortman [2] has rightly stressed out that even though many aspects of human behavior may never be entirely predictable, at least some are, at least partly, and it therefore stands to reason that social scientists, as well as archaeologists, should be able to expand knowledge of predictable behavior with appropriate efforts. He continues providing with a short list of some good examples of modern predictable events and process, such as: "demographic rates have predictable effects for tomorrow's economy, insurance companies use actuarial tables to predict payouts and adjust premiums with reasonable confidence; political scientists create models based on demographic and socioeconomic characteristics of subgroups that predict election results; the daily movements of individuals follow predictable patterns that allow our smartphones to plot the most time-efficient route of travel between two places; simple models often surpass expert judgment in predicting the outcomes of sporting events; and tech companies use browsing and posting habits to predict which products we are most likely to purchase". In the construction of such predictive models, ML techniques certainly yield a significant role.

Although ML based models are still not able to fully reproduce the irrational aspects of human behavior and are still considered a "black-box" approach that does not really push the envelope of archaeological theory, research design and inquiry by some scholars [100], it must be recognized that they can allow archaeological predictive models to describe interactions between environment and humans and highlight past human dynamics over the landscape precisely enough to obtain significant conclusions. In general, such techniques hold enormous promises. Predictive maps based on ML approaches could become a vital resource just as it is already in natural hazards management, not only with applications in cultural heritage management, but also in academic research.

This Study, starting by documenting the processes of data collation, database architecture, and modeling infrastructure with geo-spatial techniques and ML algorithms, aims in a first instance at shedding light on the digital (and non-digital) Swiss archaeological archives, highlighting issues and strengths pertaining to this system with relevance to their re-usability for quantitative and spatial analyses and research. The collation and analysis of data on a country-wide scale represents an important innovation in the domain and the information derived from this work could be further used in more specialized investigations.

With a focus on the methodology and a constant emphasis on "how to do", this Thesis primarily tries to provide a response to the need of establishing a connection between different research disciplines, as well as between academia and governmental institutions. It further aims at building a bridge between different disciplines and domains: natural sciences, archaeology and cultural heritage management. Furthermore, it contributes to advance the research in the fields above mentioned.

By providing cantonal archaeological departments with further new information and perspectives about their areas of competences, the modeling outputs can help to identify the areas that require more attention and resources (both human and financial) because of their intrinsic probability to hide archaeological evidences. Establishing these factors would both help in the organization of new excavation campaigns and, more importantly, provide a warning for any urban expansion that may be planned in these zones, thereby helping to protect cultural heritage.

By employing a ML algorithm and geo-statistics, this research investigates whether a data-driven approach is able to model the distribution of such a heterogeneous phenomenon and thus to map the strong spatial continuity/discontinuity of archaeological presences over the landscape. It further aims at gaining insight into past human behavior and its relationship with the environment in order to identify any environmental influence that could steer the site distribution. It intends to empirically discover both known and unknown spatial correlations between "events" and finally to define high and low suitability areas for event occurrence (settlement presence).

However, it must be admitted that, as Achino and Barceló (2019) remind us, "most archaeological contexts do not respect a "Pompeii premise" [101] and therefore, the preserved record cannot be considered a frozen snapshot of the last cultural deposition. Archaeological observations must be seen as the product of both human behavior and post-depositional forces that have modified the structure, the content and often the location of archaeological artefacts." Thus, the investigation of remaining human traces can be biased by several factors and in several ways. The time and expense involved in archaeological field and laboratory work as well as the changing interests of investigators over time may be enumerated as a source of bias to this effect [2]. Indeed, Ortman [2] claimed that "there is error associated with every measurement, and we cannot know, for example, the exact momentary population, or the precise rate of pottery consumption for any past settlement."

Nevertheless, it is argued the lack of small- and large-scale predictive modeling approaches covering the Swiss territory should be seen as an ideal opportunity to test (i) the goodness of archaeological data for setting up quantitative and spatial analyses, (ii) the quality of management strategies in archiving information and (iii) the possibility of data reuse in downstream studies, with a particular attention on small- and large-scale site distribution modeling.

If on the one hand, this situation represents an advantage for the conduction of innovative research projects, on the other hand, it may represent a big challenge by means of a concrete implementation of the project. As previously mentioned, the peculiar Swiss context requires more attention with respect to the archaeological heritage management. The goal of this Thesis moreover matches with the quest for suitable solutions to face the following acknowledged challenges: (i) ensure the reusability and transferability of archaeological data, (ii) record and retrieve information faster and in larger quantities than before and (iii) obtain reliable scientific models that can help revealing patterns and producing new information about site distribution while making predictions about yet unknown ones. Obtaining a reliable model at regional and supra regional scale would be important both for scientific interest both for the application of this knowledge to practical problems like the protection of archaeological sites against human and natural destruction.

With this perspective in mind, the reconstructed databases and the modeling procedure protocol performed are therefore intended to highlight geographical and environmental factors as well as a variety of criteria, not all of which immediately apparent, that could have been relevant in the regional and supra-regional settlement patterns. The applied methodology aims at overcoming modern administrative borders for establishing new lines of research that could look beyond individual sites and local or regional patterns, in order to examine wider trends both in relation to known historical events and in terms of broader settlement processes. At the same time, it aims at developing a theoretical and quantitative methodology that can be repeated or adapted for answering different archaeological questions in different geographical contexts and to contribute to the production of relevant scientific knowledge in a domain that of (digital archaeology) still at its dawn.

As stressed out by Huggett [100], it is important to push the boundaries of archaeological disciplines to welcome inputs from outside, moving the academic research into uncharted territories of quantitative and digital approaches, taking archaeology out of its comfort zones. To put it more into context, this Study likely aims at raising awareness between the Swiss public authorities (as the cantonal archaeological departments) and the academic research about the need to steer efforts towards common objectives and finally to "raise our sights to look for paradigm-shifting developments which separately or together have revolutionary potential."

1.5 Thesis Outline

The structure of this Thesis, as well as the analysis of the data and the modeling procedure have been determined by the Study's core research questions. Figure 5.1 shows the modeling workflow in more detail. Whilst the processes may appear complex at first glance, the different colored frames highlight phases which roughly equate to the actions carried out in Chap. 5.

The following section introduces the theoretical background of the methodological procedure applied within this research. Chapter 2 primarily aims at framing this study as part of landscape archaeology and quantitative studies. Chapter 3 provides with a more specific prospect of predictive modeling approaches. In doing so, a selection of recent and relevant Archaeological Predictive Modeling (APM) applications is presented, ranging from a global to a local perspective and originating from academia as well as from cultural heritage management. Since most of the case studies presented in the Chap. 3 made use of the most advanced and popular techniques currently in use in various fields of scientific research, Sects. 3.4 and 3.5 provide the reader with more details about those innovative technologies. Section 3.3 addresses uncertainty and vagueness issues in APM procedures and summarizes the state of art with respect to this research field.

In Chap. 4, the data used in this Study are outlined. Firstly, an historical framework is given. Secondly, the archaeological datasets are detailed by study area (Canton). Each dataset description is preceded by a general geographical and a more detailed historical description of the study area. Finally, the environmental aspects considered for the APM are outlined.

Chapter 5 is completely devoted to the technical step by step construction of the modeling procedure. It is organized in 5 subchapters that tackle the most relevant aspects in defining a clear and effective communicative protocol. In doing so, the first 2 subchapters in particular refer to the elaborated conceptual modeling and the solution experimented for dealing with uncertainty in archaeological data. These steps are exposed together with the quantification process of the archaeological data and the environmental factors as part of the pre-processing phase. This chapter also deals with the spatial and statistical exploration of the data ensemble as Locational Preference Analysis, and returns some of the first interesting results about site locations and their environment, at a regional and supra regional scale. The end of this chapter outlines the ML algorithm selected (Random Forest) for the development of the final predictive modeling approach.

In Chap. 6, the results obtained are presented and discussed, charting further information from a comparison between the results derived from the Locational Preference Analysis and those derived from the further modeling carried out with ML technique. Finally, the conclusions are delineated and the main research achievements and outlooks addressed in Chap. 7.

References

1. Viejo-Rose D (2018) Cultural heritage management and armed conflict. In: Smith C (ed) Encyclopedia of global archaeology. Springer, Cham. https://doi.org/10.1007/978-3-319-51726-1_1816-2
2. Ortman SG (2019) A new kind of relevance for archaeology. Front Digit Humanit 6:16. https://doi.org/10.3389/fdigh.2019.00016
3. Smith ME, Feinman GM, Drennan RD, Earle T, Morris I (2012) Archaeology as a social science. Proc Natl Acad Sci 109(20):7617–7621
4. Kintigh KW, Altschul JH, Beaudry MC et al (2014) Grand challenges for archaeology. Am Antiquity 79(01):5–24
5. Altschul JH, Kintigh KW, Klein TH, Doelle WH, Hays-Gilpin KA, Herr SA, Kohler TA, Mills BJ, Montgomery LM, Nelson MC, Ortman SG, Parker JN, Peeples MA, Sabloff JA (2017) Opinion: Fostering synthesis in archaeology to advance science and benefit society.

Proceedings of the National Academy of Sciences of the United States of America. 114: 10999-11002. PMID 29073009 https://doi.org/10.1073/Pnas.1715950114
6. Altschul JH (2016a) The Society for American Archaeology's Task Forces on Landscape Policy Issues. Advances in Archaeological Practice: A Journal of the Society for American Archaeology 4(2):102–105.
7. Altschul JH (2016b) The Role of Synthesis in American Archaeology and Cultural Resource Management as Seen Through an Arizona Lens. Journal of Arizona Archaeology 4(1):68–81.
8. Heilen M, Altschul J, Lüth F (2018) Modelling resource values and climate change impacts to set preservation and research priorities. Conserv Manag Archaeol Sites 20(4):261–284
9. Rockman M, Morgan M, Ziaja S, Hambrecht G, Meadow A (2016) Cultural resources climate change strategy. National Park Service, Washington, DC
10. Heilen M, Altschul J (2016) Cultural resources phase I efforts for modeling site location in Georgia. Report submitted to the Georgia Department of Natural Resources Historic Preservation Division. Statistical Research, Tucson (AZ)
11. Heilen M, Leckman P, Byrd A, Homburg J, Heckman R (2013) Archaeological sensitivity modeling in Southern New Mexico: automated tools and models for planning and management. Technical Report 11–26. Statistical Research, Tucson (AZ)
12. Klimeš J (2013) Landslide temporal analysis and susceptibility assessment as bases for landslide mitigation Machu Picchu, Peru. Environ Earth Sci 70:913–925
13. Hadjimitsis D, Agapiou A, Alexakis D, Sarris A (2013) Exploring natural and anthropogenic risk for cultural heritage in Cyprus using remote sensing and GIS. Int J Digit Earth 6:115–142
14. Sdao F, Lioi DS, Pascale S, Caniani D, Mancini IM (2013) Landslide susceptibility assessment by using a neuro-fuzzy model: a case study in the Rupestrian heritage area of Matera. Nat Hazards Earth Syst Sci 13:395–407
15. Heilen M, Sebastian L, Altschul J, Leckman P, Byrd A (2012) Modeling of archaeological site location and significance at White Sands Missile Range, New Mexico. Technical Report 12–06. Statistical Research, Tucson (AZ)
16. Tarragüel AA, Krol B, Westen VC (2012) Analysing the possible impact of landslides and avalanches on cultural heritage in Upper Svaneti, Georgia. J Cult Herit 13:453–461
17. Sdao F, Simeone V (2007) Mass movements affecting Goddess Mefitis sanctuary in Rossano di Vaglio (Basilicata, southern Italy). J Cult Herit 8:77–80
18. Pederson JL, Petersen PA, Dierker JL (2006) Gullying and erosion control at archaeological sites in Grand Canyon, Arizona. Earth Surf Process Landf 31:507–525
19. Bedaux R, MacDonald K, Person A, Polet J, Sanogo K, Schmidt A, Sidibé S (2001) The Dia archaeological project: rescuing cultural heritage in the Inland Niger Delta (Mali). Antiquity 75:837–848
20. Canuti P, Casagli N, Catani F, Fanti R (2000) Hydrogeological hazard and risk in archaeological sites: some case studies in Italy. J Cult Herit 1:117–125
21. Altschul J (1988) Models and the modeling process. In: Judge W, Sebastian L (eds) Quantifying the past and predicting the past: theory, method, and application of archaeological predictive modeling. US Bureau of Land Management, Denver (CO), pp 61–96
22. Hafner A (2013) Archäologische Kulturgüter in der Schweiz - eine Ressource im Spannungsfeld von Zersiedlung und Verdichtung. NIKE-Bulletin 28(4):20–23
23. Kaeser MA (2012) L'archéologie des grands travaux, Laténium, p 66
24. Kaeser MA (2013) Archéologie et Tourisme en Suisse, Archéologie et tourisme, pp 10–64
25. Kaenel G (2002) Autoroutes et archéologie en Suisse. Revue du Nord 348:33–41
26. Kaenel G (ed) (1998) 30 ans de grands travaux. Quel bilan pour la préhistoire suisse ? Actes du colloque GPS/AGUS Bâle 13–14 mars 1998. Lausanne, Documents du GPS, p 94
27. Caspari G (2020) Mapping and damage assessment of "Royal" burial mounds in the Siberian Valley of the Kings. Remote Sens 12(5):773
28. Caspari G, Crespo P (2019) Convolutional neural networks for archaeological site detection–Finding "princely" tombs. J Archaeol Sci 110:104998
29. Oonk S, Spijker J (2015) A supervised machine-learning approach towards geochemical predictive modelling in archaeology. J Archaeol Sci 59:80–88

30. Baudron P, Alono-Sarría F, García-Aróstegui JL, Cánovas-García F, Martínez-Vicente D, Moreno-Brotóns J (2013) Identifying the origin of groundwater samples in a multi-layer aquifer system with Random Forest classification. J Hydrol 499(2013):303–315. https://doi.org/10.1016/j.jhydrol.2013.07.009
31. Abedi M, Norouzi G-H (2012) Integration of various geophysical data with geological and geochemical data to determine additional drilling for copper exploration. J Appl Geophys 83:35–45
32. Abedi M, Norouzi G-H, Bahroudi A (2012) Support vector machine for multiclassification of mineral prospectivity areas. Comput Geosc 46:272–283
33. Tehrany MS, Kumar L, Jebur MN, Shabani F (2019) Evaluating the application of the statistical index method in flood susceptibility mapping and its comparison with frequency ratio and logistic regression methods. Geomat Nat Haz Risk 10(1):79–101
34. Biondi G, Campo L, D'Andrea M, Degli Esposti S, Fiorucci P, Tonini M (2018) Wildfire susceptibility mapping in Liguria (Italy): comparison of statistical driven partitioning and machine learning approach. In: Viegas DX (ed) Advances in forest fire research 2018. Chapter 1—Fire risk management., https://doi.org/10.14195/978-989-26-16-506_20
35. Reichenbach P, Rossi M, Malamud BD, Mihir M, Guzzetti F (2018) A review of statistically based landslide susceptibility models. Earth Sci Rev 2018(180):60–91
36. Deluigi N (2018) Data-driven mapping of the potential mountain permafrost distribution. PhD thesis, University of Lausanne
37. Leuenberger M, Parente J, Tonini M, Pereira MG, Kanevski M (2017) Wildfire susceptibility mapping: deterministic vs. stochastic approaches. Environ Model Softw 101:194–203 (2018). https://doi.org/10.1016/j.envsoft.2017.12.019
38. Zêzere JL, Pereira S, Melo R, Oliveira SC, Garcia RAC (2017) Mapping landslide susceptibility using data-driven methods. Sci Total Environ 2017(589):250–267
39. Pham BT, Pradhan B, Tien Bui D, Prakash I, Dholakia MB (2016) A comparative study of different machine learning methods for landslide susceptibility assessment: a case study of Uttarakhand area (India). Environ Model Softw 2016(84):240–250
40. Goetz JN, Brenning A, Petschko H, Leopold P (2015) Evaluating machine learning and statistical prediction techniques for landslide susceptibility modeling. Comput Geosci 2015(81):1–11
41. Gillings M, Hacıgüzeller P, Lock G (eds) (2020) Archaeological spatial analysis. Routledge, London. https://doi.org/10.4324/9781351243858
42. Wood D (2012) The anthropology of cartography. In: Roberts L (ed) Mapping cultures: place, practice, performance. Palgrave Macmillan, Houndmills, Basingstoke, Hampshire and New York, pp 280–303
43. Carlson D (2017) Quantitative methods in archaeology using R (Cambridge manuals in archaeology). Cambridge University Press, Cambridge. https://doi.org/10.1017/9781139628730
44. Barceló JA, Bogdanovic I (eds) (2015) Mathematics and archaeology. CRC Press. https://doi.org/10.1201/b18530
45. Hintz M, Laabs J, Castiello ME (2019) Archaeology that counts. International colloquium on digital archaeology. Pages Mag 27(1). https://doi.org/10.22498/pages.27.1.37
46. Dubbini N, Lodoen A (2014) Statistical and mathematical models for archaeological data mining: a comparison. In: Proceedings of the 42nd annual conference on computer applications and quantitative methods in archaeology, pp 509–516
47. Giligny F, Djindjian F, Costa L, Moscati P, Robert S (2010) CAA2014 proceedings. In: Proceedings of the 42nd annual conference on computer applications and quantitative methods in archaeology, pp 1–6
48. Djindjian F (2009) The golden years for mathematics and computers in archaeology (1965–1985). In: La nascita dell'informatica archeologica. Atti del Convegno Internazionale, Roma, Accademia Nazionale dei Lincei, 24 ottobre2008. Archeologia e Calcolatori 20:61–73
49. Danese M, Masini N, Biscione M, Lasaponara R (2014) Predictive modeling for preventive archaeology: overview and case study. Cent Eur J Geosci 6(1):42–55. https://doi.org/10.2478/s13533-012-0160-5

50. Rogers SR, Fischer M, Huss M (2014) Combining glaciological and archaeological methods for gauging glacial archaeological potential. J Archaeol Sci 52:410–420. https://doi.org/10.1016/j.jas.2014.09.010
51. Kvamme KL (2006) There and back again: revisiting archaeological locational modeling. In: Mehrer MW, Wescot KL (eds) GIS and archaeological site location modeling. CRC, London & New York, pp 3–38
52. Delley G (2013) Le financement de l'archéologie en Suisse dans la seconde moitié du xxe siècle'. Les nouvelles de l'archéologie 133:34–38. https://doi.org/10.4000/nda.2122
53. Luginbühl T, Monnier J, Dubois Y (ed) (2001) Vie de palais et travail d'esclave: la villa romaine d'Orbe-Boscéaz. Lausanne: Musée cantonal d'archéologie et d'histoire, pp 18–23. (Document du Musée cantonal d'archéologie et d'histoire de Lausanne). https://archive-ouverte.unige.ch/unige:116051
54. Castella D (Dir) (1998) Aux portes d'Aventicum. Dix ans d'archéologie autoroutière à Avenches (Documents du Musée Romain d'Avenches 4) Montreux
55. Demoule JP (2020) Aux origines, L'Archéologie. Une science au cœur des grands débats de notre temps. (Eds) La Découverte, Paris
56. Paunier D (2006) Sans Autoroute, Pas d'histoire? In: Arnold B, Bauermeister N, Ramseyer D (Hrsg) Archéologie Plurielle. Mélanges Offerts à Michel Egloff à l'occassion de Son 65e Anniversaire, Archéologie Neuchâteloise 34 (Hauterive 2006), pp 25–35
57. Kaenel G (2007) Les archéologies en Suisse : un regard critique. Annuaire d'Archéologie Suisse 90:37–40
58. Archäologie Schweiz (2017) Die Stellung der Archäologie in den kantonalen Gesetzen. Aktualisierung März 2017, Basel, 2017 (http://www.archaeologie-schweiz.ch/fileadmin/user_upload/customers/archaeologie_schweiz/AS/Dokumente_dt/Kommissionen_dt/KAR/20170304_RapLoi_Aktualisierung.pdf, accessed 11.11.2021)
59. Schwab H (1973) Die Vergangenheit des Seelandes in neuem Licht. Archäologische Entdeckungen und Ausgrabungen bei der 2. Juragewässerkorrektion Fribourg
60. Schwab H (1989) Archéologie de la 2e Correction des Eaux du Jura Vol. 1 ' Les Celtes sur la Broya et La Thielle. Archéologie Fribourgeoise/Freiburger Archäologie 5, Fribourg
61. Kaenel G (2012) Les archéologies en Suisse : un regard critique. Jahrbuch Archäologie Schweiz = Annuaire d'Archéologie Suisse = Annuario d'Archeologia Svizzera = Annual review of Swiss Archaeology, Band (Jahr): 90. Available at https://doi.org/10.5169/seals-117921. Accessed on 10 February 2015.
62. Hooke R (2012) Land transformation by humans: a review. GSA Today 22:4–10
63. Sandweiss DH, Kelley AR (2012) Archaeological contributions to climate change research: the archaeological record as a paleoclimatic and paleoenvironmental archive. Annu Rev Anthropol 41:371–391
64. Hambrecht G, Rockman M (2017) International Approaches to Climate Change and Cultural Heritage. American Antiquity 82(4): 627–41. https://doi.org/10.1017/aaq.2017.30
65. Erlandson JM, Rick TC (2010) Archaeology meets marine ecology: The antiquity of maritime cultures and human impacts on marine fisheries and ecosystems. Annual Review of Marine Science 2, 231–251
66. Rick TC, Sandweiss DH (2020) Archaeology, climate, and global change in the age of humans. Proc Natl Acad Sci 117(15):8250–8253. https://doi.org/10.1073/pnas.2003612117
67. Holm P, Winiwarter V (2017) Climate change studies and the human sciences. Glob Planet Change 156:115–122. https://doi.org/10.1016/j.gloplacha.2017.05.006
68. Kintigh KW, Altschul JH, Beaudry MC, et al (2014). Grand Challenges for Archaeology. American Antiquity, 79(01), 5–24.
69. Van Dieter M (2017) Living along the Limes. PhD thesis, Universiteit Utrech
70. Kamermans H, Deeben J, Hallewas D, Zoetbrood P, van Leusen M, Verhagen P (2005) Project proposal. In: van Leusen M, Kamermans H (eds) Predictive modelling for archaeological heritage management: a research agenda. Nederlandse Archeologische Rapporten 29. Rijksdienst voor het Oudheidkundig Bodemonderzoek, Amersfoort, pp 13–23

71. Ejstrud B (2003) Indicative models in landscape management: testing the methods. The archaeology of landscapes and geographic information systems. Predictive maps, settlement dynamics and space and time in prehistory. Kunow & Müller (eds) 2003:119–134
72. Parker SC (1985) Predictive modeling of site settlement systems using multivariate logistics. In: Carr C (ed) For concordance in archaeological analysis: bridging data structure, quantitative technique, and theory. Waveland Press, Prospect Heights, CO, pp 173–207
73. Graves McEwan D, Millican K (2012) In search of the middle ground: quantitative spatial techniques and experiential theory in archaeology. J. Archaeol Method Theory 19(4):491–494
74. Verhagen P, Whitley TG (2012) Integrating archaeological theory and predictive modeling: a live report from the scene. Journal of Archaeological Theory and Method 19/1, 49–100. https://doi.org/10.1007/s10816-011-9102-7
75. Kvamme KL (2006) There and back again: revisiting archaeological locational modeling. In: Mehrer MW, Wescot KL (eds) GIS and Archaeological Site Location Modeling. CRC, London & New York, 3–38
76. Deeben J, Hallewas DP, Maarlevelt TJ (2002) Predictive modelling in archaeological heritage management of the Netherlands: the indicative map of archaeological values (2nd generation), Berichten ROB 45, 9–56
77. Brouwer Burg M, Peeters H, Lovis WA (2016) Introduction to uncertainty and sensitivity analysis in archaeological computational modeling. https://doi.org/10.1007/978-3-319-27833-9_1
78. Clark JE (2000) Toward a better explanation of heredity inequality: a critical assessment of natural and historic human agents. In: Dobres MA, Robb JE (eds) Agency in archaeology. Routledge, London, pp 92–112
79. Cowgill GE (2000) "Rationality" and contexts in agency theory. In: Dobres MA, Robb JE (eds) Agency in archaeology. Routledge, London, pp 51–60
80. Shanks M, Tilley C (1987) Re-constructing archaeology: theory and practice. Routledge, London
81. Earley-Spadoni T, Harrower M (2020) Spatial archaeology: mapping the Ancient past with the humanities and the sciences. In: Bodenhamer DJ, Ell PS (eds) Int J Humanit Arts Comput 14(1–2). ISSN 1753-8548. Available Online Feb 2020
82. Lake MW (2014) Trends in Archaeological Simulation. Journal of Archaeological Method and Theory 21:258–287. https://doi.org/10.1007/s10816-013-9188-1
83. Doran JE (2000) Trajectories to complexity in artificial societies: Rationality, belief and emotions. In: Kohler TA, Gumerman GJ(eds) Dynamics in human and primate societies: Agent-based modeling of social and spatial processes. pp. 89–144. New York: Oxford University Press.
84. Verhagen P, Whitley TG (2012) Integrating archaeological theory and predictive modeling: a live report from the scene. J Archaeol Theory Method 19/1:49–100. https://doi.org/10.1007/s10816-011-9102-7
85. Kemp BM, Judd K, Monroe C, Eerkens JW, Hilldorfer L, Cordray C, Schad R, Reams R, Ortman SG, Kohler TA (2017) Prehistoric mitochondrial DNA of domesticate animals supports a 13th century Exodus from the Northern US Southwest. PLOS ONE 12(7):e0178882. https://doi.org/10.1371/journal.pone.0178882
86. Deravignone L, Blankholm HP, Pizziolo G (2015) Predictive modeling and artificial neural networks: from model to survey. In: Barceló JA, Bogdanovic I (eds) Mathematics and archaeology. CRC Press, Boca Raton, pp 335–351
87. Glowacki DM (2015) Living and leaving: a social history of regional depopulation in thirteenth-century Mesa Verde. University of Arizona Press, Tucson
88. Pizziolo G, Sarti L (2015) Predicting prehistory: predictive models and field research methods for detecting prehistoric contexts. Museo e Istituto Fiorentino di Preistoria "Paolo Graziosi", Siena
89. Verhagen P (2007) Testing archaeological predictive models: a rough guide in layers of perception. In: Proceedings of the 35th computer applications and quantitative methods in archaeology conference, Berlin, Germany, April 2–6, 2007, Bonn, pp 285–291

90. Verhagen P (2007) Case studies in archaeological predictive modelling. PhD thesis (Archaeological Studies of Leiden University 14), 224 ppages, 31 figures, 57 tables. Leiden University Press. 978-90-8728-007-9 paperback. Antiquity 83(319):232–233
91. Van Leusen PM, Kamermans H (eds) (2005) Predictive modelling for archaeological heritage management: a research agenda, Amersfoort, ROB, PlantijnCasparie Almere
92. Lloyd CD, Atkinson PM (2004) Archaeology and geostatistics. J Archaeol Sci 31:15–165
93. Verhagen P, Gili S, Micó R, Risch R (1999) Modelling prehistoric land use distribution in the Rio Aguas Valley (SE Spain). In: Dingwall L et al (eds) Archaeology in the age of the Internet. Proceedings of the CAA97 conference. BAR international series, vol 750
94. Kvamme KL (1990) The fundamental principles and practice of predictive archaeological modeling. In: Voorrips A (ed) Mathematics and information science in archaeology: a flexible framework. HOLOS-Verlag, Bonn, pp 275–295
95. Judge W, Sebastian L (eds) (1988) Quantifying the past and predicting the past: theory, method, and application of archaeological predictive modeling. USDI Bureau of Land Management, Denver (CO), pp 61–96
96. van Leusen M, Pizziolo G, Sarti L (2011) Hidden landscapes of Mediterranean Europe. Cultural and methodological biases in pre- and protohistoric landscape studies. Proceedings International Meeting, Siena 2007. BAR International Series 2320. Oxford: Archaeopress.
97. Achino F, Barceló JA (2019) Spatial Prediction: Reconstructing the Spatiality of Social Activities at the Intra-Site Scale. Journal of Archaeological Method and Theory 26:112–134. https://doi.org/10.1007/s10816-018-9367-1
98. Schwindt D, Bocinsky RK, Ortman SG, Glowacki DM, Varien MD, Kohler TA (2017) The Social Consequences of Climate Change in the Central Mesa Verde Region. American Antiquity 81(1):74–96
99. Carleton C, Connolly J, Jannone C (2012) A locally-adaptive model of archaeological potential (LAMAP). Journal of Archaeological Science 39 (2012) 3371–3385. https://doi.org/10.1016/j.jas.2012.05.022
100. Huggett J (2015) A manifesto for an introspective digital archaeology. Open Archaeol 1:86–95
101. Binford LR (1981) Behavioral archaeology and the "Pompeii premise." J Anthropol Res 37(3):195–208

Section II

Chapter 2
Space, Environment and Quantitative Approaches in Archaeology

Il n'est rien qui illustre si fidèlement une civilisation.
que la manière dont elle s'installe dans l'espace.
Jean - Pierre Cahen

This Chapter aims at highlighting the strong relationships between past human populations and their environmental space, while framing this Study within the wider context of landscape archaeological studies. An initial overview of the 'Landscape archaeology' concept is provided and an introduction to Geographic Information Systems (GIS) is given, with respect to their specific use for Archaeological Site Modeling.

Gillings et al. [1] have recently stressed out that "being human embodies space and spatial relationships within a material world and just as this applied to people living in the past, so it applies to those of us concerned with trying to understand those past lives through their remaining material residues."

Nowadays a growing acceptance for the significance of studying societies in relation to geography and environment exists, especially when the aim is to study long-ranging human impacts on the landscape and to provide archaeology with the necessary tools to meet current challenges and address the issues threatening our cultural heritage.

The landscape, according to Crumley and Marquardt [2], can be defined as "the spatial manifestation of the relations between humans and their environment" and to further quote the authors: "Included in the study of landscapes are population agglomerations of all sizes, from isolated farmsteads to metropolises, as well as the roads that link them." The research questions defined in this Study fit within this definition of landscape archaeology and pursue the same objectives.

However, this Chapter does not wish to provide a complete historical overview of archaeological landscape discipline, as this field of research is vast, very complex and based on region and country specific circumstances, as well as on world-wide trends. It would even hardly be possible to exhaustively cite the vast bibliography on the evolution of landscape archaeology, settlement research, and to thoroughly outline how earlier 'traditional' approaches were converted into a more advanced

M. E. Castiello, *Computational and Machine Learning Tools for Archaeological Site Modeling*, Springer Theses, https://doi.org/10.1007/978-3-030-88567-0_2

field of investigation by the numerous theoretical, technical and methodological developments [3]. Instead, this Chapter aims at introducing the modeling approach developed in this Thesis by outlining the events and the circumstances that led to the gradual adoption of such computer applications in the archaeological context. Defining a concurrence of events and facts that gradually has driven the discipline towards computational and modeling approaches will moreover endorse the use of Archaeological Site Modeling as a valuable instrument for site preservation and conservation purposes.

2.1 Landscape Archaeology: A Synopsis

Since the industrial revolution, humanity has witnessed changes of unprecedented scale and rapidity in almost every aspect of society. Technological advancements have allowed us to shorten distances and reach for previously inaccessible resources. But despite today's highly advanced technology, geographical and climatic factors continue to exert a fundamental influence upon social, cultural and economic activities all over the world. Mountain ranges, deserts, fertile soils, waterways, standing waters and oceans, woodlands, resources such as coal, oil or rare earths shape modern societies and economies [4].

As Forsythe [4] pointed out, it seems therefore as a matter of course that an inverse relationship between human technology and geographical determinism has long been considered. The less control people have over their natural environment, the stronger is the impact of their environment upon their organization. As a consequence, human activities have been determined by the influence of climate, environment and geography throughout prehistory and history [4]. It is thus of no surprise that historians, geographers and archaeologists begun to consider the ways in which people dealt with their environment and how their environment influenced human activities in the past.

Towards the middle of the twentieth century, historical geographers developed and applied the concept of the 'settlement chamber theory' [5, 6], according to which large areas of land delimitated by natural boundaries and hosting desirable resources sustained nucleated communities and remained occupied for almost every period. These communities would have evolved around specific topographical characteristics and soil types [3]. With their works, historical geographers, anthropologists and archaeologists first highlighted the long-term relationships between man and geographical space, and promoted the role of Geography and Environment in human behavior patterns (e.g. [7]). To quote Bintliff [8], this systemic and deterministic approach was primarily based "on the idea that a resourceful landscape unit, identified as self-sufficient, will always support a local community and even though, the housing location of this community shifts over time, it will remain within the chamber."

Under such theoretical trends, landscape and human studies have evolved, instigating the birth of a new branch of archaeological research, Landscape Archaeology indeed, which enjoyed great popularity during the 1960s and 1970s at the height of the New Archaeology movement[1] [9–16].

Chisholm's influential work [17] moreover addressed the economic aspect of settlement location by reason of community subsistence. The landscape starts to be approached as a spatially measurable entity and the occurrence of archaeological sites is explained based on environmental factors. Verhagen and Whitley [18] also pointed out that the cultural behavior is seen as a response to environmental stimuli. The general focus of these new studies lays on the relationship between culture and environment and cultural expression viewed as economic adaptation strategy to environmental opportunities, furthermore subject to technological potential. Ultimately, the environment is interpreted as a constraining or enabling force of human activity, which can be predictable because of its adaption according to ecological laws (e.g. the choice for the best settlement location).

As supported by the formulation of these new theories, the archaeological discipline has been constantly influenced through decades by developments in other disciplines (history, anthropology, geography, ecology, biology). In particular, the emergence of ecology and the species distribution theory, as well as the implementation of modeling techniques to identify metrics that define and predict species and ecosystem ranges [19, 20] had an important impact on archaeological research. In this context, archaeology borrowed concepts that shaped the evolution of the discipline. One such concept can be resumed as follow: if humans are considered to preferentially target favorable locations for their activities in a similar way to animals and plants, understanding the patterns of bio-geographical distribution is the key to apprehend the role of people in landscape history, because they form the basis of how, when, where and to what extent people shaped the course of landscape evolution [21, 22].

Furthermore, the rising awareness that no minor factors such as the terrain elevation, topography, climate or geological characteristics conditioned the successful establishment of human settlements in the landscape, reveals the strong influence of environmental disciplines over archaeology. The geo-environmental theoretical framework defined by geology, geomorphology and soil science are acknowledged to have enhanced understanding of long-term landscape changes and thus, when linked to human activity (settlements), they have contributed towards a clearer picture of man-nature interactions.

This interest in long-term men-environment relationships has always encouraged the search for patterns and models explaining human behavior. Thus, site density and site distribution studies increased their significance in archaeological research

[1] In a few words, the Processual Archaeology or New Archaeology approaches the study of archaeological artifacts from an ecological and evolutionary perspective. The artifacts are studied within their former cultural and geographical environment as legacy of the dynamic past societies that left them, rather than as static and finite typological objects. This involved a new emphasis on the natural formation processes (geological, biological, chemical, etc.) of archaeological sites and objects and the development of hypothetic-deductive methods to support rigorous archaeological interpretation [9].

considerably. As a consequence, the analysis of stimuli conditioned by environmental variables became a prime objective of intensive archaeological surveys and of methodological concerns arisen accordingly (to cite a few: [8, 23–29].

2.2 Computational and Quantitative Approaches

Along with the growing attention given to environmental determinants, the late 1960s also experienced a rising interest in the application of quantitative approaches, resulting in an intensification of site analyses and settlement patterns evaluation, which gathered even further momentum during the 1970s, when computer technology became sufficiently advanced to allow for more sophisticated computations and cartographic modeling. Such a boom is a consequence of the increasing importance of Information Technology used in data manipulation, mapping, and landscape studies [30–33], as well as of the spreading of statistical methods imported form mathematics, such as Sampling, Statistical tests, Map Algebra, Bayesian statistics, Kriging/coKriging, Regression and Correlation, Classification, Seriation, Graph theory, Signal Processing, Image Processing, and many others [24, 34–43].

Site Catchment Analysis[2] and Thiessen polygons[3] for example, represent some of the first approaches used in order to explore land resources on a micro and macro-regional level [46–53], as well as site hierarchies [54].

Archaeological techniques and supporting technology improved and so do theoretical concepts influencing methodological and interpretative frameworks. As explained by Barceló et al. [55]: "whenever we express an idea through order relations among its components, we are expressing it mathematically. The basic meaningful unit of this artificial language is the idea of quantity", and, only in the last few years archaeologists became aware "that the task of understanding the past can be done better with the help of geometry, probabilities, and equations". Nowadays, archaeologists are no longer only excavators and it appears very clearly that numerous aspects of archaeological information are numerical, and that archaeological analysis has an inevitable quantitative component. Reflecting this evolution of archaeology, the literature on quantitative analysis in archaeology has grown to prodigious size in the past few years only (to cite a few: [56–61]). Archaeologists have backed a fine tradition of employing the latest advances made in mathematics, computer and information technology.

---

[2] Vita-Finzi and Higgs [44] defined it as "the study of the relationships between technology and those natural resources lying within economic range of individual sites", which emphasizes the importance of the availability, abundance, spacing, and seasonality of plant, animal, and mineral resources in determining site location.

[3] According to Darvill [45], Thiessen Polygons is "a formal method for exploring settlement patterns based on notional polygons constructed around a series of distributed points by taking the calculated mid-line between each pair of adjacent points to form the lattice. They provide a general approximation of the extent, shape, and orientation of the spheres of influence or territory around recorded settlements or other nodes in the settlement pattern.".

2.3 Geographic Information Systems

Undoubtedly, one of the greatest innovations for archaeology has been the introduction of Geographical Information Systems (GIS), which have considerably increased the analytical potential of archaeological data [62–65].

GIS are in essence computer programs for capturing, storing, checking, integrating, analyzing and displaying data about the Earth, that are spatially referenced at different scales, allowing comparisons from continent-wide to regional, local and intra-site scale [66]. Initially, they were mostly oriented to produce land use maps for environmental studies. Nevertheless, despite the fact that they were not originally created within the Humanities discipline, some of their main characteristics have turned them into a basic and crucial tool for archaeological research, such as the ease to convey the results obtained in a clear and effective manner, the possibility to perform analyses that combine relational databases with spatial interpretations, and finally the contingency to produce visual outputs and maps. Acting as a methodological tool, GIS emphasize the spatial relevance of environmental and cultural systems and have continuously interacted with archaeological theory. They are suitable for carrying out specific analyses on socio-historic structures coupling them with natural landscape and its evolution. GIS are able to simultaneously take into account space, as well as time and form. At the same time, they allow the association of descriptive attributes to a graphic convey [67] for a better understanding of social action and sense-making [68]. Furthermore, GIS environments allow to store, manage and analyze disparate sets of data at the same time, ranging from aerial and satellite imagery, surveys reports, historical maps and 3D reconstruction models, etc. The combination of themes, images and cartography is not just a simple algebraic sum of elements, but rather represents an advantageous overlay of different informative layers, widely increasing the possibilities for data interpretation and validation and thus allowing new research perspectives [69].

Especially in the field of Cultural Heritage Management, the combination of database management with representations of image or vector entities makes GIS a powerful tool for documentation and management of archaeological collections over a variety of scales. Numerous projects across the world have proposed the creation of digital archives including map representations on a national or more modest level, demonstrating yet again the ability of GIS to serve archaeological purposes. Some years ago, Berg [70] explained in his work how GIS could successfully help to maintain a geospatial platform at a national scale. The Norwegian Cultural Heritage database consisting of archaeological sites, buildings and installations, gardens and parks as well as marine sites, facilitates the exchange and use of GIS data, both vector and raster,[4] between partners in the professional management of cultural heritage (museums, research institutions and public management with limited access rights). The use of these technologies has indeed triggered an increasing desire for better management of archaeological intra-site databases [71, 72] and as mentioned before,

[4] Vector and Raster are two different types of graphical or spatial data that can be incorporated in GIS (for further details see Chap. 5).

of cultural heritage monument and digital archives, prompting the definition of new analytical concepts [73].

Although the potential of GIS has been recognized over the years, it is important to emphasize that GIS has clearly become more than a data management tool. The literature provides infinite examples of successful applications in archaeology (to cite a few: [62, 64, 74–76]). The computational capacity of contemporary computers and GIS software, coupled with the unprecedented availability of high quality topographic and spatial environmental data, significantly improves our ability to perform accurate site modeling and simplifies the modeling process itself. By making use of these effective applications, many studies implemented GIS for various purposes. Clevis et al. [77] modeled geological processes in order to assess the chances of survival and detection of archaeological deposits. With the inclusion of cognitive-processual landscapes [78], the modeling process has been extended in order to consider not only the physical, but to incorporate the social landscape as well. GIS were further used for paleo-environmental reconstructions and the assessment of areas of human-environmental impact [79, 80]. Monti [81] performed an evaluation of the modern land use implementing new sources of information. More recently, the incorporation of cost-surfaces and least cost paths for modeling past human movement [82–86], together with viewsheds analysis [87, 88] have further broaden the field of GIS research in archaeology [68, 89].

However, despite all new developments and adjustments of the theory, a paramount question still remains topical in the archaeological research: Where did people choose to live, hunt, fish, grow crops, socialize, or carry out other activities in the past? [3, 90]. If we are able to answer this question, or at least to approximate the answer with a certain degree of reliability, we could then provide informed support in conservation and protection management of archaeological heritage, as well as leverage knowledge to contribute to the scientific archaeological research of past societies.

New perspectives to address these questions have opened up recently with the progressive falling of barriers between disciplines, even if such developments do not happen over night. As Barceló et al. [55] stated, "most mathematicians do not imagine the possibilities to develop new algorithms to solve archaeological problems because the archaeological problem has not yet been expressed in formal terms." Nevertheless, new mathematical approaches, such as neural networks, non-linear systems, multi-agent systems and others, are being explored in archaeology, but much of the efforts made in this direction seems to be out of mainstream academic archaeology.

This Thesis falls within the effort to innovate the archaeological research at the crossroads of multiple disciplines. By means of computational approaches, it answers the question about the location of past human activity, taking into account the influence of the environmental attributes and of the physical landscape. Computational and quantitative approaches have nowadays to be considered an integral part of the archaeological discipline, and this Study aims at unveiling the potential and validity of Artificial Intelligence and Machine Learning applications in this domain.

References

1. Gillings M, Hacıgüzeller P, Lock G (eds) (2020) Archaeological spatial analysis. Routledge, London. https://doi.org/10.4324/9781351243858
2. Crumley C, Marquardt WH (1990) Landscape: a unifying concept in regional analysis. In: Allen K, Green S, Zubrow E (eds) Interpreting space: GIS and archaeology. Taylor and Francis, London, pp 73–79
3. Vionis AK, Papantoniou G (eds) (2019) Central place theory reloaded and revised: political economy and landscape dynamics in the longue duree. Land 2019 8:36. https://doi.org/10.3390/land8020036
4. Forsythe G (2005) A critical history of early Rome: from prehistory to the first Punic war. University of California Press. https://doi.org/10.1525/j.ctt1ppxrv
5. Lehmann H (1939) Die Siedlungsräume Ostkretas. Geographische Zeitschrift 45:212–228
6. Philippson A (1950–1959) Die Griechischen Landschaften. Eine Landeskunde; Vittorio Klostermann: Frankfurt am Main, Germany.
7. Spaulding ACM (1953) Statistical techniques for the discovery of artifact types. Am Antiq 18:305–313
8. Bintliff J (2000) Beyond dots on the map: features directions for surface artifacts survey in Europe. Sheffield. Sheffield Academin Press, pp 3–20
9. Pelegrin J (2001) Lithics and archaeology. In: Smelser NJ, Baltes (eds). Pergamon, Oxford, pp 8984–8989. https://doi.org/10.1016/B0-08-043076-7/02061-1
10. Binford LR (1964) A consideration of archaeological research design. Am Antiq 29(4):425-441
11. Bintliff J (2000) The concept of 'site' and 'off-site' archaeology in surface artifacts survey. In: Pasquinacci M, Trément F (eds) Non destructive techniques applied to landscape archaeology. Oxford Oxbow Books, pp 200–215
12. David B, Lourandos H (1999) Landscape as mind: land use, cultural space and change in North Queensland prehistory. Quatern Int 59:107–23. https://doi.org/10.1016/S1040-6182(98)00074-3
13. Ingold T (1993) The temporality of landscape. World Archaeol 25(2):152–174. https://doi.org/10.1080/00438243.1993.9980235
14. Renfrew AC, Bahn P (1991) Archaeology: theories. Thames & Hudson, Methods and Practice, London
15. Yamin Y, Bescherer KM (eds) (1996) Landscape archaeology: reading and interpreting the American historical landscape. University of Tennessee Press
16. Kvamme K, Kohler T (1988) Geographic information systems: technical aids for data collection, analysis, and display. In: James Judge W, Sebastian L (eds) Quantifying the present and predicting the past. Theory, method, and application of archaeological predictive modeling. U.S. Department of the Interior, Bureau of Land Management Service Center, Denver, Co, p 690
17. Chisholm M (1962) Rural settlement land use. AldineTransaction, New Brunswick & London
18. Verhagen P, Whitley TG (2012) Integrating archaeological theory and predictive modelling: a live report from the scene. J Archaeol Theory Method 19/1:49–100. https://doi.org/10.1007/s10816-011-9102-7
19. Guisan A, Zimmermann NE (2000) Predictive habitat distribution models in ecology. Ecol Model 135(2):147–186. https://doi.org/10.1016/S0304-3800(00)00354-9
20. Guisan A, Tingley R, Baumgartner JB, Naujokaitis-Lewis I, Sutcliffe PR, Tulloch AIT, Regan TJ, Brotons L, Mcdonald-Madden E, Mantyka-Pringle C (2013) Predicting species distributions for conservation decisions. Ecol Lett 16(12):1424–1435. https://doi.org/10.1111/ele.12189
21. Bintliff J (1976) The plain of Western Macedonia and the Neolithic site of Nea Nikomedeia. Prooccedings of the Prehistoric Society 42:241–262
22. Jarman MR, Bailey GN, Jarman HN (1982) Early European agriculture: its foundations and development. Cambridge University Press, Cambridge
23. Fasham PJ, Shadla-Hall RT, Shennan SJ, Bates PJ (1980) Fieldworking for archaeologists. Hampshire Field Club Archaeol Soc, Winchester

24. Hodder I, Orton C (1976) Spatial analysis in archaeology. Cambridge University Press, Cambridge
25. Rupp DW (2004) Evolving strategies for investigating an extensive terra incognita in the Paphos district by the Canadian Palaipahos survey project and the western Cyprus project. In: Iacovou M (ed) Archaeological field survey in cyprus: past history, future potentials. British School at Athens, London, p 6
26. Willey G (1953) Prehistoric settlement patterns in the Viru valley, Peru. Bulletin 1. Bureau of American Ethnology, Washington, DC
27. Gamble CS, Boismier WA (eds) (1991) Ethnoarchaeological approaches to mobile campsites. Ethnoarchaeological Series 1. Int Monogr Prehist, Ann Arbor
28. Kuna M (1998) Method of surface artefact survey. In Neustpny E (ed) Space in prehistoric, Institute of Archaeology and Czech Academy of Sciences, Bohemia Praha, pp 77–83.
29. Hudak GJ, Hobbs E, Brooks A, Sersland CA, Phillips C (eds) (2002) Final report: a predictive model of precontact archaeological site location for the state of Minnesota. Minnesota Department of Transportation, St. Paul
30. Gaffney VL, Stančič Z (1991) GIS approaches to regional analysis: a case study of the Island of Hvar. Znanstveni inštitut, Filozofske fakultete
31. Hunt ED (1992) Upgrading site-catchment analyses with the use of GIS: investigating the settlement patterns of horticulturalists. World Archaeol 24:283–309
32. Kvamme KL (1990) The fundamental principles and practice of predictive archaeological modeling. In: Voorrips A (ed) Mathematics and information science in archaeology: a flexible framework. HOLOS-Verlag, Bonn, Germany, pp 275–295
33. Neiman FD (1997) Conspicuous consumption as wasteful advertising: a darwinian perspective on spatial patterns in classic Maya terminal monuments dates. In: Barton MC, Clark GA (eds) Rediscovering darwinian evolutionary theory and archaeological explanation archaeological papers of the American anthropological association, 7. Am Anthropol Assoc, Washington, DC, pp 267–290
34. Clarke D (1968) Analytical archaeology. Methuen, London
35. Djindjan F, Ducasse H (eds) (1987) Data processing and Mathematics applied to archaeology. Mathématiques et Informatique appliquées a l'Archéologie, PACT 16, Conseil de l'Europe
36. Gardin JC (1970) Archéologie et calculateurs. Problèmes sémiologiques et mathématiques. Actes du Colloque international, Marseille, 7–12 avril 1969. Paris : Editions du CNRS
37. Gardin J-C (1971) Archaeology and computers: new perspectives. In: Use of computers, documentations and the social science. Int Soc Sci J 23(2):189–203
38. Heizer RF, Cook SF (eds) (1960) The application f quantitative methods in archaeology, Viking fund publications in anthropology, 28. Quadrangle Books, Chiago
39. Hodson FR, Kendall DG, Tautu P (eds) (1971) Mathematics in the archaeological and historical sciences. In: Hodson FR, Kendall DG, Tautu P (eds) Proceedings of the Anglo-Romanian conference, Mamaia 1970. Edimburgh University Press, Endiburg
40. Hymes D (1965) The use of computers in anthropoly. Stud Gen Anthropol 2. Mouton, London
41. Lock G (ed) (2014) Using computers in archaeology, using computers in archaeology. https://doi.org/10.4324/9780203451076
42. Wheatley D, Gillings M (2000) Vision, perception and gis: developing enriched approaches to the study of archaeological visibility. In: Lock GR (ed) Beyond the map: Archaeology and spatial technologies. IOS Press, Oxford, pp 1–26
43. Kamerman H, van Leusen M, Verhagen P (eds) (2009) Archaeological prediction and risk management alternatives to current practice. Leiden University Press, The Netherlands
44. Vita-Finzi C, Higgs ES (1970) Prehistoric economies in the mount carmel area of palestine: site catchment analysis. Proc Prehist Soc 36:1–37
45. Darvill T (2019) Thiessen polygons. In: The concise Oxford dictionary of archaeology, 2nd edn. Oxford University Press
46. Cui YF, Liu YJ, Ma MM (2018) Spatiotemporal evolution of prehistoric Neolithic-bronze age settlements and influencing factors in the Guanting basin, northeast Tibetan Plateau. Sci China Earth Sci 61(2):149–162

47. Lombardo U, Prümers H (2010) Pre-Columbian human occupation patterns in the eastern plains of the Llanos de Moxos, Bolivian Amazonia. J Archaeol. Sci 37:1875–1885. https://doi.org/10.1016/j.jas.2010.02.011
48. Roper DC (1979) The method and theory of site catchment analysis: a review. Adv Archaeol Method Theory 2:119–140. Springer. http://www.jstor.org/stable/20170144
49. Zheng HB, Zhou YS, Yang Q, Hu ZJ et al (2018) Spatial and temporal distribution of Neolithic sites in coastal China: sea level changes, geomorphic evolution and human adaptation. Sci China Earth Sci 61(2):123–133
50. Bintliff J, Snodgrass A (1988) Off-site pottery distributions: a regional and interregional perspective. Current Anthropol 3(29):503–516
51. Higgs ES, Vita-Finzi C (1972) Prehistoric economies: a territorial approach. In: Higgs ES (ed) Papers in economic prehistory. Cambridge University Press, Cambridge, pp 27–36
52. Lock G, Pouncett J (2017) Spatial thinking in archaeology: Is GIS the answer? J Archaeol Sci 84:129–135. https://doi.org/10.1016/j.jas.2017.06.002
53. Li Q, Zou Q, Ma D, Wang Q, Wang S (2018) Dating ancient paintings of Mogao Grottoes using deeply learnt visual codes. http://arxiv.org/abs/1810.09168
54. Duff PR (2015) Site size hierarchy in middle-range societies. J Anthropol Archaeol 37:85–99
55. Barceló JA, Achino KF, Bogdanovic I, Capuzzo G, Del Castillo F, Moitinho de Alveida V, Negre J (2015) Measuring, counting and explaining: an introduction to mathematics in archaeology. In: Mathematics and archaeology. CRC Press. https://doi.org/10.1201/b18530
56. Baxter MJ (1994) Exploratory multivariate analysis in archaeology, 2nd edn. Edinburgh University Press, Edinburgh
57. Baxter MJ (ed) (2003) Statistics in archaeology. Arnold, London
58. Djindjian F (1991) Méthodes pour l'archéologie. Armand Colin, Paris. In: Doelle W, Barker P, Cushman D, Heilen M, Herhahn C, Rieth C (2016) Incorporating archaeological resources in landscape-level planning and management. Adv Archaeol Pract 4(2):118–131
59. Drennan RD (2010) Statistics for archaeologists: A commonsense approach, 2nd edn. Springer, New York, NY
60. Shennan S (1997) Quantifying archaeology. Edimburgh University Press, Edimburgh
61. VanPool TL, Leonard RD (2011) Quantitative analysis in archaeology. Wiley-Blackwell, Chichester
62. Allen K, Green S, Zubrow EBW (eds) (1990) Interpreting space: GIS and archaeology. Taylor & Francis, London
63. Burg MB, Howey M (2020) Unbinding Diversity measures in archaeology using GIS. J Comput Appl Archaeol 3(1):170–181. https://doi.org/10.5334/jcaa.55
64. Lock G, Stančič Z (eds) (1995) Archaeology and geographic information systems: a European perspective. Taylor and Francis, London
65. Lock GR (ed) (2000) Beyond the map. Archaeol Spat Technol. IOS Press, Amsterdam
66. Verhagen P (2007a) Testing archaeological predictive models: a rough guide in layers of perception. In: Proceedings of the 35th computer applications and quantitative methods in archaeology conference, Berlin, Germany, 2–6 April 2007, Bonn, pp 285–291
67. Rua H (2009) Geographic information systems in archaeological analysis: a predictive model in the detection of rural Roman villae. J Archaeol Sci 36(2):224–235. https://doi.org/10.1016/j.jas.2008.09.003
68. Howey MCL, Brouwer Burg M (2017) Assessing the state of archaeological GIS research: Unbinding analyses of past landscapes. J Archaeol Sci 84:1–9
69. Caracausi S, Berruti LF, Daffara S, Bertè B, Borel FR (2018) Use of a GIS predictive model for the identification of high altitude prehistoric human frequentations. Results of the Sessera valley project (Piedmont, Italy). Quatern Int 490:10–20. https://doi.org/10.1016/j.quaint.2018.05.038
70. Berg E (2007) Using a GIS-based database as a platform for cultural heritage management of sites and monuments in Norway. In: CAA2006. Digital discovery. Exploring new frontiers in human heritage. Computer applications and quantitative methods in archaeology. Proceedings of the 34th conference, Fargo, United States, April 2006, pp 345–351

71. Kantner J (2008) The archaeology of regions: from discrete analytical toolkit to ubiquitous spatial perspective. J Archaeol Res16(1):37e81
72. Montalvo Puente CE (2020) La prospección arqueológica basada en imágenes satelitales: el caso de la zona norte del país caranqui (Imbabura, Ecuador). Arqueología Iberoamericana, vol 45, pp 35–42
73. Sanjuan L, Wheatley DW (eds) (1999) Mapping the future of the past. Manag Spat Dimen Euro Archaeol Resour, Sevilla, pp 103–108
74. Aldenderfer M, Maschner HDG (eds) (1996) Anthropology, space and geographic information systems. Spatial information series. Oxford University Press, New York
75. Kvamme KL (1999) Recent directions and developments in geographical information systems. J Archaeol Res 7(2):153–201
76. Wescott K, Brandon R (eds) (2000) Practical applications of GIS for archaeologists: a predictive modeling kit. Taylor & Francis, London
77. Clevis Q, Tucker GE, Lock G, Lancaster ST, Gasparini N, Desitter A, Bras RL (2006) Geoarchaeological simulation of meandering river deposits and settlement distributions: a three-dimensional approach. Geoarchaeology 21(8):843–874
78. Trifković V (2006) Persons and landscapes: shifting scales of landscape archaeology. In: Lock G, Molyneaux BL (eds) Confronting scale in archaeology. Issues of theory and practice. Springer, New York, pp 217–324
79. Kempf M (2019) The application of GIS and satellite imagery in archaeological land-use reconstruction: a predictive model? J Archaeol Sci Rep 25:116–128. https://doi.org/10.1016/j.jasrep.2019.03.035
80. Ullah IIT (2010) A GIS method for assessing the zone of human-environmental impact around archaeological sites: a test case from the Late Neolithic of Wadi Ziqlâb, Jordan. J Archaeol Sci 38:623–632. https://doi.org/10.1016/j.jas.2010.10.015
81. Monti A (2010) Human space and disadvantage in settlement distribution a GIS analysis on the case of "ronchi" and some new considerations about the approach. In: Beyond the artifact. Digital interpretation of the past. Proceedings of CAA2004. Prato 13–17 April 2004
82. Güimil-Fariña A, Parcero- Oubiña C (2015) "Dotting the joins": a non-reconstructive use of least cost paths to approach ancient roads: the case of the Roman roads in the NW Iberian Peninsula. J Archaeol Sci 54:31–44. https://doi.org/10.1016/j.jas.2014.11.030
83. Herzog I (2017) Reconstructing pre-industrial long distance roads in a hilly region in Germany, based on historical and archaeological data. Stud Digit Herit 1(2):642–660. https://doi.org/10.14434/sdh.v1i2.23283
84. Verhagen P, Nuninger L, Bertoncello F, Castrorao Barba A (2015) Estimating memory of the landscape, In: CAA2015. Keep The revolution going. Proceedings of the 43rd annual conference on computer applications and quantitative methods in archaeology, vol 1
85. Herzog I (2018) Least-cost networks, archaeology in the digital era, (Pampus 1998), pp 237–248. https://doi.org/10.1515/9789048519590-026
86. Fonte J, Parcero-Oubiña C, Costa-García JM (2017) A GIS-based analysis of the rationale behind Roman roads. The case of the so-called via XVII (NW Iberian Peninsula). Mediterr Archaeol Archaeom 17(3):163–189. https://doi.org/10.5281/zenodo.1005562
87. Llobera M (2012) Life on a pixel: challenges in the development of digital methods within an "interpretive" landscape archaeology framework. J Archaeol Method Theory 19(4):495–509
88. Llobera M (2010) Archaeological visualization: towards an archaeological information science (AISc). J Archaeol Sci 18(3):193–223.
89. Fabrega-Alvarez P, Parcero-Oubina C (2019) Now you see me. An assessment of the visual recognition and control of individuals in archaeological landscapes. J Archaeol Sci 104:56–74. https://doi.org/10.1016/j.jas.2019.02.002
90. Graves McEwan D (2009) Predictive modelling and quantitative GIS-based analysis of ritual and settlement landscapes of Neolithic mainland Scotland, c 4000–2500 BC. PhD thesis, Department of Archaeology: University of Edinburgh

Chapter 3
Predictive Modeling

The present is pregnant with the future
and the future can be read in the past...
Leibniz, 1714

3.1 Theoretical Perspective and Model Definition

Traditionally, archaeological research uses models to explain things by integrating archaeological data with other sorts of information, such as geographical, ethno-archaeological, historical, environmental records. As emphasized by Ebert and Kohler [1]: "Explaining things in archaeology is indeed a two-way street, a progression of theory and method." Thus, as "everything starts with the model's concept and there are probably as many definitions of the term model as there are scientific disciplines" [2], it seems important to highlight the Model definition which this Thesis refers and to frame the different positions that have been taken upon archaeological predictive modeling over the years.

If we refer to Clarke [3], models are "hypotheses or sets of hypotheses which simplify complex observations whilst offering a largely accurate predictive framework structuring these observations." Clarke speaks about those observations or aspects of the real world that are selected for inclusion in a model and assumed to be significant by the person constructing that model according to his interests and set of problems. Predictions represent the way in which models are tested and if predictions are found to be successful, the model and the theories upon which it is based tend to be confirmed.

Thus, one of the most common definition for Archaeological Predictive Models (APM) is given by Kohler and Parker [4], who describe APM as "a technique that, at a minimum, tries to predict the location of archaeological sites or materials in a region, based either on a sample of that region or on fundamental notions concerning human behavior." Hence, Graves McEwan [5] reminds us that APM have been primarily

M. E. Castiello, *Computational and Machine Learning Tools for Archaeological Site Modeling*, Springer Theses, https://doi.org/10.1007/978-3-030-88567-0_3

built "to predict where people in the past chose to settle, to hunt, to bury the dead, to create or discard objects in particular locations to the exclusion of others and so on."

If we further analyze the message underlined in Clarke's statement, mentioned above, we can realize that while it indicates that there is no such thing as a truly objective model (all models reflect, to a considerable degree, the subjectivity on the part of the observer), be it inductively or deductively generated, it also alludes to the great debate that has grown over the past years upon APM, leading to strong criticism and division among scholars [6–14]. The origin of such debate has to be traced in the data-driven modeling approach, established initially by the U.S. National Environmental Protection Agency and the State Historic Preservation Office in the late 1970s and early 1980s [11, 15]. After settlement pattern studies were initiated in the 1950s [16], these federal agencies used APM in compliance with land use planning projects with the aim of generating models that estimate the probability of archaeological sites being present anywhere within the study region in a relatively cheap and efficient way. The great interest shown by U.S. archaeology in APM stems from the particular North American landscape, characterized by large tracts of land often covered only by small-scale surveys on which to base an understanding of site distribution [4, 17, 18].

Since the application of APM in North America [10, 19–23], other countries followed very quickly. The Netherlands was the first country in Europe to widely embrace APM within a Cultural Heritage Management (CHM) context [11, 12, 24]. As a result of the *European Convention for the Protection of the Archaeological Heritage* in 1992, the first predictive map at a national scale, the *Indicatieve Kaart van Archeologische Waarden* (IKAW) [25] saw the light. Dutch models are a unique example of APM currently still used for CHM as a decisional tool for an early determination whether a survey appears as necessary or not.

Besides the Netherlands, Germany developed several regional APM projects [26–29], but none represents a mandatory procedure to be followed before the beginning of any urban development procedures. France made a first attempt with the *ArcheoDyn* project [30] developed by the CNRS (Centre National de la Recherche Scientifique) and the Scientific Research Centre of the Slovenian Academy of Science laboratories, but mostly with the aim of evaluating heterogeneity in regional archaeological inventories. In the UK, the commercial archaeology has led to a predominant role of private companies and consultants. The activity of protection and preservation of archaeological sites has been seriously affected by this fragmentation [31]. A similar phenomenon can be observed in Italy, where the protection of archaeological heritage is entrusted to the Regions, which in turn delegate the obligation to elaborate APM to private societies and individuals, resulting in a multiplicity of final products, as well as of methods and techniques applied [32].

Without analyzing each Country individually, it is possible to observe that European archaeologists have utilized APM to a much lesser degree than their U.S. homologues, both in academia and CHM, which is probably due to different CHM traditions and different characteristics of the archaeological records [11]. Another reason could probably be found in the general skepticism of the academic world in this regard

[33–35]. Indeed, archaeological theorists, criticizing their lack of theoretical foundation, largely turned away from such rule-based and 'pragmatic' North American approaches, estimating that they fail in taking into account the cultural and environmental mechanisms [6, 36–38]. As a consequence, a strong dichotomy appeared between predictive models used in CHM and those used by academic researchers, which is equally reflected in a division between deductive and inductive approaches.

Kohler defines a deductive, knowledge-driven modeling approach as based on "a theory as to how people use a landscape" allowing "to deduce from that theory where archaeological materials should be located" [39]. This approach is built on expert judgment and the explanation represents the ultimate goal of such activity [40]. Models developed within a CHM framework on the other hand are generally inductive, data-driven models and have their primary purpose not in explanation, but in prediction. Kincaid [41] gives the following definition of inductive models: "Inductive models proceed from data to theory; observed correlations in the data are used to formulate general hypotheses."

The principal reason for the popularity of inductive approaches in CHM is that the majority of the data required for modeling already exists in the form of site databases and geospatial data. This significantly reduces the costs of any modeling project, making it considerably appealing to agencies funding predictive modeling exercises [42]. Verhagen and Whitley [14] highlighted that "the development of predictive modeling has veered away from mainstream archaeological thought and theory and has now become a largely self-contained activity—enjoying reasonable success as a tool for CHM." On the other hand, these models were criticized to be simply maps showing where sites were likely to be located relative to various environmental variables, without offering any further explanations about past human behavior or any new insight about the human system that created and deposited the cultural materials that form the archaeological sites. In this context, essentially two major criticisms have been raised with respect to predictive models used in CHM: (i) model data are incomplete, biased and unable to quantify or predict human behavior [9], archaeological sites are affected by chronological and functional uncertainty [22, 43] and (ii) APM, especially those used in CHM, are environmentally deterministic (as discussed by Gaffney and van Leusen [6]).

It is worth mentioning here that models, regardless of the specific domain in which they have been realized, establish a broad and inclusive understanding of where sites are known or expected to occur and if any hierarchization can be discerned among sites. Indeed, a major strength of predictive models is their ability to identify sensitivities and risks to which an archaeological site is exposed, for example in the context of large-scale planning purposes or due to natural agents. Such maps showing the distribution of sites and indicating areas with high probabilities to make new archaeological discoveries respond well to the immediate needs of CHM, and thus have always represented a useful support tool in decision making processes by providing an esteem of the archaeological risk [44].

According to Danese et al. [45], archaeological risk can be defined as the product of hazard, vulnerability and exposure with respect to any archaeological evidence, where the hazard represents the probability that an event causes any damage, be it due to anthropic hazard such as urban sprawl, construction works, looting, or to natural risk factors such as landslides, fire, earthquakes, etc. The exposure expresses the value of the element at risk (the archaeological evidence). This is particularly difficult to evaluate in an objective way in the case of archaeological heritage, as the value of heritage arguably reaches far beyond the simple material or market value of an object. Finally, the vulnerability is linked to the elements at risk represented by the archaeological site.

By providing useful information and hence reducing archaeological risk linked both to anthropic and natural factors, APM can decisively contribute to the preservation of archaeological evidence and their valorization.

It is outside the scope of this Thesis to present a comprehensive history or review of the development of the processual/post-processual debate in archaeology. Useful synopses can be found in Parker [46], Judge and Sebastion [19], Kvamme [21], Preucel [47], Whitley [48], Hodder [49] and Westcott and Brandon [50]. Nevertheless, it appears to be useful in the context of this Thesis to address the place of data-driven modeling approaches in archaeological research.

Despite the large variety of different approaches applied in CHM and in academic research, hundreds of articles and books published on this topic strengthen the fact that often a same construction process is to be followed, a same methodology as well as the same final purpose is to be achieved by both types of APM, that is trying to locate the past human activity. Despite the criticism addressed to inductive, data-driven approaches, these are more and more integrated not only in the field of CHM, but also in the academic research. As shown in the following paragraphs, recent and extensive research widely supports inductive data-driven modeling approaches as valid and scientifically significant methodologies.

The next pages are particularly dedicated to stress APM as an integral part of archaeological applications, whether they follow deductive or inductive approaches, and to demonstrate that, supported with an increasing number of advanced methodologies and practical examples, they attempt to fulfill similar purposes. APM aims at providing support in testing hypothesis and answering traditional questions related to past human activities and land-use. By doing so, they place the accent on applicability for preservation and conservation activities, which are the core principles of CHM, but also of vital interest for academic research.

3.2 From Global to Local Scale: Indicative Case Studies and Experiences

In a first instance, this Chapter aims at giving a general theoretical framework for statistical learning and classification and regression modeling. In a second, several APM applications are presented and their strengths and weaknesses addressed. The evolution of simple statistical models towards more complex, non-linear Machine Learning approaches show the progressive integration of data-driven approaches in archaeology in order to cope with methodological limitations and obtain more accurate results. In a third instance, the incorporation of uncertainty in APM is tackled. Finally, a case study of APM in Switzerland is presented.

3.2.1 Theoretical Framework

As discussed in the previous Chapters, predictive models are based on the first assumption that the location of archaeological remains in the landscape is not random, but is related to certain characteristics of the natural environment. Now, these repeating patterns can be identified through statistical methods and models, which can further be applied to un-surveyed areas, in order to identify new locations that may have also been occupied by human activities [7, 51].

Generally speaking, statistical learning aims at classifying a target variable, as a function of a series of predictors:

$$Y = f(X) + \varepsilon \tag{3.1}$$

where Y denotes the target variable (archaeological sites), X denotes the predictors (environmental data) and ε is a random error component, which depends on various factors such as measurement errors and represents the irreducible part of the model [52].

The pioneering work of Kvamme [15, 17, 20] has been the first and most significant with regard to a consistent analytical and procedural study of predictive modeling methods. He significantly advanced the technique's potential by creating several models in Colorado in the early 1980s. A lot of energy has been spent by then for the further development of statistical and spatial techniques (where statistical tests are applied to see if a relationship can be found between a sample of known archaeological sites and a selection of landscape characteristics or 'environmental factors').

The basic methodology of these models makes use of multivariate statistical techniques based on logistic regression [22, 53, 54]. Binary Logistic Regression (BLR) identifies predictive variables that correlate with site presence or absence. Once predictive correlations have been defined, they can be projected onto areas with no existing site presence/absence data in order to indicate areas of archaeological potential [55].

Espa et al. [56] sketch the principles of how logistic regression works in modeling procedures: "The study area is divided into M contiguous squared cells called pixels. Assuming that in N of such cells ($N < M$) the site survey was able to assess the presence or absence of an archaeological site, we can define a variable Y_i such that:

$$Y_i = \begin{cases} 1 & \text{if the cell } i \text{ contains an archaeological site} \\ 0 & \text{if otherwise} \end{cases}$$

$$i = 1, 2, \ldots, N. \tag{3.2}$$

Assuming further that for all M cells a set of auxiliary information about k variables (qualitative and quantitative)

$$X_i = (X_{1i}, \ldots, X_{ki})$$
$$i = 1, 2, \ldots, M \tag{3.3}$$

is available (by direct observation or by interpolation), these variables are used to forecast the probability that in each pixel an archaeological site is located. [...] After collecting data, we can define a model by choosing a subset of the N cells containing n training sites ($n < N$) where the information "absence" or "presence" of site is known. The model chosen must be estimated and cross-validated by contrasting the results with the $N - \mathrm{n}$ observed sites." Thus constructed, the model can be employed to forecast the probability of archaeological site location on the unobserved cells and the visual output will be a probability map of archaeological site location with high, medium and low value zones.

3.2.2 From Spatial Analysis to Machine Learning Applications: Case Studies

The work of Carrer [43] provides an example of locational predictive model based on logistic regression using 10 independent variables (or geo-environmental features: elevation, slope, aspect, cost surface rivers, cost surface lakes, geology, avalanche risk areas, morphometry, topographic index, profile curvature) and a sample data of 83 modern alpine pastoral sites (defined *malghe*), collected in Trentino province (eastern Italian Alps). Likewise, a dataset of absences, 83 non-sites, was randomly created to balance the model. The predictive model was tested on two different areas using two archaeological datasets containing moreover different types of upland settlements (dry-stone huts, rock shelters and dry-stone enclosures). The validation through Kvamme's Gain [22] obtained 0.59 Gain value, showing that the model successfully predicted the location of the modern *malghe*, but, as the author argues himself, failed in predicting the location of other types of archaeological seasonal pastoral and upland sites.

Next to the presence of different distribution patterns mentioned in the study, the issues encountered by the authors can perhaps be ascribed to one of the main problems related to the logistic regression approach: the spatial interdependence of the dependent variable [56]. According to Carrer [43], ethnoarchaeological interpretation is furthermore necessary in order to achieve a better understanding of the distribution of the other types of sites. This approach represents an attempt to define a suitable methodology for discovering new archaeological sites in mountainous regions and at the same time to combine the preservation and conservation purpose with the interest in developing a theoretical background for discerning patterns in the behavioral reasoning underlying pastoral settlement strategies.

On the other hand, the APM developed by Nicu et al. [57] stresses the benefits that such instruments could bring to CHM policies if they were comprehensively integrated. Their APM for the northeastern part of Romania addresses specific issues of prevention and conservation of Eneolithic sites threatened by modern development. The methodology applied relies on Frequency Ratio,[1] already well known in geosciences for erosion susceptibility mapping, flood and landslide mapping. The model used a set of only three environmental factors: soil type, heat load index and slope position classification, along with a set of 100 archaeological sites, to demonstrate that Eneolithic site locations are strongly dependent and influenced by the soil type as conditioning factor, immediately followed by the heat load index and the slope position classification factors in the ranking of variable importance.

The good model accuracy, attested by a Receiver Operating Curve (ROC) with an Area Under the Curve (AUC) showing a value of 0.72, made it an extremely useful supporting tool for CHM in its efforts to avoid destructions in conjunction with future infrastructure works intersecting with some of the areas falling within the high and very high probability classes of the predictive map obtained by Nicu et al. [57].

Within a similar CHM framework, Caraucasi et al. [59] have proposed an APM for the study of the distribution of high altitude prehistoric settlements in the Piedmont region (western Italian Alps). This model, unlike Carrer's and Nicu's, made use of 'weights' (and of reclassification) in the definition of different parameters or variables (slope, aspect, water resources). Looking across the GIS and APM literature, it is easy to find numerous examples and applications of reclassification, overlaying, weighting, and summation of map themes, often referred to as 'map algebra' [18, 60–62].

In general, the procedure followed here is the Bayesian Weight of Evidence—WofE[2] [63, 64], where odds are used to determine the likelihood that a site will be found in a particular region given a set of spatial variables. As Canning [40] pointed out, WofE techniques substantially make use of the combination of various forms of evidence to support a hypothesis or hypotheses. These forms of evidence can be binary (i.e. presence or absence of sites) or might integrate other non-binary variables. These can be difficult to introduce into models because of their non-binary values (e.g. distance to water).

---

[1] Frequency Ratio (FR) is used to calculate the ratio of cells with archaeological occurrences in each class for a reclassified factor or categorical factor (i.e., geology and land cover), and the ratio is assigned to each factor class again. The FR is the ratio of archaeological occurrences in a desired class as a percentage of all archaeological sites to the area of the class as a percentage of the entire map [58].

[2] The WofE model in particular uses statistical associations between known sites called training points and different map themes (such as geology, aspect and slope) in order to calculate a set of weights. These weights are defined on the expert knowledge-basis and used to evaluate every possible combination of the different map layers in order to produce a single map (a unique conditions grid) showing probability of the presence of a site.

This kind of methodology can be applied to different landscapes and for modeling archaeological sites belonging to more recent epochs. Kay and Witcher [65] for example, attempted to model the distribution of early imperial Roman villas and the agricultural productivity of the middle Tiber valley (Italy), while [66] applied this technique in her Ph.D. studies, presenting a simplified GIS framework for the development of an automated APM of the Roman rural villas in Portugal (for other uses of WofE in APM see for example: [30, 67–69]).

However, a significative issue of this weighted map-layers approach lies in the selection of weights. The subjectivity in the choice of odds may alter the final result of the model when based exclusively on expert judgment. "By simply changing the size or ordering of weights, it is possible to achieve profoundly different results" [62]. Moreover, the spatial correlation problem between the environmental factors is inevitably reflected in the final predictive map, where the areas of very low probability sit next to areas of high probability [62].

The APMs presented above [43, 57, 59, 66, 65], as many others present in the literature (to cite only a few: [27, 32, 44, 70–72]), show a clear significance of the environmental 'themes' used for APM procedure in terms of both past human behavior (e.g. site location preference) and archaeological recovery. The consolidated use of modern environmental themes can be furthermore generalized to different landscapes, be it an alluvial valley or a mountainous area (geology, topography and slope, proximity to water resources, etc.). Furthermore, if modeling is treated as an iterative process, with new results progressively added in, models can be constantly refined to attain a better performance [66].

An increased attention in quantitative and mathematical methods results in an extensive literature and numerous case studies, which illustrate how archaeological research is progressively taking advantage of the rapid increase in computation capabilities and improvements of methods, and finally realizing the great potential hold by Machine Learning (ML) and Artificial intelligence (AI) techniques.

Numerous studies performed in this domain support the technological predominance and the better accuracy of ML algorithms compared to classical statistical techniques such as discriminant analysis or logistic regression. This appears to be especially the case when modeling complex spaces or areas (i.e. when the study area is expected to present multi-dimensional features and the relationship between the dependent variable (i.e. archaeological sites) and the independent input variables (i.e. environmental features) is expected to be non-linear). ML techniques have also performed particularly well when the input datasets are expected to show different statistical distributions [73–77]. Thus, ML algorithms have a great potential to identify and model complex and non-linear relationships between archaeological occurrences and environmental variables [78].

Espa et al. [56] developed one of the first examples of APM using a classification algorithm (Classification and Regression Tree CART) as described by Breiman [79], focused on archaeological sites of the Roman period in the Cures Sabini area (Tiber Valley—Italy). The model is presented as a first "good" alternative to the classic logistic regression, because (i) it is not affected by outliers; (ii) it is easily converted into an automatic tool; (iii) it can be used for modeling different type of settlements; (iv) it provides graphical outputs easily interpretable by non-statisticians; (v) it does not conflict with spatial models but is useful to select predictors and to detect significant interactions.

However, at the current state of knowledge, the methodology presents some critical points that alter the effectiveness of this APM: (i) there is no clear distinction between "no data" and "negative occurrences" (no site) used in the training and testing phase; (ii) the edge-borders effect, e.g. the probability to find archaeological site is homogeneous in the areas of wide extension. Such absence of trends can be entirely ascribed to CART, which recursively partition all data in rectangular regions; (iii) high probability areas on the final predictive map are concentrated in zones more intensively sampled, whereas they are not in the parts of territory far from the sample points; (iv) the variable that represents the proximity to river water curiously retains very low importance in the final prediction.

Finally, if the example of Espa et al. [56] provided a first attempt of APM using a classification algorithm, the results obtained demonstrate that the methodology of that time was still in need of major refinements.

A step forward in this direction may have been made by Märker and Bolus [80], who carried out an explorative spatial analysis and created an APM of Neanderthal fossil sites across Europe (resolution of the analysis: 250×250 m cell size).

Once again, the importance of modern environmental factors in determining site location preferences is confirmed. The outstanding performance of ML algorithm—Boosted Regression Tree (BRT) [81, 82] compared to other methods (MaxEnt, [83]) is demonstrated as well.

Elith et al. [84] describe BRT as follow: "Boosted Regression Trees combines the strengths of two algorithms: regression trees (models that relate a response to their predictors by recursive binary splits) and boosting (an adaptive method for combining many simple models to give improved predictive performance). The final BRT model can be understood as an additive regression model in which individual terms are simple trees, fitted in a forward, stagewise fashion."

Although the archaeological dataset used in this procedure is very skimpy in relation to the large-scale analysis (250 m $\times$ 250 m cell size) encompassing 219 sites from 29 different countries, the selected methodology demonstrates to perform well even on a reduced archaeological sample. The environmental variables originally consisting in 61 climate related indices (aspect, hillshade, insolation, etc.), water related indices (e.g. distance to the major water courses) and strategic indices (visibility analysis and terrain ruggedness index) were manually reduced to 18 based on their importance for the prediction procedure, as suggested by the classic procedure for Decision Trees (DT) algorithms.

In general terms, it as argued that BTR and DT suffer from overfitting and do not generalize well, while the variable selection procedure is not random but based on the best results obtained in trees construction ("only the first tree is estimated for the training data; all successive trees are grown on the residuals of the preceding tree" [80]). Thus, the variables considered to be more influential in the predictive procedures have to be chosen and selected in advance.

Such issues have drew the interest of archaeologists toward ensemble methods and deep learning, like Random Forest (RF), Convolutional Neural Networks (CNN), Support Vector Machine (SVM), etc. [85]. The application of ML techniques has evolved considerably since statistical approaches were employed for the very first time.

Märker and Heydari-Guran [86] developed probably one of the first examples of APM using Random Forest (RF) algorithm [79] for the creation of a distribution map of Paleolithic sites in the Zagros Mountains (Iran). A great attention was reserved to the selection of geo-environmental variables (15 topographic indices) and to the dataset containing all archaeological information related to presences and absences (site occurrence).

The authors see several advantages in the RF method applied, namely: "(i) that predictors [variables influencing the site location] are automatically selected, (ii) that the method is more accurate than a single tree approach, (iii) that the data do not have to be rescaled, (iv) that it is resistant to overtraining and (v) that it provides an internal cross validation using 'out of bag' (OBB) data [79, 87]."

The model indeed reached an accuracy of 99% in predicting the location of new sites (according to the Receiver Operating Curve (ROC)), showing that Paleolithic sites were strongly related to terrain characteristics and processes and could be assessed by topographic indices.

ML applications in APM stretch from the wider, large-scale example of Neanderthal sites in Europe, described before, to a very local and confined application of RF in Tasmania, Australia. With the aim of disclosing the bio-geographical patterns behind the distribution and locational preferences of Aboriginal archaeological sites, Jones et al. [88] developed a RF habitat suitability model focused on the island of Tasmania.

Though habitat suitability models were originally designed for Ecology studies [89, 90], the authors showed that such models, by taking advantage of ML techniques, perform well in a wide range of applications, among which archaeological predictive modeling. For their model, 31 environmental variables were initially chosen (e.g. climate, topography, proximity to inland water, proximity to certain vegetation species, proximity to roads) to characterize the landscape that people have likely occupied in the past. The total number of variables was then reduced to 13 following a standard procedure designed for variable selection in RF in order to avoid the inclusion of useless variables and highly correlated groups of predictors that could affect the model output [91].

A set of isolated artifacts, artifact scatters and mixed site types with artifacts constitutes the archaeological dataset used along with an equal number of pseudo absences randomly generated. Although the authors particularly point at the limited quality of the archaeological data and the bias that may have derived by the survey methods, the RF model reached 92% of accuracy. The final suitability map shows which types of landscape had the greatest and smallest concentration of archaeological sites. The landscape was characterized primarily based on the use of natural resources by Tasmanian Aboriginal people.

To date, the globally shared vision in Australia and Tasmania is that Aboriginal people did not make use of rainy forests, woodlands and grassland areas. The output of this model provides relevant evidence to support an inverse hypothesis. According to the map, the highest probability of human activities is to be found in the Midlands around several major river valleys and along the coast, while the lowest probabilities are located along mountain ridges, high alpine plateaus, etc.

From a research point of view, this model represents an analytical support at a wide landscape-scale for reinterpretation of previous historical theories. At the same time, it can serve a more pragmatic purpose related to CHM, such as the protection of archaeological sites against lightning-ignite fires in the framework of a more effective fire management regime [88].

The outstanding results obtained with RF in many disciplines (for RF applications in biology see for example: Mi et al. [92], for applications in geosciences see [93, 94]) are driving the future trends in archaeological site detection techniques at an undiminished rate.

Despite such undeniable advancements, the use of RF in APM is still in its infancy. The work of Rodriguez-Galiano et al. [78] and Oonk and Spijker [95] has confirmed the great added value of using a mixed methodology in obtaining good predictive maps by incorporating geological, geochemical and geophysical data and testing different ML algorithms at once (weighted k-nearest neighbors analysis (kNN), support vector machines (SVM) and artificial neural networks (ANN)).

As a consequence of the good results obtained with ML algorithms, combined with the growing availability and accessibility of large quantities of satellite and aerial imagery, researchers have recently started to combine ML and remote sensing techniques, using Archaeological Site Modeling to accelerate surveys and protect site locations. Chen et al. [96] gives a glimpse of the potential of RF combined with remote sensing to create an 'enhanced' APM capable of automating the pixel classification process in thousands of satellite images from Airborne Laser Scanner and tagging evidences of archaeological artifacts in Ft. Irwin, California.

3.2.3 Delivering Uncertainty with Archaeological Predictive Models

From the first conceptualized Archaeological Predictive Models (APM), developed as allocation—location analysis, where the aim was to allocate suitable locations to specific types of human activity and their archaeological remains [37], the research tends to focus today on more sophisticated models based on statistical algorithms that at the best try to automate the process of human patterns recognition and to maximize their applicability and efficiency on complex and incomplete datasets by further taking into account the delivering of uncertainty information until the very final output.

The delivering of uncertainty is an issue widely addressed by APM. Catering uncertainty related to the information contained in the data used is not an easy task, but the ability to manage it in APM procedures can further strengthen the model accuracy. The uncertainty or the bias encompassed in the archaeological datasets often trace back to the survey methods and information digitalization methods. Over the past years, several attempts were made to address these kinds of issues. The Archäoprognose Brandenburg project [26] and the related APM represent one of the steps made in this direction. Born from a joint collaboration between the Brandenburg State Authority for Heritage Management and the University of Bamberg, this APM is a good example of how archaeological research and CHM can efficiently work together to meet common purposes, addressing the protection of archaeological monuments in planning procedures and generating new information to gain useful insights into large-scale processes of past settlement strategies of Neolithic and Bronze Age populations of a German region.

As said, a key concept of this APM is the management of uncertainty as introduced by missing data, incomplete datasets, errors, and the diverse sources of information. Moreover, in that particular region, the problem related to the accelerated soil erosion provoked by human land use has been identified as another important source of uncertainty in the study of archaeological sites and their distribution over the landscape. The procedure selected for addressing these issues mainly relies on Dempster Shafer Theory of Evidence (DST), which is also the basis of WofE.

DST, as defined by Dempster [97] and Shafer [98], is a generalization of the Bayesian Theory [99, 100] under consideration of all possible outcomes and is built around the central concept of 'belief' (or confidence, truth) [28, 29, 34, 101, 102]. 'Belief' refers to the fact that the researcher can directly make statements about the uncertainty of the used data and explicitly incorporate such subjective uncertainty in the modeling procedure [103]. Ducke [26] describes DTS as "a flexible mathematical framework that allows pooling of data from a variety of sources in a natural, straight-forward manner, explicitly representing uncertainty and producing a range of interesting output metrics that can be used in decision making processes."

In general terms, unlike Bayesian Theory, DST assigns an explicit set of values to the data e.g. by weighting the data or variables based on expert judgment or subjective knowledge. Thus, these values do not have to be mathematical probabilities. The uncertainty in the model is then calculated applying the *belief interval* between the *plausibility* (the probability that all the evidences would be true) and the *belief* (the degree of confidence in the hypothesis: 'site' or 'no site'). The resulted *weight of conflict* map shows the places in the map where there is contrast between the evidences. Thus, "a high weight of conflict might indicate a serious flaw in the model design or disagreement of evidences supplied by different data sources" [26].

Finally, this APM may represent a concrete way to handle the uncertainty issue by (subjectively) providing a quantitative degree of 'belief' for those variables included in the model (rainfall intensity, soil erosion, soil types, etc.) or by assigning a Basic Probability Number (BPN) to them [26].

3.2.4 A Swiss Case Study

Finally, the work carried out by Ebersbach [104] provides one of the few examples of APM pursuing CHM objectives in Switzerland. The study entitled "Eine Potenzialkarte für den Kanton Bern" aimed at creating a map of the archaeological potential in the Canton of Bern with a focus on Roman villas. The choice of the type of site to be mapped was motivated by the ease of detecting the structures in the modern landscape and their better representativeness and recording compared to other archaeological sites stored in the Cantonal inventory (in total 103 villas were considered).

The environmental variables considered in this study consisted mainly in the altitude, the slope, the distance to the nearest water sources and the proximity between sites. The altitude and the slope were derived from the digital elevation terrain model, while the distance to main rivers and lakes was obtained by including the modern water network and by digitizing the historical water network from the Dufour map.[3] The proximity to the other sites was calculated by assigning a buffer area around each villa. For this specific areal calculation, the author referred directly to a comprehensive work conducted by Schucany [105] on the spatial extension of Roman villas. Schucany [105] considers the constructed areas of a villa to occupy between 1 and 5 ha and the *fundus* (the cultivated surrounding and the lands left for grazing) to reach up 1000–8000 ha for the largest domains. According to this source, a 0.01 km^2 (1 ha) buffer area was assigned to each villa to represent the constructed part (*pars urbana and pars rustica* annexes) and an additional 1 km^2 (100 ha) buffer was computed for the *fundus*. However, different values were assigned to the villas classified as "possible" in the Cantonal database: 5000 m^2 (0.5 ha) for the structures and 0.5 km^2 (50 ha) for the fundus.

Starting from an overlay of the sites and the environmental factors, a statistical analysis of the distribution of the villas over altitude, slope and terrain orientation was performed, resulting in the following conclusions: most of the villas lie at relatively low altitude (within 550 m a.s.l.) with irrelevant slope values (between 3° and 5°). No site was found above 900 m a.s.l. and on a slope above 20°. The sites seem to be located at a distance of 2 km from each other and within 250 m from the nearest water source. The water is intended here as source for cultivation, supply for the livestock and for the structures proper to the *pars urbana*.

The model procedure chosen relies essentially on a WofE methodology, which, as seen previously in this Chapter, is based on the addition of subjectively assigned weights for each of the variable considered as an influencing factor in the location of Roman villas (30% each to the slope and the proximity to sites; 20% to the proximity to water; 10% each to the orientation and altitude) and a subjective reclassification of the value ranges for each different environmental variable.

According to the results obtained from different weighted models, around 1800 pixels with a surface of 1 ha each are assigned to the highest probability class to host

[3] https://www.swisstopo.admin.ch/en/knowledge-facts/histcoll/historical-maps/dufour-map.html.

a Roman villa. As a villa is considered to occupy 100 ha, 18 new villas might be waiting to be rediscovered in the Canton of Bern.

With the aim to test the accuracy and precision of the model defined for the Roman villas, a two-folds cross-validation was performed: (i) once with the entire dataset embedding all the pre-medieval settlements and graves (including the Roman villas used to train the model) and (ii) once with a dataset containing villages firstly mentioned as own parishes dating to the twelfth and the sixteenth centuries. Assuming that a parish was only assigned to a settlement that knew a certain continuity over time, the parish villages were considered to be located at the same place of Roman villas. By doing so, the list of parish villages was used to evaluate the presence of Roman villas over the landscape.

An additional potential map was produced for Neolithic settlements (36 sites) by using only the altitude, the slope and the distance to water sources. Finally, a third map drawing a general archaeological potential of the region without differentiation between epochs was created by using the altitude, the slope, the proximity to water sources and an additional map representing potentially marshy areas derived from the slope and including all areas with a slope of less than 1°. In this case, the altitude and the slope were weighted 5 times more than the distance to water sources, and the potentially marshy areas 2.5 times more than distance to water sources.

The exploratory spatial-statistical analysis performed in Ebersbach [104] provides a useful framework for evaluating locational criteria for Roman villas and Neolithic sites. The study furthermore offers an interesting comparison between the potential map for Roman villas and the location of villages known as parishes from historical sources. The models created here however might suffer from a lack of validation on an independent testing dataset kept aside during the modeling procedure. The Roman villas used to train the model were included in the dataset used for the cross-validation, which may cause an overvaluation of the model performance. Also, the assumption of continuity between Roman villas and parish villages introduces additional uncertainty in the model validation. The subjective weight assignment may lead to the overrepresentation of some variables in the final results (altitude and slope are weighted 5 times more than proximity to water). Nevertheless, this study provides a significant first analysis of site distribution and an innovative approach to APM in the Canton of Bern, which can inspire further developments of APM in Switzerland, considering further environmental factors, e.g. agriculture suitability.

3.3 Uncertainty and Vagueness in Archaeological Predictive Modelling

There are known knowns.

These are things we know that we know.

There are known unknowns.

That is to say, there are things that we know we don't know.

But there are also unknown unknowns.
There are things we don't know we don't know.
Donald Rumsfeld: February 12, 2002

The issue of delivering uncertain information in Archaeological Predictive Modeling was only briefly addressed in the previous Chapter while exploring relevant worldwide case studies. In this Chapter instead, different notions of uncertainty that modelers in archaeology cope with are disentangled. It is furthermore pointed out that dealing with uncertainty and incorporating it in exploratory predictive models is central to a robust analysis, while sketching how the approach selected in this Thesis may lead to clear modeling improvements.

Technically, when working with models, simulations and methodological experiments, it is important to address and incorporate uncertainty, bias and model validation issues, especially when presenting APM as exploratory data tool. The integration of uncertainty can moreover allow for questioning how models can be compared and provide more information about the data and the model itself [106]. Unfortunately, this is a difficult task to accomplish for archaeologists, as the reality studied in this domain consists often not only of human artifacts but also of a myriad of natural and anthropogenic processes that may have altered initial patterns in terms of composition and distribution of observable phenomena. On the other hand, the archaeological reality deals with theoretical conceptions and assumptions about the exactly definition of an archaeological evidence [107–113].

Concretely, when elaborating a database that will further be used to construct an APM, archaeologists use maps (sometimes in the traditional paper format, which needs to be digitized), photos (that need to be adjusted with photogrammetry programs) and drawings (both hand and computer made) and by doing so perform some kind of simplification, that in some cases may also result in an over-simplification of the past reality, as pointed out by Niccolucci [114].

Already at the source of the data, it is a rather common issue in geospatial databases to have to deal with imprecise archaeological site borders [115, 116]. Even if in most cases it is possible to determine whether a point is located inside a particular site or not, the site border itself cannot be represented by a sharp line. Likewise, when typologically classifying an archaeological object, several aspects may affect the accuracy of the archiving procedure. Indeed, if we look at the excavation activity, the uncovered objects or structures are often only the remains of what were probably the foundations of ancient buildings. This scarce information needs to be completed by analyzing sparse material, using ethnographic parallels, textual descriptions or by comparing with better preserved sites (if they exist). The rest of the interpretation has to be completed by the researcher based on his/her knowledge or, ultimately left incomplete [114].

With regard to the case studies considered in this Thesis and in general, the creation of an excavation database, the interpretation of the analytical results are always prone to an unknown degree of uncertainty. Similarly, many uncertainties surround the interpretive possibilities of an archaeological record as well as the subsequent contextualization and modeling of human behavior. These issues linked to uncertainty have been largely acknowledged within the field and have led to many discussions (see [117, 118]).

Recently, Martin-Rodilla et al. [119], Martin-Rodilla and Gonzalez Perez [120] determined how, in digital humanities as well as in archaeology, uncertainty may derive from the degree of ignorance of certain aspects of the information by the researchers (which is properly defined as epistemic vagueness) or how on the other hand it may be due to the nature of the entities studied, which are intrinsically vague and unclear (defined as ontological vagueness [119, 121]). As in archaeology the events studied are often difficult to assess because they happened in the past, we refer more often to epistemic vagueness [121]. Epistemic vagueness occurs when we deal with the problem of the definition of "what exactly is a settlement" or with the difficulty to ascribe an object to an epoch.

In this research, two case studies present this kind of epistemic vagueness in a tangible manner: the Cantons of Aargau and Geneva, where the uncertainty is mainly expressed through linguistic statements, in the sense that archaeologists qualitatively evaluated and inventoried the archaeological objects by means of a degree of subjective reliability: "sure", "unsure", "unknown", "undefined", "possible", "potential." The difficulty to reconstruct the information related to a discovery with precision, and to give a defined interpretation of the evidence, causes several issues for APM development. Moreover, the predictive maps accumulate a lack of standard both in the quality of information both in the final result, by limiting any critical analysis if uncertainty and vagueness are not defined and considered [121].

Unfortunately, the uncertainty and vagueness may not only be held by the archaeological datasets as input data, but also in different stages of building the model. Evans [118] frames that the uncertainty can have the following sources, occurring throughout the life cycle of any model: uncertainty associated with input data, model choice, model parameters, and finally model outputs. Input data can introduce uncertainty by various means, such as errors in data measurement, overlooking data, or through the choice of inappropriate sample sizes or inappropriate discretization measures. Choices in model settings might also introduce uncertainty in different ways, in particular by the selection of variables, scale, parameters, and algorithms or mathematical transformations [122].

From its starting point, an APM can accumulate an unknown degree of uncertainty and reliability. Such risk has casted this Thesis towards tackling uncertainty and vagueness as part of the process of Archaeological Site Modeling. The advantage of incorporating, not only textually but also numerically (calculating an index of reliability, for example one that goes from "totally sure"—this is what we believe—to pure uncertainty—that is "we have no proof whatsoever") and visually, such qualitative adjective/predication into the final model is further emphasized [121]. Finally, looking at the literature, the issue of data uncertainty is hardly discussed in archaeological predictive models based on Machine Learning techniques, but according to [121], "revealing all doubts as well as the general model limitations to the reader can be seen as a matter of scientific ethics, at least as important as compiling a convincing story."

In this Study, the problem of uncertainty is tackled by statistical means. Basically, the procedure established follows the steps previously made by Martin-Rodilla et al. [119] and Gonzalez-Perez [121], which proposed a solution to deal with uncertainty and vagueness in archaeology, through the use of a conceptual model (ConML)[4] and a fuzzy logic approach.

To simplify, fuzzy logic (a branch of mathematics evolved out of "fuzzy set" theory) is a technique that allows considering uncertainty by ranking the "truth" or accuracy of the modeled data by degree or percentage rather than seeing it as a binary (true/false) information [123–126]. Zadeh [123] introduced the so-called fuzzy sets, characterized by a function that may vary between 0 and 1, not only assuming the two extreme values as for ordinary sets (crisp sets) [122]. This technique has been applied in many research fields [127–130] as well as in archaeology [12, 131–133].

As Machine Learning algorithms like RF are based on a binary logic, they can offer a suitable solution if combined with a fuzzy logic approach. Binary logic, in the case of archaeological sites, theoretically considers a location either to be a site or not, even though in reality a location has a potential to be a site to some degree and not to be a site to another degree. Only by excavating one can be 'sure' of a site's identity, and even then, uncertainty may persist [134].

Thus, accordingly to the method suggested by Martin-Rodilla et al. [119] and Gonzalez-Perez [121], uncertainty contained in the archaeological databases of Aargau and Geneva was incorporated in the model procedure from the beginning of the work and transmitted to the final output. A fuzzy truth function f was defined for the single finds database of Aargau, varying between 0 (absence) and 3 (single find-sure) and assuming intermediate values (unknown, unsure) and a similar function computed for the database of Geneva, varying between 0 (absence) and 2 (sure), 1 being probable or potential settlements.

Furthermore, as Hatzinikolaou et al. [133] pointed out, "by using fuzzy logic, this multiple ability of an area to belong in more than one classes can be properly treated. This gives archaeological prediction a new potential, as by defining different types of sites this method can contribute more efficiently in cultural resource management and in an hierarchically driven archaeological survey as well." Consequently, fuzzy logic can be considered as a form of probabilistic reasoning that can lend a great deal of freedom of choice to a decision-making model, even more if it is combined with binary classification algorithms [135]. In the particular case of RF regression, an algorithm capable of handling discrete values, the modeling procedure has been extended to compute the numerical confidence values assigned to the presences in the Cantons of Aargau and Geneva according to the procedure detailed in Chap. 6.

---

[4] Incipit ConML Technical Specification. ConML 1.4.4.2015 http://www.conml.org/Resources_TechSpech.aspx.

3.4 Data Mining and Machine Learning Techniques

There is an extensive literature for data mining and machine learning applications in several research fields, like bio-computing, business and finance, socio-economic sciences and more. Recently, they gained a great popularity in geosciences and environmental sciences, which have successfully applied such techniques for geological prospection, geological and mineral mapping [76, 80, 95, 136, 137], and natural hazard susceptibility mapping [93, 94, 138–145].

The environmental research domain deals with problems that are similar to archaeological ones in various ways. In geoscientific modeling, a lot of work is concerned with the prediction and assessment of resource potential (*what? where? how much?*) or the effects of geophysical dynamics on landforms (*what happens under changing conditions?*). Such questions are very similar to those rising in archaeology and archaeological predictive modeling (APM), where the analysis refer to a different subject: the human activity and human impact, but still try to answer the same questions: what? where? how much? what happens under changing conditions? As stated by Earley-Spadoni and Harrower [146], this similarity is moreover corroborated by a wide diversity of advanced methods and practices applied in archaeological research, framing it as a science with a well-defined scientific perspective.

Just like in environmental sciences, in the recent years alone, the use of cutting edge techniques had a surge in archaeological research. According to Davis [147], the explosion of machine learning applications within the environmental domain in the past several years has also effected archaeological research around the world, allowing for comprehensive studies on settlement and mobility, effects of environmental change on human societies, ecological effects of anthropogenic land use and other significant topics. Such machine learning-based research has demonstrated to be extremely valuable for archaeological site detection (to cite a few: [148–154] and can make all the difference in areas where cultural heritage is increasingly at risk.

Moreover, due to the large amount of complex data that became available and freely accessible, several studies have been conducted to develop and incorporate more robust and efficient methods for data analysis, modeling and visualization [155]. In this regard, assessing the relationships between settlement choices and landscape elements, whether they are natural (rivers and lakes) or anthropogenic (roads and towns), on a large scale requires an extensive amount of data processing and computational power. More specifically, methods explored to this end come from a combination of computational disciplines, including Statistics, Mathematics, Artificial Intelligence and Information Technology [145].

Data Mining serves this purpose well as a computational process often applied to the analysis of large datasets, to the discovery of patterns or to predict outcomes of future or unknown events (to cite a few: [85, 156–162]). In the same way, Machine learning can provide a wide array of scalable and reliable methods, which can be automatic or semiautomatic, as powerful mechanisms for collecting more complete and systematic information about the archaeological records, allowing for more comprehensive and reproducible research analysis and results [163, 164].

It is possible to distinguish between two types of classification: Supervised and unsupervised. Supervised classification is based on the selection of some representative samples for each input variables, called "training data." It requires pre-existing labels to be applied to the training outcomes. The computer uses the numerical information contained in all pixels of the training dataset to create an inferred function that can be applied to recognize similar information in other areas and to label (or classify) the other pixels of the study area. This method gained particular popularity especially for image recognition [157, 165–168].

Unsupervised classification is a method that aims at finding the structure and regularity of an unlabeled dataset for the purpose of extracting useful representations [169–171]. It tests a large number of unknown pixels and groups them into a number of classes based on natural groupings in the image values, these groupings are called "clusters".

In a broad sense, it is possible to determine two major paradigms of Machine Learning (ML) algorithms: prediction and knowledge discovery. Within these, there are four sub-categories: (1) Classification and regression, (2) Clustering, (3) Association Rule Mining, (4) Outlier/Anomaly Detection [161]. For example, TreeNet (boosting; [172]), Random Forest (bagging; [79]), CART [173] and Maxent [83] are considered to be among the most powerful ML algorithms both for prediction usage [84, 174, 175] and for obtaining powerful ensemble models [92, 176, 177].

Fundamentally, the modeling procedure defined in this Study can be described as a supervised ML technique. The Random Forest algorithm set up in this Study belongs to the first group of classification and regression, where the objective is to induce a predictive model, capable of classifying samples as belonging to a particular category or label such as site—no site.

As seen, many recent studies, carried out in several research fields, suggest that ML techniques, and RF in particular, perform better than traditional and individual models [84, 178], which is fundamentally the reason for choosing this specific algorithm within this research.

3.5 Random Forest: Classification and Regression Trees

It is essential here to provide more in depth knowledge about the functioning of Random Forest (RF) and its strengths. RF is an ensemble learning method based on decision trees, capable of learning from data and making predictions based on the acquired knowledge through the modeling of hidden relationships between a set of input variables (i.e. geo-environmental features prone to influence the site location) and output variables (i.e. the archaeological settlements/single finds) [79, 87, 173, 179]. Thus, to understand Random Forest, it is necessary to prior know what a single decision tree is.

A decision tree is a decision support system that breaks down a dataset into smaller subsets. The tree is composed of internal nodes and leaf nodes. A decision node has two or more branches and represents a test on an attribute (e.g. elevation below or above a certain value). Each branch of the tree represents the outcome of the previous test (separating between two main groups by maximizing the difference among them in terms of presences and absences). Leaf nodes represent a class label or decision, after computing all attributes (if a site is present or absent at pixel level). The topmost decision node in a tree, which corresponds to the best predictor, is called root node. This path from root to leaf represents the rules of the model [180–183].

Decision tree based models are in general very good at handling tabular data with numerical features, or categorical features with fewer than hundreds of categories [181, 184, 185]. As highlighted by Bellinger et al. [161], decision tree is often preferred over the more sophisticated models because humans can easily understand the decisions leading to the predictions. The decision trees are simpler to understand and interpret than more complex algorithms. There is no need to select training data with a unimodal distribution. Furthermore, decision trees can handle both numerical and categorical variables, a decisive advantage when considering various forms of environmental variables, and the classification is fast once the rules are developed. On the other hand, two major limitations can be attributed to a decision tree. It is prone to overfitting and tends to be non-robust, meaning that a small change in the training data can result in a very different tree.

As comprehensively outlined by Lotfian [157], Random Forest overcomes these two drawbacks of a decision tree by generating multiple decision trees trained on different parts of the training set and by averaging and aggregating the predictions of each individual tree. Each tree creates a classification and records a vote for the resulting class. The class with most votes among all the trees in the forest is finally chosen and a single prediction produced. Each tree in the forest is grown according to the following steps [157, 179, 186].

Assuming that the number of classes found in the training data set is N, it randomly samples N classes, but with replacement from the original data (bootstrapping) rather than collecting values on a continuous surface. This random sample will be used as training set for growing the tree.

If there are M input variables, a number $m \ll M$ is specified, so that at each node, m variables are selected randomly out of M and the best split on these m variables is used to split the node. The value of m is held constant during the entire process of forest growing.

Each tree is grown to the largest possible extent. There is no pruning.

It is therefore possible to define two major steps while creating the tree in the forest. The first step is a random selection of the training data (2/3 of data) to build each tree, whereby for each tree a different subset of training data is being used to develop the decision tree model. The remaining 1/3 of data will be used in a second step to test the accuracy of the model and is called "out-of-bag."

Theoretically, RF doesn't need to be cross-validated using a separate testing dataset, since the classification error is estimated internally in the run. As suggested by the literature (in particular in [157]) however, an external cross-validation using a separate testing set allows for an additional accuracy measurement.

Nevertheless, a number of important parameters must be specified when using Random Forest [160]:

- The Input training data including predictor variables and response variables.
- The number of trees which must be developed.
- The number of predictor variables for creating the binary rules for each split.
- Parameters for calculation of error information and variable significance.

Finally, two key features differentiate RF from other ML methods. The first is the novel measure of variable importance, which is not affected by some of the shortcomings of traditional variable selection methods, such as selecting only one or two variables among a group of equally good but highly correlated predictors.

The second key feature distinguishing RF from other methods is the array of analyses that can be carried out, such as cluster analysis and multidimensional scaling, and their graphical representations [187].

Ultimately, RF gives the possibility to visualize the influence of each class of a predictor variable through Partial Dependence Plots. These plots, computed for one or two predictor variables at a time, may be constructed for any classifier model [181].

3.6 Towards New Perspectives in Archaeological Practices

Since its origin as a descriptive, documentary discipline, archaeology has evolved in recent years into a science that attempts to understand certain aspects of human behavior by explicitly taking into account independent events and environmental variables. This effort to understand human behavior has orientated archaeology in a new direction and given it a new sense of purpose, pushing archaeology towards Natural Sciences. By introducing recent and concrete applications, this Chapter aimed at providing an evolutionary framework of the techniques and methods used in Archaeological Predictive Modeling (APM), both in Cultural Heritage Management (CHM) and in academic research. It attempted to stress the greatest and most impressive advancements made in the discipline thanks to the use of ML algorithms in recent years. Moreover, it was highlighted how all models, regardless of the methods used, can be constructed at almost any scale. APM can be constructed on study areas ranging from a few hectares up to very large, Europe-wide areas. All models presented here share a number of common attributes and simple rules and obtained significant results and accuracy. The choice of the approach and style of modeling depends as much upon the required outcomes and the available data, as it does on the methods applied [40].

Models may be constructed to predict the location of archaeological sites from any temporal period or archaeological class. For example, APM have been developed to model Eneolithic and Neolithic sites across Europe and the Near East, to analyze land use strategies by the Aboriginal population in Tasmania, to assess locational preferences of Roman populations in the Tiber valley, as well as in rural contexts in Portugal.

From mountainous regions to alluvial plains, the role of environmental variables in modeling procedures is paramount. The only limitation for predictive modeling is the availability of the appropriate data upon which to base the models.

Nevertheless, some significant and useful guidelines may be drawn to highlight what components, methods and steps are relevant in the design and development of archaeological site models or data-driven predictive models. The description of the modeling approach is the core of the next chapter.

To summarize, all kinds of models should be testable and simple enough to be useful [4, 15, 39]. Either way, 'there is nothing inherently unscientific about either approach' [55] and each method has its champions and its critics [7, 188].

References

1. Ebert JI, Kohler TA (1988) The theoretical basis of predictive modeling and a consideration of appropriate data-collection methods. In: Judge WJ, Sebastian L (eds) Quantifying the present and predicting the past. Theory, method, and application of archaeological predictive modeling. US Department of the Interior, Bureau of Land Management Service Center, Denver, pp 97–171
2. Judge W, Sebastian L (eds) (1988) Quantifying the past and predicting the past: theory, method, and application of archaeological predictive modeling. USDI Bureau of Land Management, Denver, CO, pp 61–96
3. Clarke D (1968) Analytical archaeology. Methuen, London
4. Kohler TA, Parker SC (1986) Predictive models for archaeological resources location. In: Schiffer MB (ed) Advances in archaeological methods and theory, vol 9. Academic Press, New York, pp 397–452
5. Graves McEwan D (2012) Qualitative landscape theories and archaeological predictive modelling—a journey through no man's land? J Archaeol Method Theory 19:526–547. https://doi.org/10.1007/s10816-012-9143-6
6. Gaffney V, Van Leusen M (1995) Postscript-GIS, environmental determinism and archaeology: a parallel text. In: Lock GR, Stancic Z (eds) Archaeology and geographical information systems: a European perspective. Taylor and Francis, London, pp 367–383
7. Ebert JI (2000) The state of the art in "Inductive" predictive modeling: seven big mistakes (and lots of smaller ones). In: Wescott KL, Brandon RJ (eds) Practical applications of GIS for archaeologists. A predictive modeling kit. Taylor and Francis, London, pp 129–134
8. Verhagen P, Wansleeben M, van Leusen M (2000) Predictive modelling in the Netherlands. The prediction of archaeological values in cultural resource management and academic research. In: Harl O (ed) Archäeologie und Computer 1999. Forschungsgeselschaft Wiener Stadtarchäeologie 4:66–82
9. Wheatley D (2004) Making space for an archaeology of place. Internet Archaeol 15. http://eprints.soton.ac.uk/28800
10. Kvamme KL (2006) There and back again: revisiting archaeological locational modeling. In: Mehrer MW, Wescot KL (eds) GIS and archaeological site location modeling. CRC, London & New York, pp 3–38
11. Verhagen P (2007a) Testing archaeological predictive models: a rough guide in layers of perception. In: Proceedings of the 35th computer applications and quantitative methods in archaeology conference, Berlin, Germany. April 2–6, 2007, Bonn, pp 285–291
12. Verhagen P (2007b) Case studies in archaeological predictive modelling, PhD thesis, Archaeological Studies of Leiden University 14. 224 pages, 31 figures, 57 tables. 2007. Leiden University Press; 978-90-8728-007-9 paperback. Antiquity 83(319):232–233

13. Kamermans H (2008) Smashing the crystal ball: a critical evaluation of the Dutch national archaeological predictive model (IKAW). Int J Human Arts Comput 1:71–84
14. Verhagen P, Whitley TG (2012) Integrating archaeological theory and predictive modeling: a live report from the scene. J Archaeol Theory Method 19/1:49–100. https://doi.org/10.1007/s10816-011-9102-7
15. Kvamme K, Kohler T (1988) Geographic information systems: technical aids for data collection, analysis, and display. In: James Judge W, Sebastian L (eds) Quantifying the present and predicting the past. Theory, method, and application of archaeological predictive modeling. U.S. Department of the Interior, Bureau of Land Management Service Center, Denver, Co, p 690
16. Willey G (1953) Prehistoric settlement patterns in the Viru Valley, Peru. Bulletin, 1. Bureau of American Ethnology, Washington, DC
17. Kvamme KL (1983) Computer processing techniques for regional modeling of archaeological site locations. Adv Comput Archaeol 1:26–52
18. Dalla Bona L (1994) Ontario ministry of natural resources archaeological predictive modelling project. Center for Archaeological Resource Prediction, Lakehead University, Thunder Bay (Ontario)
19. Judge W, Sebastian L (eds) (1988) Quantifying the past and predicting the past: theory, method, and application of archaeological predictive modeling. USDI Bureau of Land Management, Denver (CO), pp 61–96
20. Kvamme KL (1989) Geographic information systems in regional archaeological research and data management. In: Schiffer M (ed) Archaeological method and theory, vol 1. University of Arizona Press, Tucson (AZ), pp 139–203
21. Kvamme KL (1990) The fundamental principles and practice of predictive archaeological modeling. In: Voorrips A (ed) Mathematics and information science in archaeology: a flexible framework. HOLOS-Verlag, Bonn, Germany, pp 275–295
22. Kvamme KL (1999) Recent directions and developments in geographical information systems. J Archaeol Res 7(2):153–201
23. Hudak GJ, Hobbs E, Brooks A, Sersland CA, Phillips C (eds) (2002) Final report: a predictive model of precontact archaeological site location for the state of Minnesota. Minnesota Department of Transportation, St. Paul
24. Van Leusen PM, Kamermans H (eds) (2005) Predictive modelling for archaeological heritage management: a research agenda. Amersfoort, ROB, PlantijnCasparie Almere
25. Deeben J, Hallewas DP, Maarlevelt ThJ (2002) Predictive modelling in archaeological heritage management of the Netherlands: the indicative map of archaeological values (2nd generation). Berichten ROB 45:9–56
26. Ducke B (2014) An integrative approach to archaeological landscape evaluation: locational preferences, site preservation and uncertainty mapping. Archaeol Eros Eros Archaeol 1:13–22
27. Herzog I (2014) A review of case studies in archaeological least-cost analysis. Archeologia e Calcolatori 25:223–239
28. Ducke B, Münch U (2005) Predictive modelling and the archaeological heritage of Brandenburg (Germany). In: van Leusen, Kamermans (eds) pp 93–107
29. Ducke B (2003) Archaeological predictive modelling in intelligent network structure. In: Doerr M, Sarris A (ed) The proceedings of the 29th conference of the computer applications in archaeology. Hellenic Ministry of Culture, Crete, pp 267–273
30. Oštir K, Kokalj Ž, Saligny L, Tolle F, Nunninger L, avec la collaboration de F. Pennors et K. Zaksek (2007) Confidence maps: a tool to evaluate archaeological data's relevance in spatial analysis. In: Layers of perception. Proceedings of the 35th computer applications and quantitative methods in archaeology conference, Berlin, Germany, April 2–6, 2007, Bonn, pp 272–277
31. Demoule JP (2019) Aux origines, l'archéologie : une science au coeur des grands débats de notre temps. La Découverte (Ed.)
32. Cecamore C, Castiello ME (2014) Un modello speditivo per la carta del Rischio Relativo nei Beni Culturali, in Atti della 15a Conferenza Italiana Utenti Esri. GEOmedia, [S.l.], v. 18, n.

2, giugno 2014. ISSN 2283-5687. https://www.mediageo.it/ojs/index.php/GEOmedia/article/view/873/801
33. Woodman PE, Woodward M (2002) The use and abuse of statistical methods in archaeological site location modelling. In: Wheatley D, Earl G, Poppy S (eds) Contemporary themes in archaeological computing, Oxford, pp 22–27
34. Ejstrud B (2003) Indicative models in landscape management: testing the methods. The archaeology of landscapes and geographic information systems. In: Kunow, Müller (eds) Predictive maps, settlement dynamics and space and time in prehistory,pp 119–134
35. Kamermans H, Deeben J, Hallewas D, Zoetbrood P, van Leusen M, Verhagen P (2005) Project proposal. In: van Leusen M, Kamermans H (eds) Predictive modelling for archaeological heritage management: a research agenda. Nederlandse Archeologische Rapporten 29. Rijksdienst voor het Oudheidkundig Bodemonderzoek, Amersfoort, pp 13–23
36. Van Leusen M (1993) Cartographic modelling in a cell-based GiS. In: Andresen et al (2003), pp 105–124
37. Van Leusen PM (2002) Pattern to process: methodological investigations into the formation and interpretation of spatial patterns in archaeological landscapes. PhD thesis, Faculty of arts. http://dissertations.ub.rug.nl/faculties/arts/2002/
38. Verhagen P, Nuninger L, Bertoncello F, Castrorao Barba A (2015) Estimating memory of the landscape, In: CAA2015. Keep the revolution going. Proceedings of the 43rd annual conference on computer applications and quantitative methods in archaeology, vol 1
39. Kohler TA (1988) Predictive locational modeling: history and current practice. In: Judge WL, Sebastian L (eds) Quantifying the present and predicting the past: theory, method and application of archaeological predictive modeling. US Bureau of Land Management, Denver, pp 19–59
40. Canning S (2003) Site unseen: archaeology, cultural resource management, planning and predictive modelling in the Melbourne metropolitan area. PhD thesis, La Trobe University, Australia
41. Kincaid C (1988) Predictive modeling and its relationships to cultural resource management, pp 549–569. In: Sebastian, Judge (eds) (1988) Quantifying the preset and predicting the past: theory, method and application of archaeological predictive modeling. U.S. Department of the Interior, Bureau of Land Management Service Center. Denver, Co. xx, p 690
42. Church T, Brandon RJ, Burgett G (2000) GIS applications in archaeology: method in search of theory. In: Wescott K, Brandon R (eds) Practical applications of GIS for archaeologists: a predictive modeling kit. Taylor & Francis, London, pp 135–156
43. Carrer F (2013) An ethnoarchaeological inductive model for predicting archaeological site location. A case-study of pastoral settlement patterns in the Val di Fiemme and Val di Sole (Trentino, Italian Alps). J Anthropol Archaeol 32(1):54–62. https://doi.org/10.1016/j.jaa.2012.10.001
44. Danese M, Masini N, Biscione M, Lasaponara R (2014) Predictive modeling for preventive archaeology: overview and case study. Central Eur J Geosci 6(1):42–55. https://doi.org/10.2478/s13533-012-0160-5
45. Danese M, Masini N, Biscione M, Lasaponara R (2014) Predictive modeling for preventive archaeology: overview and case study. Cent Eur J Geosci 6(1):42–55. https://doi.org/10.2478/s13533-012-0160-5
46. Parker SC (1985) Predictive modeling of site settlement systems using multivariate logistics. In: Carr C (ed) For concordance in archaeological analysis: bridging data structure, quantitative technique, and theory. Waveland Press, Prospect Heights, CO, pp 173–207
47. Preucel RW (1995) The postprocessual condition. J Archaeol Res 3(2):147–175
48. Whitley DS (ed) (1998) Reader in archaeological theory: postprocessual and cognitive approaches (Routledge readers in archaeology). Routledge, New York
49. Hodder I (1999) The archaeological process: an introduction. Blackwell, Oxford
50. Wescott K, Brandon R (eds) (2000) Practical applications of GIS for archaeologists: a predictive modeling kit. Taylor & Francis, London

51. Balla A, Pavlogeorgatos G, Tsiafakis D, Pavlidis G (2014) Recent advances in archaeological predictive modeling for archeological research and cultural heritage management. Mediter Archaeol Archaeom 14(4):143–153
52. James G, Witten D, Hastie T, Tibshirani R (2013) An introduction to statistical learning: with applications in R. Springer, New York
53. Vaughn S, Crawford T (2009) A predictive model of archaeological potential: an example from northwestern Belize. Appl Geogr 29(4):542–555
54. Wachtel I, Zidon R, Garti S, Shelch-Lavi G (2018) Predictive modeling for archaeological site locations: comparing logistic regression and maximal entropy in north Israel and north-east China. J Archaeol Sci 92:22–36. https://doi.org/10.1016/j.jas.2018.02.001
55. Warren RE (1990) Predictive modelling of archaeological site location: a primer. In: Allen KMS, Green SW, Zubrow EBW Interpreting space: GIS and archaeology. Taylor & Francis, London
56. Espa G, Benedetti R, Meo AD, Ricci U, Espa S (2006) GIS based models and estimation methods for the probability of archaeological site location. J Cult Herit 7:147–155
57. Nicu I, Mihu-Pintilie A, Williamson J (2019) GIS-based and statistical approaches in archaeological predictive modelling (NE Romania). Sustainability 11(21):5969. https://doi.org/10.3390/su11215969
58. Silalahi F, Yukni Arifianti P, Hidayat F (2019) Landslide susceptibility assessment using frequency ratio model in Bogor, West Java, Indonesia. Geosci Lett 6(10). https://doi.org/10.1186/s40562-019-0140-4
59. Caracausi S, Berruti LF, Daffara S, Bertè B, Borel FR (2018) Use of a GIS predictive model for the identification of high altitude prehistoric human frequentations. Results of the Sessera valley project (Piedmont, Italy). Quatern Int 490:10–20. https://doi.org/10.1016/j.quaint.2018.05.038
60. Brown PE, Rubin BH (1981) Patterns of desert resource use: an integrated approach to settlement analysis. In: Brown PE, Stone CL (eds)Granite Reef: a study in desert archaeology. Arizona State University Anthropological Research Papers No. 28, pp 267–305
61. Berry JK (1987) Fundamental operations in computer-assisted map analysis. Int J Geogr Inf Syst 1:119–136
62. Brandt RW, Groenewoudt BJ, Kvamme KL (1992) An experiment in archaeological site location: modelling in the Netherlands using GIS techniques. World Archaeol 24:268–282
63. Agterberg F, Bonham-Carter G, Cheng Q, Wright D (1993) Weights of evidence modeling and weighted logistic regression for mineral potential mapping. In: Davis J, Herzfeld U (eds) Computers in geology, 25 years of progress. Oxford University Press, Oxford, pp 13–32
64. Bonham Carter G, Agterberg F, Wright D (1989) Weights of evidence modelling: a new approach to mapping mineral potential. In: Statistical applications in the earth sciences. No. 89-9. Geological Survey of Canada Paper, pp 171–183
65. Kay SJ, Witcher RE (2009) Predictive modelling of Roman settlement in the middle Tiber valley. Archeologia e calcolatori 20:277–290. Available at http://www.archcalc.cnr.it/indice/PDF20/21_Kay.pdf. Accessed on 25 February 2020
66. Rua H (2009) Geographic information systems in archaeological analysis: a predictive model in the detection of rural Roman villae. J Archaeol Sci 36(2):224–235. https://doi.org/10.1016/j.jas.2008.09.003
67. Ford A, Clarke KC, Raines G (2009) Modeling settlement patterns of the late classic Maya civilization with Bayesian methods and geographic information systems. Ann Assoc Am Geogr 99(3):496–520. https://doi.org/10.1080/00045600902931785
68. De Vries P (2008) Archaeological Predictive Models for the Elbe Valley around Dresden, Saxony, Germany in CAA2007. In: Layers of perception. Proceedings of the 35th international conference on computer applications and quantitative methods in archaeology, Berlin, Germany, April 2–6, 2007 (Kolloquien zur Vor- und Frühgeschichte, Vol. 10)
69. Goodchild H (2009) Modelling roman agricultural production in the middle tiber valley, central Italy. PhD Thesis, University of Birmingham. Available at https://core.ac.uk/download/pdf/40062866.pdf

70. Baena J, Blasco C, Recuero V (1995) The spatial analysis of Bell Beaker sites in the Madrid region of Spain. In: Lock G, Stančič Z (eds), pp 101–116
71. Ford A, Clarke K (2000) Modeling settlement patterns of the late classic maya civilization with Bayesian methods and geographic information systems. Ann Assoc Am Geogr 99(3):496–520. https://doi.org/10.1080/00045600902931785
72. Anichini F et al (2011) MAPPA project methodologies applied to archaeological potential predictivity. In: MapPapers 1en-I, pp 23–43. https://doi.org/10.4456/MAPPA.2011.02
73. Brown WM, Gedeon TD, Groves DI, Barnes RG (2000) Artificial neural networks: a new method for mineral prospectivity mapping. Aust J Earth Sci 47:757–770
74. Harris D, Zurcher L, Stanley M, Marlow J, Pan G (2003) A comparative analysis of favorability mappings by weights of evidence, probabilistic neural networks, discriminant analysis, and logistic regression. Nat Resour Res 12:241–255
75. Piccini C, Marchetti A, Farina R, Francaviglia R (2012) Application of indicator kriging to evaluate the probability of exceeding nitrate contamination thresholds. Int J Environ Res 6:853–862
76. Abedi M, Norouzi G-H, Bahroudi A (2012) Support vector machine for multiclassification of mineral prospectivity areas. Comput Geosci 46:272–283
77. Zuo R, Carranza EJM (2011) Support vector machine: a tool for mapping mineral prospectivity. Comput Geosci 37:1967–1975
78. Rodriguez-Galiano V, Sanchez-Castillo M, Chica-Olmo M, Chica-Rivas M (2015) Machine learning predictive models for mineral prospectivity: an evaluation of neural networks, random forest, regression trees and support vector machines. Ore Geol Rev 71:804–818. https://doi.org/10.1016/j.oregeorev.2015.01.001
79. Breiman L (2001) Random forests. Mach Learn 45:5–32. https://doi.org/10.1023/A:1010933404324
80. Märker M, Bolus M (2018) Explorative spatial analysis of neandertal sites using terrain analysis and stochastic environmental modelling. GI_Forum 2018, Issue 2, pp 21–38. https://doi.org/10.1553/giscience2018_02_s21
81. Schapire R (2003) The boosting approach to machine learning—an overview. In: Denison DD, Hansen MH, Holmes C, Mallick B, Yu B (eds) MSRI workshop on nonlinear estimation and classification, 2002. Springer, New York
82. Friedman JH, Hastie T, Tibshirani R (2000) Additive logistic regression: a statistical view of boosting. Ann Stat 28:337–407
83. Phillips SJ, Dudík M, Schapire RE (2004) A maximum entropy approach to species distribution modeling. In: Proceedings of the twenty-first international conference
84. Elith J, Leathwick JR, Hastie T (2008) A working guide to boosted regression trees. J Anim Ecol 77:802–813
85. Klassen S, Weed J, Evans D (2018) Semi-supervised machine learning approaches for predicting the chronology of archaeological sites: a case study of temples from medieval Angkor, Cambodia. PLoS One 13(11). https://doi.org/10.1371/journal.pone.0205649
86. Märker M, Heydari-guran S (2009) Application of datamining technologies to predict Paleolithic site locations in the Zagros Mountains of Iran. In: Proceedings of computer applications in archaeology, Williamsburg, Virginia, USA. March 22–26, 2009, pp 1–7
87. Breiman L (1996) Bagging predictors. Mach Learn 24:123–140
88. Jones PJ, Williamson GJ, Bowman DMJS, Lefroy EC (2019) Mapping Tasmania's cultural landscapes: using habitat suitability modelling of archaeological sites as a landscape history tool. J Biogeogr 46(11):2570–2582. https://doi.org/10.1111/jbi.13684
89. Guisan A, Tingley R, Baumgartner JB, Naujokaitis-Lewis I, Sutcliffe PR, Tulloch AIT, Regan TJ, Brotons L, Mcdonald-Madden E, Mantyka-Pringle C (2013) Predicting species distributions for conservation decisions. Ecol Lett 16(12):1424–1435. https://doi.org/10.1111/ele.12189
90. Guisan A, Zimmermann NE (2000) Predictive habitat distribution models in ecology. Ecol Model 135(2):147–186. https://doi.org/10.1016/S0304-3800(00)00354-9

91. Genuer R, Poggi JM, Tuleau-Malot C (2010) Variable selection using random forests. Pattern Recognit Lett 31(14):2225–2236. https://doi.org/10.1016/j.patrec.2010.03.014
92. Mi C, Huettmann F, Guo Y, Han X, Wen L (2017) Why choose random forest to predict rare species distribution with few samples in large undersampled areas? Three Asian crane species models provide supporting evidence. Peer J. https://doi.org/10.7717/peerj.2849
93. Tonini M, D'Andrea M, Biondi G, Degli Esposti S, Trucchia A, Fiorucci P (2020) A machine learning-based approach for wildfire susceptibility mapping. The case study of the Liguria region in Italy. Geosciences 10:105
94. Goetz JN, Brenning A, Petschko H, Leopold P (2015) Evaluating machine learning and statistical prediction techniques for landslide susceptibility modeling. Comput Geosci 81:1–11
95. Oonk S, Spijker J (2015) A supervised machine-learning approach towards geochemical predictive modelling in archaeology. J Archaeol Sci 59:80–88
96. Chen F, Lasaponara R, Masini N (2017). An overview of satellite synthetic aperture radar remote sensing in archaeology: from site detection to monitoring. J Cult Herit 23:5–11. https://doi.org/10.1016/j.culher.2015.05.003
97. Dempster AP (1967) Upper and lower probabilities induced by a multi-valued mapping. Ann Math Stat 38:325–339
98. Shafer G (1976) A mathematical theory of evidence. Princeton University Press, Princeton
99. Millard AR (2005) What can Bayesian statistics do for predictive modelling? In: van Leusen M, Kamermans H (eds) Predictive modelling for archaeological heritage management: a research agenda. Rijksdienst voor het Oudheidkundig Bodemonderzoek, Amersfoort, pp 169–182
100. Finke PA, Meylemans E, Van De Wauw J (2008) Mapping the possible occurrence of archaeological sites by Bayesian inference. J Archaeol Sci 35:2786–2796
101. Ducke B (2010) Regional scale predictive modelling in North-Eastern Germany, CAA2004. Beyond the artifact. Digital interpretation of the past. In: Proceedings of CAA2004, Prato 13–17 April 2004, pp 296–301
102. Ejstrud B (2005) Taphonomic models. Using Dempster-Shafer theory to assess the quality of archaeological data and indicative models. In: Kamermans/van Leusen 2005, pp 189–198
103. Verhagen Ph, Kamermans H, van Leusen M (2008) The future of archaeological predictive modelling. In: Proceedings of symposium the protection and development of the Dutch archaeological historical landscape: the European dimension, 20–23 May 2008, Lunteren
104. Ebersbach R (2015) Eine Potentialkarte Archäologie für den Kanton Bern. Archäologie Bern/Archéologie Bernoise 2015:212–233
105. Schucany C (2006) Die römische Villa von Biberist- Spitalhof/SO (Grabungen 1982, 1983, 1986–89). Ausgrabungen und Forschungen 4. Remshalden
106. Morrison MS (2015) Reconstructing reality: models, mathematics, and simulations. Oxford University Press, New York
107. Taheri SM, Ghadim FI, Kabirian M (2019) Application of fuzzy inference systems in archaeology
108. Ramos-Soto A, Alonso JM, Reiter E, van Deemter K, Gatt A (2017) An empirical approach for modeling fuzzy geographical descriptors. IEEE
109. Barceló JA, Bogdanovic I (eds) (2015) Mathematics and archaeology. CRC Press. https://doi.org/10.1201/b18530
110. Lieskovský T, Ďuračiová R, Karell L (2013) Selected mathematical principles of archaeological predictive models creation and validation in the GIS environment. Interdisciplinaria Archaeologica Nat Sci Archaeol 4(2):33–46
111. Mink PB, Ripy J, Bailey K, Grossardt TH (2009) Predictive archaeological modeling using GIS-based fuzzy set estimation: a case study in Woodford County, Kentucky
112. De Runz C, Desjardin E, Piantoni F, Herbin M (2007) Using fuzzy logic to manage uncertain multi-modal data in an archaeological GIS. http://citeseerx.ist.psu.edu/viewdoc/summary?doi=10.1.1.108.7063
113. Nicolucci F, Hermon S (2004) A fuzzy logic approach to reliability in archaeological virtual reconstruction. In: CAA2004: Beyond the artifact—digital interpretation of the past. Proceedings of CAA2004, Prato 13–17 April 2004

114. Niccolucci F (2006) Managing uncertainty in archaeological GIS applications, reading historical spatial information from around the world studies of culture and civilization based on geographic information systems data
115. Rivett P (1997) Conceptual data modelling in an archaeological GIS. In: Proceedings of the 2nd annual conference of geocomputation'97 & SIRC'97. University of Otago, New Zealand, 26–29 August 1997.
116. Benvenuti A, Niccolucci F (1996) Historical objects with indeterminate boundaries. In: Paper presented at the IX international conference of the AHC, Moscow 1996
117. Refsgaard JC, van der Sluijs JP, Etejberg AL, Vanrollegham PA (2007) Uncertainty in the environmental modeling process—a framework and guidance. Environ Model Softw 22:1543–1556
118. Evans A (2012) Uncertainty and error. In: Heppenstall AJ, Crooks AT, See LM, Batty M (eds) Agent-based models of geographical systems. Springer, New York, pp 309–346
119. Martin-Rodilla P, Pereira-Fariña M, Gonzalez-Perez C (2019) Qualifying and quantifying uncertainty in digital humanities: a fuzzy-logic approach. In: ACM international conference proceeding series, pp 788–794. https://doi.org/10.1145/3362789.3362833
120. Martin-Rodilla P, Gonzalez-Perez C (2019) Conceptualization and non-relational implementation of ontological and epistemic vagueness of information in digital humanities. Informatics 6(2). https://doi.org/10.3390/informatics6020020
121. Gonzalez-Perez C (2018) Information modelling for archaeology and anthropology. https://doi.org/10.1007/978-3-319-72652-6
122. Brouwer Burg M, Howey M (2017) Assessing the state of archaeological GIS research: unbinding analyses of past landscapes. J Archaeol Sci 84:1–9. https://doi.org/10.1016/j.jas.2017.05.002
123. Zadeh LA (1965) Fuzzy Sets. Inf Control 8:338–353
124. Hájek P (1998) Metamathematics of fuzzy logic. Kluwer, Dordrecht, The Netherlands
125. Halpern JY (2003) Reasoning about uncertainty. MIT Press, Cambridge, MA
126. Yager RR, Kacprzyk J, Fedrizzi M (eds) (1995) Advances in Dempster-Shafer theory of evidence. Wiley, New York, pp 5–34
127. Ragin C (2000) Fuzzy-set social science. University of Chicago Press, Chicago, IL
128. Roberts DW (1986) Ordination on the basis of fuzzy set theory. Vegetatio 66:123–131
129. Moraczewski IR (1993) Fuzzy logic for phytosociology II. Generalizations and predictions. Vegetatio 106(1):13–20
130. Sattler R (1996) Classical morphology and continuum morphology: Opposition and continuum. Ann Bot 78:577–581
131. Crescioli M, D'Andrea A, Niccolucci F (2000) A GIS-based analysis of the Etruscan cemetery of Pontecagnano using fuzzy logic. In: Lock GR (ed) Beyond the map: archaeology and spatial technologies. IOS Press, Amsterdam, pp 157–179
132. Barceló JA, Pallarés M (1998) Beyond GIS: the archaeology of social spaces. Archaeologia e Calcolatori 1:47–80
133. Hatzinikolaou EG, Hatzichristos T, Siolas A, Mantzourani E (2003) Predicting archaeological site locations using GIS and fuzzy logic. In: Doerr M, Sarris A (eds) The digital heritage of archaeology. Computer applications and quantitative methods in archaeology. Archive of Monuments and Publications, Hellenic Ministry of Culture, Heraklion (Greece), pp 169–178
134. Hermon S, Niccolucci F (2003) A fuzzy logic approach to typology in archaeological research. In: Doerr M, Sarris A (eds) The digital heritage in archaeology: computer applications and quantitative methods in archaeology. Archive of Monuments and Publications, Hellenic Ministry of Culture, Heraklion, pp 169–178
135. Bashir Musa A (2014) Logistic regression classification for uncertain data. Res J Math Stat Anal 2(2):1–6. ISSN 2320–6047
136. Baudron P, Alono-Sarría F, García-Aróstegui JL, Cánovas-García F, Martínez-Vicente D, Moreno-Brotóns J (2013) Identifying the origin of groundwater samples in a multi-layer aquifer system with random forest classification. J Hydrol 499:303–315. https://doi.org/10.1016/j.jhydrol.2013.07.009

137. Abedi M, Norouzi G-H (2012) Integration of various geophysical data with geological and geochemical data to determine additional drilling for copper exploration. J Appl Geophys 83:35–45
138. Tonini M, Cama M (2019) Spatio-temporal pattern distribution of landslides causing damage in Switzerland. Landslides 16:2103–2113. https://doi.org/10.1007/s10346-019-01236-1
139. Tehrany MS, Kumar L, Jebur MN, Shabani F (2019) Evaluating the application of the statistical index method in flood susceptibility mapping and its comparison with frequency ratio and logistic regression methods. Geomat Nat Hazards Risk 10(1):79–101
140. Biondi G, Campo L, D'Andrea M, Degli Esposti S, Fiorucci P, Tonini M (2018) Wildfire susceptibility mapping in Liguria (Italy): comparison of statistical driven partitioning and machine learning approach. In: Viegas DX (ed) Advances in forest fire research 2018. Chapter 1—fire risk management. https://doi.org/10.14195/978-989-26-16-506_20
141. Reichenbach P, Rossi M, Malamud BD, Mihir M, Guzzetti F (2018) A review of statistically based landslide susceptibility models. Earth Sci Rev 180:60–91
142. Deluigi N (2018) Data-driven mapping of the potential mountain permafrost distribution. PhD thesis, University of Lausanne
143. Zêzere JL, Pereira S, Melo R, Oliveira SC, Garcia RAC (2017) Mapping landslide susceptibility using data-driven methods. Sci Total Environ 589:250–267
144. Pham BT, Pradhan B, Tien Bui D, Prakash I, Dholakia MB (2016) A comparative study of different machine learning methods for landslide susceptibility assessment: a case study of Uttarakhand area (India). Environ Model Softw 84:240–250
145. Leuenberger M, Parente J, Tonini M, Pereira MG, Kanevski M (2017) Wildfire susceptibility mapping: deterministic vs. stochastic approaches. Environ Model Softw 101:194–203. https://doi.org/10.1016/j.envsoft.2017.12.019
146. Earley-Spadoni T, Harrower M (2020) Spatial archaeology: mapping the ancient past with the humanities and the sciences. In: Bodenhamer DJ, Ell PS (eds) Int J Human Arts Comput 14(Issue 1–2). ISSN: 1753-8548 Available Online Feb 2020
147. Davis D (2020) Geographic disparity in machine intelligence approaches for archaeological remote sensing research. Rem Sens 12(6). https://doi.org/10.3390/rs12060921
148. Caspari G, Crespo P (2019) Convolutional neural networks for archaeological site detection—finding "princely" tombs. J Archaeol Sci 110. https://doi.org/10.1016/j.jas.2019.104998
149. Klehm C, Follett F, Simon K, Kiahtipesc C, Mothulatshipi S (2019) Toward archaeological predictive modeling in the Bosutswe region of Botswana: utilizing multispectral satellite imagery to conceptualize ancient landscapes. J Anthropol Archaeol 54:68–83. https://doi.org/10.1016/j.jaa.2019.02.002
150. Menze BH, and Ur JA (2014) Multitemporal fusion for the detection of static spatial patterns in multispectral satellite images—with application to archaeological survey. IEEE J Selec Topics Appl Earth Observ Remote Sens 7(8):3513–3524. https://doi.org/10.1109/jstars.2014.2332492
151. Wernke S, VanValkenburghb P, Saito A (2020) Interregional archaeology in the age of big data: building online collaborative platforms for virtual survey in the Andes. J Field Archaeol 45(S1):S61–S74. https://doi.org/10.1080/00934690.2020.1713286
152. Verschoof-van der Vaart WB, Lambers K (2019) Learning to look at LiDAR: the use of R-CNN in the automated detection of archaeological objects in lidar data from the Netherlands. J Comput Appl Archaeol 2:31–40
153. Thabeng OL, Merlo S, Adam E (2019) High-resolution remote sensing and advanced classification techniques for the prospection of archaeological sites' markers: the case of dung deposits in the Shashi-Limpopo Confluence area (southern Africa). J Archaeol Sci 102:48–60. https://doi.org/10.1016/j.jas.2018.12.003
154. Trier ØD, Cowley DC, Waldeland AU (2019) Using deep neural networks on airborne laser scanning data: results from a case study of semi-automatic mapping of archaeological topography on Arran, Scotland. Archaeol Prospect 26:165–175
155. Mantovan L, Nanni L (2020) The computerization of archaeology : survey on AI techniques

156. Barone G, Mazzolenia P, Spagnolo GV, Raneric S (2019) Artificial neural network for the provenance study of archaeological ceramics using clay sediment database. J Cult Herit 38:147–157. https://doi.org/10.1016/j.culher.2019.02.004
157. Lotfian M (2016) Urban climate modeling, case study of Milan city. Master thesis, Politecnico di Milano
158. Shalev-Shwartz S, Ben-David S (2014) Understanding machine learning: from theory to algorithms. Cambridge University Press, New York, NY, USA
159. Pal M (2005) Random forest classifier for remote sensing classification. Int J Remote Sens 26(1):217–222
160. Liaw A, Wiener M (2002) Classification and regression by random forest. R News 2(3):18–22
161. Bellinger C, Jabbar MSM, Zaiane S, Osornio-Vargas A (2017) A systematic review of data mining and machine learning for air pollution epidemiology. BMC Public Health 17:907. https://doi.org/10.1186/s12889-017-4914-3
162. Seong H, Son H, Kim C (2018) A comparative study of machine learning classification for color-based safety vest detection on construction-site images. KSCE J Civ Eng 22(2018):4254–4262
163. Huggett J (2015) A manifesto for an introspective digital archaeology. Open Archaeol 1:86–95
164. Monna F, Magailb J, Rollanda T, Navarroc N, Wilczek J et al (2020) Machine learning for rapid mapping of archaeological structures made of dry stones—example of burial monuments from the Khirgisuur culture, Mongolia. J Cult Herit. Elsevier Masson SAS, pp 1–11. https://doi.org/10.1016/j.culher.2020.01.002
165. Gualandi ML, Scopingo R et al (2016) ArchAIDE-archaeological automatic interpretation and documentation of ceramics. In: Catalana CE, De Luca L (eds) Eurographics workshop on graphics and cultural heritage. https://doi.org/10.2312/gch.2016140
166. Kalayci T (2015) Data integration in archaeological prospection. In: Apostolos Sarris (ed) Best practices of geoinformatic technologies for the mapping of archaeolandscapes. Archeopress, Oxford
167. Assael Y, Sommerschield T, Prag J (2019) Restoring ancient text using deep learning: a case study on Greek epigraphy. https://arxiv.org/abs/1910.06262
168. Gattiglia G (2018) Classificare le ceramiche: dai metodi tradizionali all'intelligenza artificiale. L'esperienza del progetto europeo ArchAIDE. In: ARCHEOLOGIA QUO VADIS? Riflessioni metodologiche sul futuro di una disciplina. Atti del Workshop Internazionale Catania, 18–19 Gennaio 2018
169. Gultepe E, Conturo ET, Makrehchi M (2018) Predicting and grouping digitized paintings by style using unsupervised feature learning. J Cult Herit 31:13–23. https://doi.org/10.1016/j.culher.2017.11.008
170. Bishop CM (2006) Pattern recognition and machine learning. Information science and statistics. Springer, Berlin, Heidelberg
171. Fiorucci M, Khoroshiltseva M, Pontil M, Traviglia A, Del Bue A, James S (2020) Machine learning for cultural heritage: a survey. Pattern Recognit Lett 133:102–108
172. Friedman JH (2002) Stochastic gradient boosting. Comput Stat Data Anal 38(4):367–378. https://doi.org/10.1016/S0167-9473(01)00065-2
173. Breiman L, Friedman JH, Olshen RA, Stone CJ (1984) Classification and regression trees. Wadsworth and Brooks/Cole, Monterey, California, USA
174. Williams JN, Seo C, Thorne J, Nelson JK, Erwin S, O'Brien JM, Schwartz MW (2009) Using species distribution models to predict new occurrences for rare plants. Divers Distrib 15(4):565–576. https://doi.org/10.1111/j.1472-4642.2009.00567.x
175. Wisz MS, Hijmans RJ, Li J, Peterson AT, Graham CH, Guisan A (2008) Effects of sample size on the performance of species distribution models. Divers Distrib 14(5):763–773. https://doi.org/10.1111/j.1472-4642.2008.00482.x
176. Araújo MB, New M (2007) Ensemble forecasting of species distributions. Trends Ecol Evol 22(1):42–47. https://doi.org/10.1016/j.tree.2006.09.010
177. Hardy SM, Lindgren M, Konakanchi H, Huettmann F (2011) Predicting the distribution and ecological niche of unexploited snow crab (Chionoecetes opilio) populations in Alaskan

waters: a first open-access ensemble model. Integr Comp Biol 51(4):608–622. https://doi.org/10.1093/icb/icr102
178. Dahinden C (2009) An improved random forests approach with application to the performance prediction challenge datasets. Hands on pattern recognition. Microtome
179. Breiman L, Cutler A (2010) Random forests. http://www.stat.berkeley.edu/~breiman/RandomForests/
180. Nilsson A (2016) Predicting the archaeological landscape. Archeological density estimation around the Ostlänken railroad corridor. Dissertation, Uppsala University. http://urn.kb.se/resolve?urn=urn:nbn:se:uu:diva-303949
181. Hastie T, Tibshirani R, Friedman J (2008) The elements of statistical learning, 2nd edn. Springer, p 745
182. Alpaydin E (2014) Introduction to machine learning, 3rd edn. In: Alpaydin E (ed) MIT Press, Cambridge, Mass
183. Sammut C, Webb GI (2010) Encyclopaedia of machine learning. Springer, Boston MA USA
184. Charalambous E, Dikomitou-Eliadou M, Milis GM, Mitsis G, Eliades DG (2016) An experimental design for the classification of archaeological ceramic data from Cyprus, and the tracing of inter-class relationships. J Archaeol Sci Rep 7:465–471. https://doi.org/10.1016/j.jasrep.2015.08.010
185. Lipo CP, Madsen M, Dunnel R, Hunt T (1997) Population structure, cultural transmission, and frequency seriation. J Anthropol Archaeol 16:301–333
186. Kulkarni VY, Sinha PK (2012) Pruning of random forest classifiers: a survey and future directions. In: International conference on data science & engineering (ICDSE), Cochin, Kerala, 2012, pp 64–68. https://doi.org/10.1109/ICDSE.2012.6282329
187. Cutler DR, Edwards Jr TC, Beard KH, Cutler A, Hess KT, Gibson J, Lawler JJ (2007) Random forests for classification in ecology. Ecology 88(11):2783–2792. https://doi.org/10.1890/07-0539.1
188. Westcott KL, Kuiper J (2000) Using a GIS to model prehistoric site distributions in the upper Chesapeake Bay. In: Wescott KL, Brandon RJ (eds) Practical applications of GIS for archaeologists: a predictive modelling kit. Taylor & Francis, London, pp 59–72

Section III

Chapter 4
Materials and Data

Archaeological data is always incomplete, frequently unreliable, often replete with unknown unknowns, but we nevertheless make the best of what we have and use it to build our theories and extrapolations about past events.

Jeremy Huggett, 2019

This Chapter introduces the data used in the research. In order to put the analyzed data in their temporal context and to allow a better understanding of Roman site distribution in Switzerland, a brief historical framework is outlined. The data provided by the Cantonal archaeological departments, representing the regions of interest, are then listed and discussed, followed by the description of the environmental variables assumed to have influenced site location choices of the population during Roman times. In a first step, the archaeological data is introduced and a comprehensive description of their specific characteristics, strengths, limitations and constraints provided. The following part is concerned with a general description of each study area and with a detailed presentation of the datasets gathered and their restructuring. Finally, the geographical and environmental variables considered in the modeling procedure are outlined.

4.1 Premise

The archaeological records refer to the evidences discovered in six Swiss administrative units (Zurich, Aargau, Graubünden, Vaud, Geneva and Fribourg) collected by the respective Cantonal Archaeological departments over decades and until 2015.

As mentioned above, the archaeological databases belong to a precise time span. The idea to use a temporally bounded dataset lies in the fact that the research results aim at being testable. Using data covering a period ending five years ago gives the possibility to further validate the modeling results either by means of back-end analysis (by checking the survey documentation created after 2015) or by conducting direct fieldwork (by planning archaeological surveys). Thus, the predictive maps

M. E. Castiello, *Computational and Machine Learning Tools for Archaeological Site Modeling*, Springer Theses, https://doi.org/10.1007/978-3-030-88567-0_4

could be used to check whether the zones defined as of 'high probability' hide other still unknown sites, confirming or infirming the model prediction.

As it is mostly the case, the recorded archaeological information exist in a variety of forms. As it has been discussed in Sect. 1.2, this is even more true in Switzerland, where site records are stored and dispersed among different institutions. The Cantonal institutions conduct individual archives and records that are not duplicated at the level of a federal state repository.

Due to the decentralized political organization, the extant archaeological data varies considerably in structure and quantity from one region to another. This variety represents a particular challenge for modeling procedures and spatial statistical analysis, due to the introduction of data biases and the absence of a standardized supra structure. Such limitations and constraints are moreover tangible when aiming at developing predictive maps with a standardized and interoperable procedure.

This premise further aims at tackling one paramount challenge of this research, namely the difficult task of integrating data, collected for cultural heritage management purposes, in an academic or research oriented question or series of questions. In general, Cultural Heritage Management (CHM) data is a collection of inventoried regional archaeological site information, not primarily designed for in depth analyses or interregional comparison.

When archaeological sites or finds are recorded by the CHM departments, different protocols of inventorying have to be followed depending on the administrative unit of reference, and the data is stored within different database systems. Moreover, the same type of artifact can be defined and stored differently, depending on the rules applied overhead and the officer creating the entry in the local system. For data belonging to certain periods, the information is stored in different digital and non-digital formats as well as on different supports (individual and local storage), which means that spatial and statistical analyses for the overall site distribution in a precise region may require additional time and efforts to be accomplished.

Resulting from the necessity of the regional administrations to respond to localized needs, the absence of a single, country-wide database with standardized definition for all the archaeological records (encompassing all the (pre)historical epochs) represents a major challenge for conducting research on the past human occupation of the entire territory. As a consequence, this research was first kept at regional scale and in a second stage, the regional data were merged in a unique geospatial database.

Nevertheless, the archaeological records kept locally by the cantonal departments represent an extremely valuable source of data and information. These databases contain very useful information on an exhaustive list of excavations carried out since the early institutional activities in Switzerland. Surveys concerning all historical epochs and all types of archaeological evidence are included in these collections, allowing for an overview of the archaeological heritage out of the museums in Switzerland. The availability of the cantonal departments to provide their data for research purposes demonstrates a praiseworthy disposition to support research with their exhaustive collections. In fact, the combination of Cantonal databases and archives holds an enormous scientific potential.

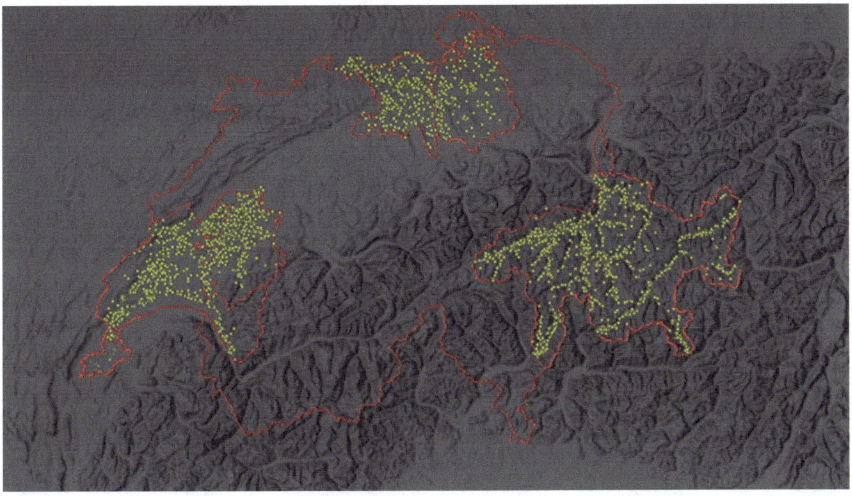

Fig. 4.1 The Swiss Cantons analyzed (From west to east: Geneva, Vaud, Fribourg, Aargau, Zurich, Grisons) are delimited by a red contour line and the presence of archaeological sites is highlighted by yellow dots

In the following pages, solutions are explored and tested to deal with the intrinsic archaeological data bias and uncertainty and a standard procedure is established to propose a reproducible approach for managing it. In order to analyze the data in a GIS (Geographic Information System) environment, the data had to match a specific data structure. Hence, a reconstructed series of geo-spatial tables was created in ESRI ArcGIS 10.6, and in a second instance in ArcGIS 10.7 (Fig. 4.1). The original digital tables consisted of hundreds of thousands of cells of information with several attribute fields, and very detailed information about the discoveries made.

First, the choice of archaeological evidence to be modeled was limited to two types of categories: 'settlements', often referred to as building-housing-living spaces, and 'single finds', which, if not otherwise declared, refer to ceramic shreds and coins found outside a settlement area.

The choice of the Roman period as historical epoch to be modeled is based first on its relative temporal proximity to our times. In fact, 2000 years are a very short time period on a geological scale, thus allowing a certain confidence in using modern geo-environmental variables as determining parameters in the model. Secondly, the fact that Roman housing structures and artifacts are generally considered to be the best preserved and easily distinguishable in the field will be an advantage in case of further field assessments and model validation tests. As shown in Fig. 4.2, the entries defined as belonging to the Roman epoch account for roughly 10–30% of all data points in the Cantons of Aargau, Zurich, Geneva and Vaud, averaging around 20% for all Cantons. Grisons shows only a very small percentage of Roman evidences, as this region is principally characterized by the discovery of pre-historical and medieval sites. For the Canton of Fribourg, the database obtained did not include any medieval site, thus explaining the particularly high percentage of Roman entries.

Fig. 4.2 Percentage of entries labeled as belonging to the Roman epoch in the original cantonal databases. *Other epochs exclude the medieval epoch in the database used for the Canton of Fribourg

Within the frame of this Thesis, no detailed typological or comparative analysis was carried out on the sites due to the already very broad context in which this research is embedded. A deeper study of patterns within the distribution of other site types can be listed among the outlooks of this research.

Thus, the use of the terms 'settlement', 'single finds' simply follows the guidelines and descriptions provided in the original databases. In order to discern locational preferences and distribution patterns among the data, a first explorative and descriptive data analysis was carried out which led to a final Locational Preference Analysis.

In the following pages, a short digression is made about the 'site' concept and its relative issue for archaeological predictive modeling, before to introduce each case study individually. This aims at better contextualizing some of the choices made concerning the database architecture and to step back from the pure theoretical interpretation of the results produced by Archaeological Site Modeling techniques.

Finally the explorative Locational Preference Analysis represents a first attempt to unveil patterns and to compare the data at regional and supra-regional level. This specific approach aims at providing with a bigger picture of the archaeological phenomena behind any individual regional study.

4.2 The Concept of 'Archaeological Site' and Relative Issues

> *...[L]ike a pair of worn suspenders, the site concept can be stretched so far. that it fails to carry any weight at all'.*
>
> Thomas [92]

Though the definition of 'Site' does not really represent one of the topics tackled and discussed in this Thesis, it seems important to underline what McCoy has recently argued and have been resumed here: 'In this digital era, geospatial technologies allows us an unprecedented level of precision and accuracy in detection and mapping at a number of different spatial scales. This pushed the research and studies in archaeology to shift their attention from the defense of the "site" seen as the core of the study (which led to ontological pragmatic issues), to the "site" seen as the locations where artifacts, human remains, and architecture have been found" [56].

However the topic of site definition is as old as the discipline of archaeology itself and it has been explored since 70s [6, 24, 60, 80, 92] there are still many definitions for the term. As stressed out by Thomas [92], these definitions are at times arbitrary, at times precise. At the date, 'site' definition is no longer universal and it is commonly accepted that several problems contribute to the difficulty in interpreting and defining what exactly is an archaeological 'site'. Nevertheless, the general notion of archaeological 'site' and the definition of what is exactly seem to be a central topic for many research and cultural heritage management activities.

In more recent years, Canning [17] emphasized the importance of the definitions of the 'site' for cultural heritage management in order to better manage the resource in question, while in archaeological research defining the 'site' refer to the actions undertaken when typological, taxonomic or functional attributes or occurrences are to be extracted, analyzed and compared at inter or intra-site levels. However, the range of definitions can be almost limitless and determining the unit of analysis is one of the fundamental steps in designing any kind of database [17]. Sites definition may be period or area dependent and for some, sites are even nothing more than archaeological constructs [7].

According to McCoy [56], several issues may contribute to the difficult task of site definition for example, that (i) we rely on contemporary observations and not on the original past context, (ii) our ability to make those observation is depending by the visibility—or not—of the archaeological evidences; (iii) post depositional processes, such as human and natural agents, continuously change the evidences and their relations between artifacts and features; (iv) a formal definition of site must rely on an subjective description, which furthermore is difficult to reproduce and transfer between regions and time periods.

Ultimately, the lack of a universal definition of what constitutes a site has been amplified by the use of geospatial technologies, which are able to handle many different data categories and formats (points, polygons, rasters, etc.). Thus, this site problem is so pervasive that it would be impossible to give a full accounting in this Thesis, which moreover focuses on different regions and deals with different categories-of evidence at the same time. As various authors have already pointed out [16, 39, 40, 106] in the end, the definition of the term archaeological 'site' is often depending on the constraints of the individual project or on the researcher conducting the project. In CHM applications for example, definitions of what is or is not a 'site' are often prescribed by the authorities or legal context.

Consequently to these shared difficulties arising from the definition of archaeological 'site', the archaeological literature provide with many methods and theories designed to overcome the discerned shortcomings of the 'site' concept. Since the early 1970s, the usefulness of the 'site' notion had come to be questioned and the general perception evolved. Archaeological record progressively appeared to be considered as a more or less continuous phenomena across landscapes, rather than as discrete or precisely outlined 'site' [16].

These pages aimed at highlighting well-known complications that the 'site' concept presents and that are specific to each geographic region. Thus, the term "site" as used in this context refers primarily to where community structure is to be located, the locations where things have been found, a shorthand for where archaeologists have made field observations or collected objects. As mentioned above, these have been furthermore divided in two main categories: 'settlements' intended as permanent installation and 'single finds' as isolated discoveries.

4.3 Historical Framework of the Research Areas

Caes., Gall. 1, 3, 1 et 1, 5, 2–3.

> … His rebus adducti et auctoritate Orgetorigis permoti constituerunt ea, quae ad proficiscendum pertinerent, comparare, iumentorum et carrorum quam maximum numerum coemere, sementes quam maximas facere, ut in itinere copia frumenti suppeteret, cum proximis civitatibus pacem et amicitiam confirmare…
>
> …ubi iam se ad eam rem paratos esse arbitrati sunt, oppida sua omnia numero ad duodecim, vicos ad quadringentos, reliqua privata aedificia incendunt, frumentum omne, praeter quod secum portaturi erant, comburunt, ut domum reditionis spe sublata paratiores ad omnia pericula subeunda essent; trium mensum molita cibaria sibi quemque domo efferre iubent….

> …Induced by these circumstances, and influenced by the authority of Orgetorix, they decided to prepare what they needed for their journey-they bought as many horses and wagons as possible-they sowed all the grain that they could sow, in order to have enough of it during the journey-and to establish peace and friendship with the neighboring populations. They believed that two years were enough to complete the preparations: they fixed with a law their departure for the third year…

> …As soon as they estimated they had made the necessary preparations, they set fire to all their *oppida* (cities), a dozen in number, and to all their four hundred *vici* (villages) along with the rest of the private buildings and all the grain, except what they wanted to take with them on the journey, in order to deprive themselves of any hope of returning home and to be all the more determined in all dangerous undertakings. Then the order was given that each of them should take along a three-month supply of flour and bread from home …

The Roman period in Switzerland is the best-recorded period [18, 31, 43] for which archaeological evidences, especially the masonry structures of private, public and religious buildings or the remains of cultural life (coins, ceramic, etc.) are abundant and often relatively easy to detect through modern technologies (i.e. aerial photography, satellite imagery, landscape surveys). This observation has motivated the choice of modeling archaeological information belonging to the Roman period. At this point, an excursus on the Roman conquest of the Swiss Plateau and the social organization of the *Helvetii* appears to be useful to frame the socio-political and socio-cultural phenomena behind the modelization performed in this study.

The territory of the *Helvetii*,[1] who settled on the Swiss plateau at the latest around the end of the second century BC, was delimited by the Rhone, Lake Leman, the Jura range and the Rhine, bordered on the other side by the Pre-Alps and the Alps (where it is difficult to establish a precise border), and stretched up to Lake Constance [34, 41]. The incorporation of the Swiss Plateau into the Roman Empire is well documented by Roman written sources. Hence, we have most of the information through Roman eyes, a point of view that may have tended to dominate and influence the interpretation of archaeological records for several years [10]. The well-known extract of *De Bello Gallico* (cited above) describes the migration of the *Helvetii* in the year 58 BC with the intention of settling in the southwest of France. This migration had an abrupt end as they were defeated by Caesar in the region of Bibracte (Mont Beuvray, west of Autun). They surrendered unconditionally to the Roman general (*par deditio*), who restored their rights as long as they would return to their country [36, 46]. This excerpt of Cesar' text is particularly interesting in the light of this research approach, since it is relaying, indirectly, some interesting details on the number of the local Celtic cities (a dozen of *oppida*), villages (approximately 400 *vici*) and of the presence of other private buildings (probably farms) existing at that precise moment: a proof of the highly hierarchized Celtic organization. This kind of information, as well as other details we can derive from historical sources, but also from environmental features, can ultimately shed new lights on the urban development of the *Helvetii* since the

[1] The Helvetii/Helvetians, one of the great Celtic populations, are mentioned by ancient authors around 100 BC. They are said to have gradually settled in Switzerland from southern Germany [111].

very dawn of the imperial age. Though, as the historical sources as well as the archaeological evidences are not sufficiently clear about the episodes following the *Helvetii*'s return to their lands, a question arises about how they were able to rebuild all their *oppida*[2] and *vici*[3] after such military defeat and whether they decided to choose new suitable locations for settling down permanently [93]. Archaeology tries to help to overcome the lack of information and explain chronological gaps by taking advantage of digital innovations in technology, hand in hand with more traditional methods such as numismatics and archaeometry analysis.

It will be sketched out in this paragraph how the Country went through the romanization process and to what extent it experienced a relative continuity in the settlement distribution between the 50–40 to 20–10 BC. More detailed historical contextualization is further provided for each canton in the corresponding following subchapter. Scholars as Tarpin [90, 113] or Vitali and Kaenel [107] draw attention on the numerous evidences concerning the first contacts with the Romans well before Caesar's war.[4] After the defeat of Bibracte, around 45–44 BC, and following a decision taken by Caesar, the veteran's colony of Nyon (*Colonia lulia Equestris*) was established on the shores of Lake Leman in the modern Canton of Vaud, and a year later Augst (*Colonia Augusta Raurica*) was founded near Basel [36]. Despite the Gallic origin of the native name of the city, Nyon has not revealed any remains of settlements dating back to times prior to the establishment of the roman colony. During the Augustan period, it presents an orthogonal grid and a first *forum*, of which only the *basilica* with two naves is known. The *Macellum*, the amphitheater and some private *domus*, as well as the remains of important hydraulic constructions probably connected with Lake Leman are later constructions. Meanwhile, Augusta Raurica was probably established on the remains of an ancient *oppidum* to control the routes leading from the Plateau to the Rhine. The particular site morphology, furthermore situated on important communication routes explains the strategic location [36]. Both colonies were installed at the opposite (southern and northern) limits of the Swiss Plateau to respond to the need to protect the access to Gaul and to strengthen the Roman presence on the local trade and transalpine commercial routes [62], Kilcher and Zaugg [53, 54]. During the same period, several military settlements were established: *Tenedo* (Zurzach), *Turicum* (Zurich Lindenhof), Uetliberg, *Vitudurum* (Oberwintertur), *Vindonissa* (Windisch), usually also based on ancient *oppida* [30, 53, 54, 62].

Frei-Stolba et al. [36] have pointed out that people living in the territory of modern Switzerland were conquered by the Romans in different stages of history and that the territorial distribution of these heterogeneous peoples exceeded the modern boundaries. Roman rule did not introduce a unity in the territory of modern Switzerland,

---

[2] The *oppida* were developed at the end of La Tène (200–15 BC). These first fortified "towns" played the role of political, economic and religious centers.

[3] The *vici* describe the smallest administrative unit of a provincial tow, a settlement.

[4] The first Roman attempts to take control of the mountain passes and the accesses to mines, e.g. the Great St Bernard Pass, as well as to control the commercial routes crossing the Alps were undertaken during the second century BC.

instead this continued to be marked by a number of administrative borders between the different Roman provinces. Similarly, it did not result in an immediate integration of the *Helvetii* in the Roman Empire. The *Helvetii* were effectively subject to Rome only after the conquest of Rhaetia and after the Romans took over a certain number of strategic sites to plan the military attack against *Germania* and the conquest of the Alps (finally led by Augustus between 25 and 7–6 BC) [31, 64].

As mentioned, the romanization process is often presented as a rapid and voluntary adoption of Roman culture by the local Gauls [33, 38, 44, 67] but in reality it might have been somewhat more of a progressive process. Romans brought several new products, new cultivated plants and livestock, new building techniques (stone buildings, canalizations, hypocaust heaters, etc.), built up a well-developed road network across the country, a uniform currency and balanced weights which favored the lively long-distance trade on the rivers and lakes as well as new cultural traditions, including a new pantheon of gods [32, 61, 62]. Nevertheless, we can observe that innovations brought by the roman conquerors were partly assimilated and rather mixed with traditional indigenous elements to create an independent *gallo*-roman culture [54]. The numerous evidences arising from excavations make it possible to draw new conclusions about everyday life, for example, while the masonry was adopted for the main part of the foundations of new buildings, mud and wood continued to be used for construction walls, illustrating the mixture of Celtic and Roman traditions. Thanks to the results of the excavations carried out during the last decades, this classical interpretation has changed [61, 114]. Today, we assume that the roman presence in Switzerland resulted in a complex process of acceptation and assimilation of new inputs on both sides: indigenous and Roman.

According to the sources, from the first decades of our era, we can assist to an intensified urban development, geographically and qualitatively, as a result of a general reorganization of Gaul (subdivision into provinces), with reuse and restoration of a series of military camps concentrated along the Rhine [115]. Punctual military settlements established from Caesar's time were extended to legionary camps to serve in the conquer/defense of the German regions. *Vindonissa* and *Aventicum* were emblematic, counting around 22,000 inhabitants [34]. They were connected by a network of high-quality roads, allowing them to control the main commercial routes that converged through the Swiss plateau towards the Mediterranean area. There are evidences that some of these main axes connected Italy to the Rhine and the Danube plain [12, 67, 96]. The installation of the military settlements had an impact on the expansion of urban agglomerations. As security brought more favorable conditions for artisanal and trade activities around these settlements, most of the *vici* were located in the proximity of those military settlements [31, 123] . The first evidences of this new phenomenon appeared in the western part of the Swiss plateau. *Aventicum* (Avenches) was designed as the capital of *Helvetii* [68] and elevated to the status of colony—*Colonia Pia Flavia Constans Emerita Helvetiorum Foederata* [3, 47] in 70 AD. *Aventicum* was built close to a preexistent *oppidum* on the hill of Bois de Châtel. As the archaeological research has shown, indigenous traditions and religious places coexisted with roman institutions and temples [11]. The city was moreover the leading center for the surrounding *vici*: Lousonna (Lausanne),

Minnodunum (Moudon), Uromagus (Oron) or Eburodunum (Yverdon), as well as for some structures in the proximity identified as: villa[5] "En Russalet", the villas "Boscéaz" in Orbe, "Mordagne" in Yvonand, the villa of Crissier, the villa "Baugy" in Montreux or the villa of Commugny (discovered during the highway construction works in 1986) [59]. All these discoveries represent another indication pointing at that the countryside was likewise highly populated from the middle of the first century AD onwards [64].

Similarly, in the northeastern part of the Country, *vici* like *Turicum* (Zurich) and *Vitudurum* (Oberwintertur) or *Vindonissa* and *Tenedo* [47, 105] were mainly located at the bends of rivers (Limmat, Reuss, Aare, Rhine) in a position easy to defend and suitable for the trading, on headlands or on the shores of main lakes like Kempraten on lake Zurich, and surrounded by trenches and fortifications [31, 38]. Similar to those in the western part of the Country, they were probably city-markets, provided by a *forum, tabernae, thermae* and one or more temples [34, 51]. Many manufacturing industries may have supplied the surrounding area with pottery, tools and other consumer goods. Crafts and trades, such as ceramics and tannery industries, were probably located on the city borders [30]. The long rectangular housing structures, often realized in wood, flanked the main road. They were decorated on both sides with narrow facades, while the cemeteries were located outside the cities, at the crossing of main roads [10, 19]. The countryside also experienced a big expansion e.g. in the Limmat and the Aare valley. In the area of modern Dietikon, the remains of an important villa has been largely studied [26]. This construction responds to the standard of the classical Mediterranean structure (*pars urbana* and *pars rustica*). In the light of its extension, researchers assumed that it could have been at the center of a bigger agglomeration of *villae*, probably a system where surrounding villas were located within a few kilometers (for example connected with the structures found in Urdorf, Zurich-Albisrieden) [31].

Generally speaking, the romanization of the countryside went hand in hand with urbanization and the enrichment of parts of the local population, resulting in a rapid evolution of the structures of rural establishments during the first centuries of our era, shifting from the first indigenous structures, essentially made out of wood, to the model of the roman Mediterranean *villae* in masonry [27, 64, 108]. Centers of a domain (*fundus*), usually respecting an axial symmetry, these *villae* generally included a residential part (*pars urbana*) with the master's house, surrounded by courtyards and gardens and equipped with central heating (hypocausts), thermal baths, frescoes, mosaics or sculpted decorations, and an area reserved for rural activities (*pars rustica*), where farm buildings (sheds, stables, barns, granaries, warehouses, workshops) and staff accommodations were scattered [31, 47]. Current research [83] is trying to clarify the extent, the economic organization and environment of these settlements (e.g. Pomy-Cuarny, Messen, Geneve Parc la Grange,

---

[5] *Villae* in the rural environment are buildings erected in the Roman era. They are self-sufficient in that they secure the basic needs of their users. They may be agricultural, industrial or administrative, depending on whether they are basically farms, produce garum or are engaged in mining. The nature of the activity would influence the type of construction, the size and the location. Their usage time is also important to how these structures were adapted to new demands.

Yvonard, Vallon). Similar to the urban agglomerations in the region of Vaud, developed at the crossroads of the routes connecting the Rhone and the Rhine basins with a commercial and craft vocation, the rural establishments played a link-role between the capital and the countryside [15]. The spatial mapping of the settlements supports the assumption that the large villas are equally distributed, giving the impression that they were the centers of larger domains to which belonged middle and smaller farmsteads [83].

Recent studies based on archaeological research tend to confirm that they were located at a distance of 30 km each other, corresponding to about one walking day along communication and commercial routes [26, 27, 31, 82].

In the Alps and the mountainous regions of the south, between the Canton of Valais and the Canton of Grisons, the romanization has left fewer visible traces still detectable today. On the one hand, the landscape morphology favored the reuse and the continuous occupation of the same settlements over the epochs. For this reason, the same urban agglomerations or alpine villages have been populated from the Iron Age or even earlier to the Late Antiquity (e.g. Scharans-Spundas, Maladers, Castel) [20, 70]. On the other hand, these regions lying outside of the most strongly urbanized areas, were interested by less intensive construction activities and consequently less extensive archaeological excavations. For this reason, most findings collected here are related to graves and coins. Generally, these alpine villages are in proximity to the transalpine paths and at the crossroads that led to the principal cities located in the central plateau (e.g. Bondo, Riam-Cadra, Schiers, Zernez or Zillis). Most of the known roman villages are located on hillsides up to 1600 m a.s.l. with a gentle to a strong slope, well exposed and sheltered from floods and landslides [70– 72]. The buildings were originally characterized by a small elongate structure with a single room (up to $40m^2$). Furthermore, the alpine regions were exploited for clay, iron and stone quarries used for artisanal production (e.g. Coira, Riom).

Modern Switzerland can be divided into 3 topological regions: the Swiss Alps in the south, the central Plateau and the Jura range and the Rhine valley in the north. The archaeological database created for this study covers almost all types of landscapes found and stretches from the westernmost to the easternmost tip of the country. It incorporates the diversity and heterogeneity of the settlement landscape resulting from the historically diverse occupation process and the various local and regional specificities. As mentioned previously, the analyzed areas (6 Cantons) were first processed individually and then as an ensemble, but from a historical point of view, these areas never experienced the Roman conquest and the romanization process in the same way.

4.4 Canton of Zurich

4.4.1 General Framework of the Region

The modern territory of the Canton of Zurich extends over 1729 km^2 and nowadays is considered productive for about 80% of its extension. Forests cover 505 km^2 and lakes 73 km^2. Most of the Canton consists of narrow river valleys that flow towards the Aare and finally the Rhine in the north. The main lake, the Lake Zurich (25 km long and max. 5 km wide), is one of the medium sized lake of the Alpine forelands that have formed after the last glaciation [25]. The hinterland is mostly shaped by moraines, also derived from the last glaciation. Along with the lakes Zurich, Greifensee and Pfäffikersee, the rivers crossing the landscape played an important role for commercial and communication purposes since the antiquity [25]. The Limmat river flows out of the north shores of lake Zurich and, in the Limmat valley, ancient populations started to settle. The *vicus* of *Turicum* (the ancient city of Zurich), the first military settlement of the Canton that occupied both sides of the valley [51], was also founded in this region [97]. Before the construction of the artificial Linth channel during the eighteenth century, the southeast region was occupied mostly by marshy and swamp areas that probably hindered any kind of land exploitation. Although the southwestern banks of the lake are very narrow and can limit the agriculture activity, the northeastern hills are quite suitable for agriculture and have been intensively exploited for their richness in soils mineral. Parts of the southwest are densely occupied by coniferous woodlands [25].

During Roman times, the landscape was particularly dense of roads and settlements. As attested by the Amt fur Raumentwicklung, Kantonsarchäologie (Archaeological department) of Zurich, around 900 archaeological sites have been discovered all over the administrative area. Among the known ancient settlements, a significant role was played by *Turicum* and *Vitudurum*, which were primarily military settlements in a wider networking context [38]. *Turicum* is named as for a first time in a funerary stele, and along with *Vitudurum* as *Obfelden-Lunnern* are described by the sources as *vici* (civil settlements) [44, 108].

Based on the state of the current knowledge and according to the literature, these *vici* were mainly located at the bends of rivers, in a position easy to defend and suitable for the trades, usually established on headlands or on the shores of main lakes and surrounded by trenches and fortifications [31, 38]. Probably also playing a role of marketplaces, the ancient *vici* were equipped with a *forum*, *tabernae*, *thermae* and one or more temples. Numerous manufacturing industries produces pottery, supplies and other materials for the surrounding areas. During the *Pax Augusti*, the settlements constantly grew, becoming more important urban centers [19, 33, 73].

Next to the *vici*, a certain number of urban *domus* and hundreds of rural *villae* of varying sizes occupied the countryside, e.g. the villas of Dietikon, Neftenbach, Buchs, etc. [26, 27, 73, 74]. Agriculture assumed a very important role as main subsistence activity. Thus, the environmental factors such as the suitability of the terrain for agriculture, the proximity to water resources, the topographic indices etc. acquire a predominant relevance for the research developed in this context.

4.4.2 Archaeological Dataset

The source data was provided by the cantonal archaeological service of Zurich[6] as a digital table containing a list of surveys carried out over the last decades (5812 rows catalogued until October 2015) and embedding artifacts ascribed to different epochs (from Mesolithic to Middle Age). This table is structured in several fields with a set of geographic coordinates in the local reference system (CH-1903 LV95), e.g.

- FS Suchbegriff (identifier of the discovery)
- FS Gemeinde (municipality)
- FS Name (name)
- FS Beschreibung (short description)
- FS Fundstellenart (type of evidence)
- FS Epoche (assigned epoch)
- FS Nord-Koordinate (X coordinate)
- FS Ost-Koordinate (Y coordinate).

For the purpose of this research, only the fields: FS Epoche, FS Beschreibung, FS Fundstellenart, FS Nord-Koordinate and FS Ost-Koordinate were considered. By taking a closer look at FS Epoche, it is possible to observe several categories for the epoch definition:

- BZ (Bronze Age)
- Eisenzeit (Iron Age)
- FBZ (Early Bronze Age)
- FLT (Early La tène)
- FrhNZ (Early Modern age)
- frhM (Early Middle Age)
- HA (Hallstatt)
- hochM (High Middle Age)
- jüngereHA (early Hallstatt)
- LT (La tène)
- MA (Middle Age)
- MBZ (Middle Bronze Age)
- ME (Mesolithic)
- mittlereNZ (Middle Modern Age)
- NE (Neolithic)
- NZ (Modern Age)
- PA (Paleolithic)

[6] Kanton Zürich, Baudirektion, Amt für Raumentwicklung, Archäologie & Denkmalpflege.

- Prähistorische (Prehistoric)
- Rezent (Recent)
- RZ (Roman age)
- SBZ (Late Bronze Age)
- Unbestimmt (undetermined)
- *blank*.

Thus, the analysis focused only on the rows classified as RZ (Roman age), for which the corresponded findings could also have had different specific socio-economic functions listed in the FS Fundstellenart and further specified in FS Beschreibung. Sites are inventoried either as a permanent settlement (e.g. housing, villa, vici), a place of worship and of religious identity (e.g. funum, temple, etc.), burial graves, necropolis, artisanal production centers, street, etc. or as a non-permanent settlement (scatter finds, ceramic shreds, coins etc.). Figure 4.3 shows the amount of sites attributed in each class for the roman period. Of particular interest for this study are the roman settlements (266 cases) and the single finds (416 cases). However, a certain degree of uncertainty exists at this stage, as not all entries possess doubtless coordinates. In fact, the coordinates of some entries are reported as imprecise or unsure. Other entries do not possess any coordinates and had to be deleted.

Some of the rows contain subjective interpretations concerning the evidences discovered, e.g. see the FS Beschreibung field (short description). This descriptive field may also contain short notes or just suggestions. An accurate screening and implementation of this original table was thus performed (Table A.6) (further details about the Exploratory Locational Analysis and Pre-Processing are given in the Chap. 5).

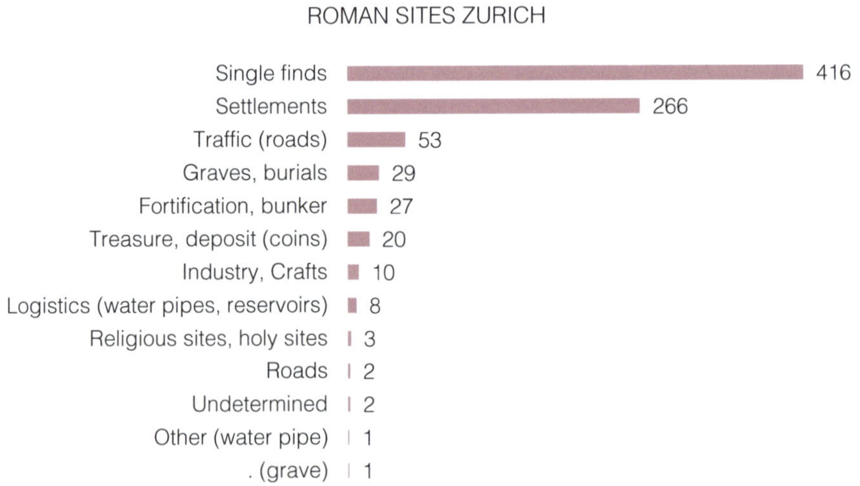

Fig. 4.3 Site typological classes for the roman sites in the original database of the Canton of Zurich

4.5 Canton of Aargau

4.5.1 General Framework of the Region

Today, the Canton of Aargau is one of the most densely populated regions of Switzerland. Its total area is 1,404 km^2. Located in the north central part of the Country, this is one of the least mountainous regions; flat and arable lands are predominant as in the Canton of Zurich and together they form part of the Swiss Plateau (Mittelland) situated between the Alps to the south and the Jura to the northeast [98]. This part of the Country is characterized by undulating landscapes and forested hills, as well as fertile valleys drained principally by the river Aare and its tributaries. Most of the lands are used for farming, while one-third of the region is wooded and the rest is unproductive because of the presence of lakes and rivers [89, 98]. The geological conformation and the geographical position of the area, crossed by the main rivers of the Swiss Plateau, explain the strategic importance of the region of Aargau since antiquity. The Rhine forms a natural boundary to the north, while the Aare crosses the Canton in a west-northeast direction from Aarau at the foot of the Jura mountains to join the Rhine in Koblenz. The Reuss and its tributaries played an important role in long-distance trade, as a south-north traffic axis [14, 89].

Situated at the crossing point of the Aare, Reuss and the Limmat rivers, the region of Aargau was chosen as settlement area by the *Helvetii* after the Battle of Bibracte (58 BC). They decided to set up a series of fortified settlements on the cliff of Windisch (i.e. Mellingen and Baden-Kappelerhof), from which the lower Aare valley could have been easily controlled [36, 48]. However, the knowledge of this period is still scarce. Similar considerations probably also led the Romans in their choice to settle on the plateau of Windisch, overlooking the confluence of the Aare and the Reuss, as a location for the military *castra* of Vindonissa, Kaiseraugst and Zurzach in order to protect the main commercial routes and the Rhine [36, 38].

In the years following 15 AD, the Plateau and hence also the territory of Aargau, were incorporated as a *civitas* of the *Helvetii* in the Roman province of *Belgica* [36]. During this period, many villages around the *castra* and several farms were built in the countryside as rural villas primarily to supply the roman legionaries. The villas were connected by a dense road network, on the main axes of two military roads: Strasbourg(F) - Augst - Windisch - Zurich and Hüfingen(D) - Zurzach - Windisch - Solothurn - Avenches, which led to the Rhaetian passes and the Great St. Bernard Pass [12, 47, 61, 105]. As for the Canton of Zurich, the Roman military expansion to the north and the shift of the limes in the second half of the first century marked the beginning of a long period of peace (150 years) on the Swiss Plateau in general, which from a border region became an internal region of the Empire, thus assuming a completely new role [38, 73, 81].

There is little information about the presence of *vici* in this area, apart from those of *Salodurum*, Olten and Holderbank, which can be classed as small towns according to Schucany [83]. As research highlighted, three different types of rural settlements on the Swiss Plateau are known at the date: large villas with domains exceeding

5 ha, middle-sized villas with domains reaching up to 3 ha, and small farmsteads cultivating areas of up to 1 ha. A number of settlements cannot be attributed to any of these three categories with any certainty. These correspond to the majority of small farmsteads with small areas lying above 500 m a.s.l., while larger villas were hardly found above this altitude [83]. Moreover, several find spots located in the Jura, lying at higher altitude have been classified by scholars as "forest dwellings" [31]. They were probably used in connection with the nearby forests for timber and coal or for cattle rearing and pastoral economy. A few of them could be interpreted as sanctuaries. A number of large and middle-sized villas have been identified, as they yield clearly visible traces unlike the small farmsteads for which only a small number have been identified up to date [31, 38, 73]. According to the natural borders, like the Jura mountains and rivers, and according to distance to the next large villas, these domains seem to extend over a surface of 2000 ha. [83].

Unfortunately, the current research state about the development of rural settlements in the area of the northeastern Swiss Plateau is incomplete and much has still to be done in order to have a more comprehensive knowledge of past human landscape occupation.

4.5.2 Archaeological Dataset

The archaeological evidences were collected by the cantonal archaeological department of Aargau.[7] As for Zurich, the source data was provided by the Kantonsarchäologie Aargau as a digital table containing a list of the surveys carried out over the last decades (3101 rows catalogued until 2015) and embedding the objects belonging to different epochs (from Mesolithic to Middle Age). The table was originally structured as follows:

- Fst-Signatur (Identifier),
- Gemeinde (municipality of the discovery),
- Nähere Ortsbezeichnung (detailed name of the location, toponym),
- Text für Gemeindeliste (Text for the municipality list),
- X Koordinate (X coordinate),
- Y Koordinate (Y coordinate),
- Befund/Qualität (type of evidence/quality of the interpretation),
- Datierung/Qualität (assigned epoch/quality of the interpretation).

Compared to Zurich, the dataset of Aargau presents two more fields annexed in the table, where the reliability and uncertainty of the information has been incorporated and stated clearly (see Table 4.1).

Uncertainty regarding the recognition of the evidences is reported in the 'Text for municipality list', in the 'Type of evidence/quality' and in the 'Assigned epoch/quality' fields.

[7] Kanton Aargau, Departement Bildung, Kultur und Sport, Abteilung Kultur, Kantonsarchäologie.

Table 4.1 Incorporation of uncertainty in the original database of Aargau (with English translation on the right)

Befund/Qualität	Datierung/Qualität	Evidence/Quality	Epoch/Quality
Siedlungsstelle - unsicher	römisch - sicher	Settlement - unsure	Roman - sure
Siedlungsstelle - unsicher	römisch - unsicher	Settlement - unsure	Roman - unsure

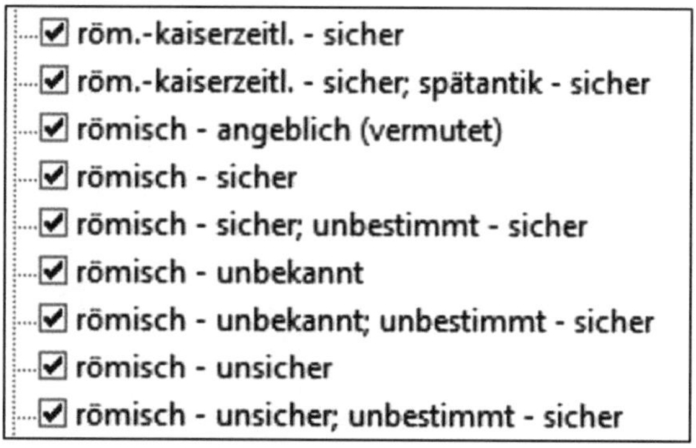

Fig. 4.4 Extract from the Befund/Qualität field of Aargau DB

Sometimes, we encounter a double uncertainty, as for example in the second row shown in Table 4.1. This entry points at a discovery located in the locality of Altenberg. The description field for this survey states: 'Fragments of Roman (?) ceramics and tiles; from a Roman settlement site?' (Römische (?) Keramik- und Ziegelfragmente; von einer römischen Siedlungsstelle?). Then, in the 'Type of evidence' field, the result of the survey is classified as 'Settlement - uncertain' (Siedlungsstelle - unischer) and the quality of the epoch assigned as: 'Roman - unsure' (römisch unsicher). Figure 4.4 shows the different categories for uncertainty implemented in the database for the roman period.

Many fields carry uncertain information and numerous entries lack geographic coordinates (often referencing to old findings) and several lines have blank cells. Another example of uncertainty is the following: The place where the archaeological survey occurred is not specified: 'Not uncommon in Aarau' (in Aarau nicht selten) and the description of the survey results is recorded as follow: 'very old finding report; not exactly localized: repeatedly graves, sometimes with offerings; dating unsure, possibly alemannic/early medieval; belonging to a grave group?' (sehr alte Fundmeldung; nicht genau lokalisiert): wiederholt Gräber, zum Teil mit Beigaben; Datierung unklar, möglicherweise alamannisch/frühmittelalterlich; zu einer Gräbergruppe gehörig?). Under the type of evidence and the epoch assigned we find: 'Flat grave group/field cemetery – uncertain' (Flach-Gräbergruppe/-Feld-Friedhof – unsicher) and 'early medieval – uncertain' (frühmittelalterlich - unsicher).

Although the new GIS database is constructed on the information provided within this digital table, for the best of the modeling procedure and in order to obtain a more reliable output, a more consistent corpus of archaeological and spatial information referring to the Roman period was elaborated (further details in the Chap. 5). The database of the Canton of Aargau contains 53 different classes of evidence (see Fig. 4.5). Amongst others, the different classes were reorganized in order to match the categories of the other Cantons (Table A.1). Finally, an ad hoc procedure was

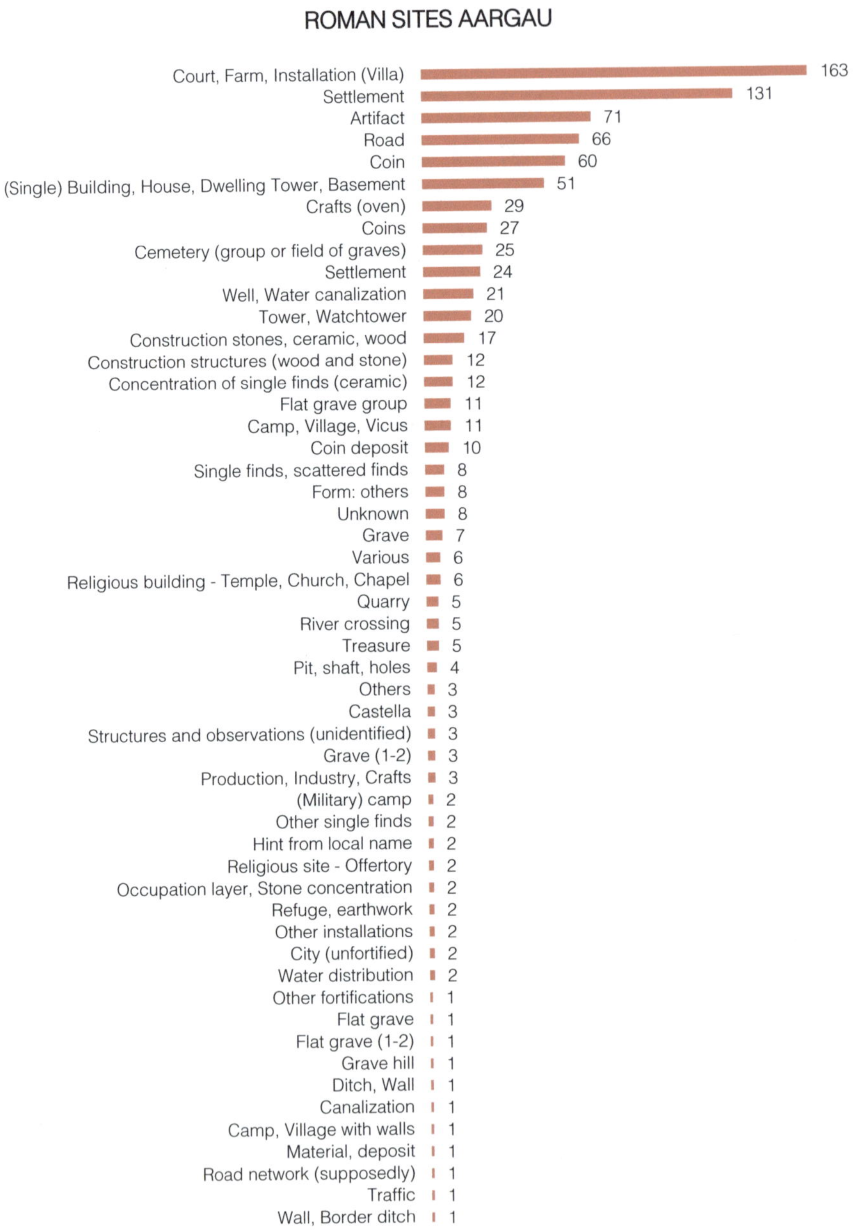

Fig. 4.5 Site typological classes for the roman sites in the original database of the Canton of Aargau

elaborated to deal with the uncertainty information contained in this database and to embed it in the final predictive map.

4.6 Canton of Grisons

4.6.1 General Framework of the Region

Grisons is the biggest Swiss Canton with a surface area of 7,105.39 km^2 and the most mountainous region of the Country (elevation between 260 and 4,049 m), where only one third of its land is considered productive because of the steep mountainous ridges. The population is concentrated in the valleys (almost 150), mainly the Vorder- and Hinterrhein valley (Surselva and Sutselva) and Engadin valley. The modern Cantonal administrative limits include international borders with Austria to the east, with Italy to the south and with Lichtenstein to the north [49]. The diversity of neighboring cultures has influenced the development of the local cultures since ancient periods. Moreover, the geographical and geomorphological position of the Canton can explain the development both of the local culture and of the settlement distribution over the centuries, subdivided into valleys where regional traditions and external influences coexisted [55].

Roman historical sources describing the situation in the region are very scarce. In general, the local populations are perceived negatively by Roman authors probably because the region was inhabited by different Celtic tribes (*Rhaetii* and *Lepontii*) [20, 55]. When the region was embedded in the *Provincia Rhaetia* after the Roman conquer of the Alps, several main towns were built in the western part of the province, including Chur (*Curia*), the current administrative capital of Grisons [38, 71, 77].

So far, there are no known written sources describing the legal status of Chur at that time. It is assumed that it was a *vicus*. The remains of the Roman city of Chur were located where the roads leading to the Rhaetian passes branched off and where navigation on the Rhine to the north probably began [50]. The important role of this city as a regional center is also documented by the remains of public and private buildings in Welschdörfli, on the left bank of the Plessur: *Thermae, officinas, tabernae, hospitia* and typical long buildings containing imported goods. There is also evidence of a military post for the control of transit routes [61, 70–72, 77].

Other settlements developed along the main routes in the valleys, or at higher altitudes reaching up to 1500 m a.s.l. as Zernez [72]. Generally, these settlements preferred south, southwest orientation and a certain distance from the nearest river. An example is the farming settlement of Riom, the best studied so far, that consisted of at least two comfortable masonry buildings, which also served as a *hospitia* for travelers. Other similar buildings seem to have had specific functions related to trading traffic [70, 72, 77]. Since the Bronze Age until the Roman epoch, the main subsistence economies in the Alpine regions are agriculture, livestock farming, and copper ore mining (e.g. Oberhalbstein, Upper Engadin, Lumnezia, etc.). Trade over the Alpine passes has also been documented [38, 61, 70].

4.6.2 Archaeological Dataset

The archaeological dataset was provided by the Archaeological department of the Canton of Grisons.[8] The source database contains a list of the surveys carried out over the last decades (3472 rows catalogued until 2015), embedding the objects belonging to different epochs (from Mesolithic to Middle Age). The table was originally structured as follows:

- FS Suchbegriff (identifier of the discovery)
- Gemeinde (municipality)
- FS Name (name)
- FS Beschreibung (short description)
- FS Fundstellenart (type of evidence)
- FS Epoche (assigned epoch)
- FS Nord-Koordinate (X coordinate)
- FS Ost-Koordinate (Y coordinate).

The organization of this dataset is similar to that of the dataset of Zurich regarding the fields and the categories. With regard to the fields considered for this research, the epoch assignment FS Epoche contains the following definition in addition to those found in the dataset presented in the Sect. 4.4.2 (Zurich):

- Geologie (Geology)
- Historische Zeit (Historical Age)
- Kaiserzeit (Imperial Age)
- FrüheKZ (Early Imperial Age)
- Spätantike (Late Antiquity)
- SpätM (Late M)
- SpätME (Late Mesolithic)
- SpätNE (Late Neolithic)
- SLT (Late La Tène)
- MittlereKZ (Middle Imperial Age).

Here too, some rows present blank cells or missing geographic coordinates. Hence, only those rows referring to the Roman epoch or the Roman Imperial Age and with consistent information were retained for the new database and the further modeling procedure (Table A.4). The total number of Roman settlements corresponds to 15 while the single finds to 57 (Fig. 4.6).

[8] Kanton Graubünden, Archäologischer Dienst Graubünden.

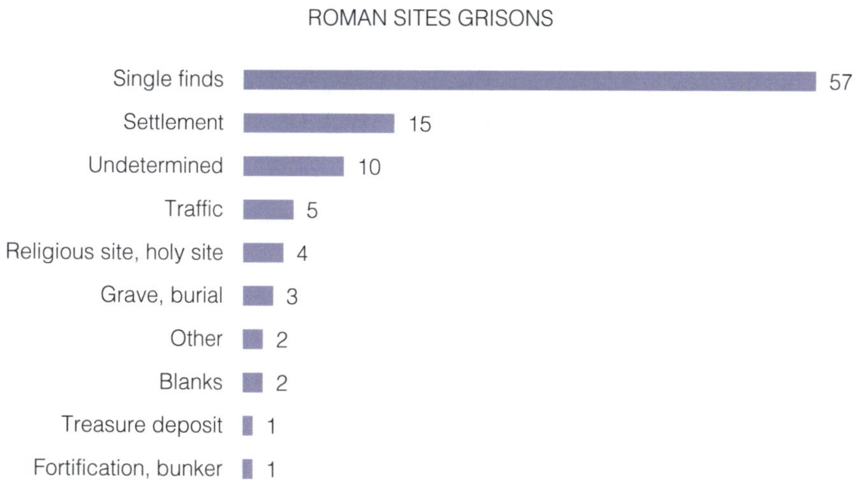

Fig. 4.6 Site typological classes for the Roman sites in the original database of the Canton of Grisons

4.7 Canton of Fribourg/Freiburg

4.7.1 General Framework of the Region

The Canton of Fribourg lies on the Swiss Plateau, between lake Neuchatel in the northwest and the peaks of the Prealps in the southeast, with an altitude ranging from 429 m to 2,389 m a.s.l. and a surface of 1,671.42 km^2. The areas to the west and to the north are mostly slightly hilly or flat and intensively cultivated while the areas to the southeast rising up towards the Alps are left used for grazing and pasture. The main river, the Sarine, crosses the Canton from south to north. In the northwest, the lakes of Morat and Neuchâtel determine the landscape. The main rivers next to the Sarine are the Broye and the Glâne in the west and the Sense in the east, all three part of the Aare basin [2]. The geomorphology of the region is generally attributed to postglacial erosion. The *Plateau Fribourgeois - Freiburg Mittelland* is composed of sandstone and moraine deposits and lake and river sediments have filled an ancient depression around the lower valleys of the Broye and the Grand-Marais. Erosion has carved sinuous gorges in the terrain left bare by the retreat of the Rhône and Sarine glaciers [13].

The first Roman settlements favored the Broye valley, the nearest Plateau and the Sarine valley, which constituted a commercial axis connecting the south to the north. The importance of the region is confirmed by its position. It acts as a transit zone for the military legions and for goods trade (silk, *sigillata*, wine, olives, oil, etc.). Several principal and secondary roads were constructed and numerous finds of coins and ceramic shreds testify of movements from Italy towards the Rhine passing through the region of Fribourg/Freiburg. A main route led from Martigny to

Vindonissa (*via* Vevey, Oron, Payerne, Avenches and Solothurn), and a secondary route led from Avenches to Yverdon [38, 84].

Along these routes, several villas were built as well as the only known *vicus* of Marsens-Riaz (I-III AC), along the Sarine river [109]. The construction of rural villas with the typical division in *pars urbana* and *pars rustica* in the countryside were favored by the well-organized route network and the highly developed economy of the region. The Broye river and lake Morat were intensively crossed by boats transporting limestone blocks from the Jura alps to *Aventicum* or to other places. They could belong to a Roman or to an aristocratic Helvetian that had some privileges [18]. As testified by the surveys, several hundreds of these villas were located around *Aventicum*, but others popped up all across the landscape, like in Ferpicloz, Chablais, Morat, Bösingen, Ursy, etc. [1]. In some cases, the luxurious furnishings like the well-known mosaics of Vallon and Cormérod were preserved (scenes of hunting in an amphitheater and Bacchus discovering Ariadne asleep in Vallon, Theseus against the Minotaur in Cormérod) [4, 37]. As for the other Swiss regions analyzed in this context, the decline of the Roman empire from the middle third century AD signed a decrease in the number of Gallo Roman establishments. Socio-economic changes and the insecurity caused by the Germanic incursions explain the decline of most of the sites [67].

4.7.2 Archaeological Dataset

The archaeological data was provided by the Service Archéologique de l'Etat de Fribourg[9] in the form of two different digital formats containing the information concerning the Roman or Gallo Roman period and the Prehistoric settlements. In the case of Fribourg, the data concerning the Roman period could be extracted directly from the local database system already inventoried in a GIS format. The database contains 798 rows organized in the following classes:

- Point (sites)
- Lines (routes)
- Polygons (remains of ancient aqueducts).

The Points class contains the sites classified in the following categories:

- Localité (Locality)
- Liue-dit (Name of the locality)
- Y (Y coordinate)
- X (X coordinate)
- Z (Z coordinate)
- Fonction (Function)
- Remarques (Comments).

[9] Etat de Fribourg, Direction de l'instruction publique, de la culture et du sport, Service archéologique de l'Etat de Fribourg.

The data for the other periods (except Middle Age) containing 823 rows was provided in a digital table and classified in following categories:

- ID (Identifier)
- Localité (Locality)
- X (X coordinate)
- Y (Y coordinate)
- Nature du site (type of evidence)
- Epoque (Epoch).

For the sake of completeness, the *Epoque* field in this table is divided in the following categories:

- Protohistorique
- Paléolithique
- Préhistorique
- Néolithic
- Mésolithic
- Bronze
- Indeterminé
- Mixte
- Hallstatt
- La Tène.

Thanks to the systematic structure of this original database, very few preprocessing maneuvers were necessary to adapt this data for the modeling procedure. As for the other Cantons, the research focuses only on the sites belonging to the Roman period (Table A.2). These are originally divided in 39 different classes shown in Fig. 4.7. Single finds and various types of settlements represent the large majority of cases.

4.8 Canton of Vaud

4.8.1 General Framework of the Region

The Canton of Vaud stretches over 3212,03 km^2 in the west of the country at the separation of the Rhone and Rhein basins and between the lakes Leman and Neuchâtel. Its lands lie between 372 and 3210 m a.s.l., mostly on the Swiss Plateau. The territory is generally very diversified with high mountain peaks and fertile lake shores. The diversity of the shoreline landscapes is one of the characteristics of this region along with that of the bordering Canton of Geneva [117]. The shores of lake Leman are hilly in the north and in the southwest. Though in some places, the slope makes the shores unstable or prone to flooding and exposed to waves, lake Leman has always been an attractive environment due to its climate, natural irrigation, fishing opportunities and easiness of travelling [104]. Everything suggests that it played an important

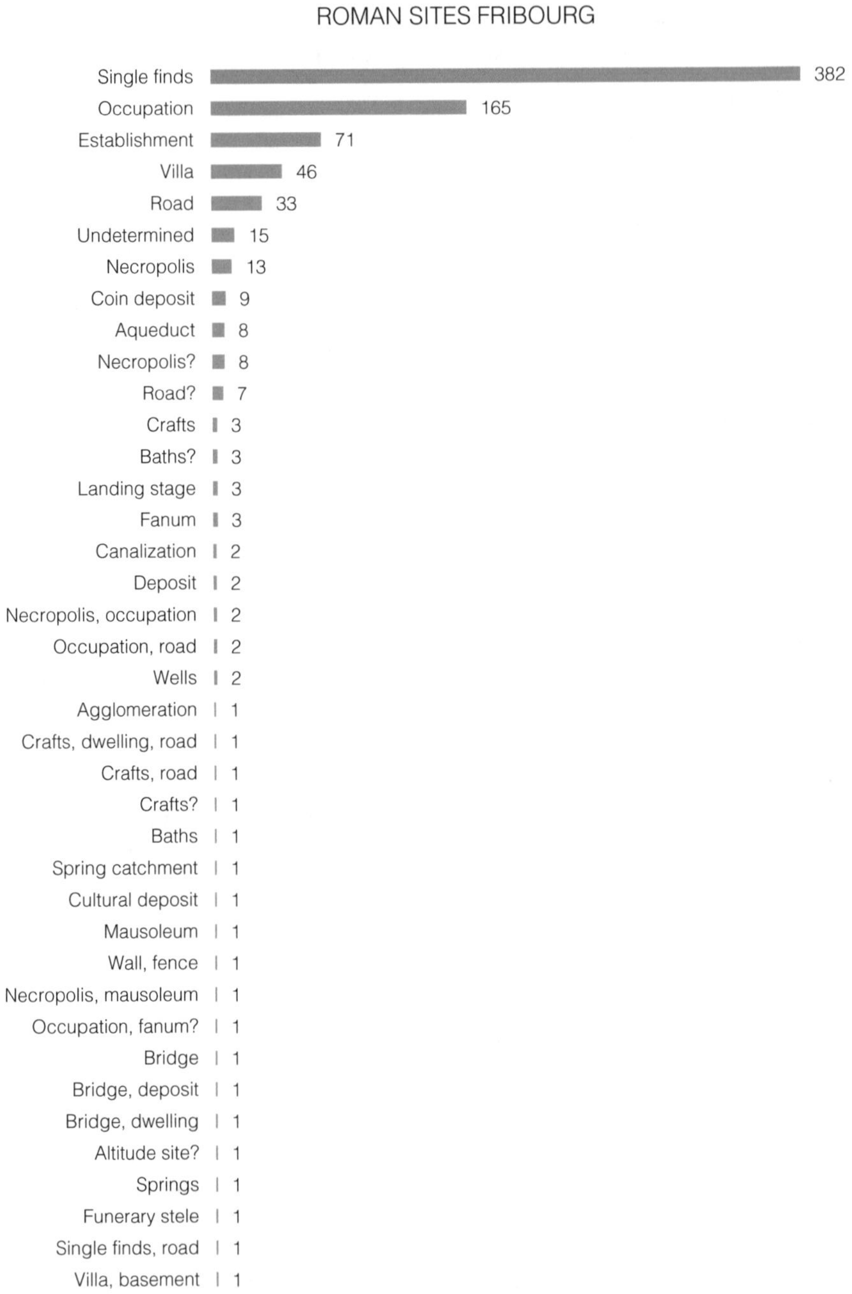

Fig. 4.7 Site typological classes for the Roman sites in the original database of the Canton of Fribourg

role in the human distribution trends over the actual regions of Vaud and Geneva. Indeed, lake Leman and the course of the Rhône river are known since Hellenistic times, although through imprecise information for most of the time. Unfortunately, no historical source makes any mention about the navigability of the lake or of the river for example, nor do they mention fishing or local products exported to Rome [10]. In general, given its vast area (2822 km^2 without lakes) and its geographical and environmental heterogeneity, the Canton of Vaud is particularly rich in prehistoric sites and single finds as well as in Roman settlements, as confirmed by more than 3500 discoveries made by the local archaeological service until 2015.

As mentioned previously in this Chapter, the *Colonia Iulia Equestris* of Nyon can be considered as the "capital" of this lake front region. Remains of epigraphs and medieval texts mention an ancient *pagus Equestris* extending from the right bank of the Rhone to Aubonne and from the Jura to the shores of the Leman, bordering with *Helvetii*, *Sequanii* and *Allobroges* territories (Buchsenschutz and Curdy [15, 35]. Nyon benefited of a programmed and regular urban planning. The city is equipped with public buildings and monumental setting, experiencing an important phase of progressive monumentalization: towards the middle of the first century, the center was constituted of around eight *insulae* with an amphitheater, commercial areas and rich suburban villas (only partially known). Evidence of water infrastructures suggest the presence of an ancient harbor (not yet identified). As Tarpin [113] pointed out, since Nyon was located on the mid-way between Geneva and Lausanne, it is likely that it mainly played the role of a showcase city, highlighting the presence of Rome in the region and its position while allowing for controlling possible southward migrations, enclosed between the Jura and the lake [57].

The Roman presence in the region is testified by the increasing volume of commercial traffic of goods from southern to northern Europe passing through the main rivers and the lakes Leman and Neuchâtel [30]. This intensive activity is moreover confirmed by the presence of the *Nautes* corporation (ship owners) of Geneva based in Lausanne-Vidy, which ensured the transportation of olive oil, wine, garum, pottery, and raw materials from Lake Leman to Yverdon [63–65].

Next to the flourishing trade economy, the agriculture and the craft production largely contributed to the growth of the region. This growth allowed for the extension and remodeling of many pre-existing indigenous settlements (e.g. Lausanne, Avenches), while others were created ex *nihilo*. Around the second century AD, villages were located roughly every 30 km along the road axes. As expression of the Roman power, these regional centers favored the romanization of the local population and provided a favorable framework for the local aristocracy to maintain its prerogatives. The main *vici* were indeed equipped with masonry and monumental buildings (thermal baths, theatres, amphitheaters as in Vevey (*Viviscum*) or Lousonna-Vidy [63, 65].

The *vicus* of Lousonna-Vidy developed on a slightly terraced terrain between the lake and the moraine hillside of Bois-de-Vaux. Similar to the majority of the areas in front of Lake Leman, the landscape is shaped by the presence of terraces, which were formed at different times. Remains of dwellings, dating between 100 and 50 BC, have been identified on the hill of the cathedral of Lausanne [63, 112, 113]. Lousonna-Vidy was an important station on the commercial routes between the Rhone and the Rhine basins, at the crossroads of the main routes linking Italy to *Gallia* and the Rhine border through the Great St. Bernard pass, and the Mediterranean sea to Germany through the Rhone valley [28, 69]. The *vicus* of Lousonna-Vidy is also the westernmost agglomeration of the territory of the *Helvetii* and was organized in length, experiencing an intensive development around 40–20 BC [5, 63]. Many public buildings have been excavated and structures dedicated to artisanal activities such as metallurgy and pottery, as well as an inscription with a mention to the *nautes* of the Leman reveal the main vocation of the ancient city of Lausanne [28, 52, 113]. At the end of the first century, the city was equipped with an important harbor. The remains of a pier were found in 1935–1940. The shore of the lake was stabilized by a rocky bank supported by oak piles. This sort of dam was used to protect the shore against the effects of large seasonal variations in lake levels and to break storm waves. Lousonna was probably an active harbor capable of handling large volumes of traffic [52, 69].

Other sites in the Leman region probably participated in establishing commercial activities. The ancient site of Vevey-Viviscus, known through incidental discoveries and mentions in geographical sources, has been partially excavated. The habitat shows a *facies* similar to the houses of Lousonna-Vidy. The presence of a milestone marking the distance from Martigny suggests that Vevey was an outermost site in the *Pennine* Alps. Villeneuve has been assimilated to the ancient site of Pennelocus. The position of Villeneuve, between Vevey and Messongex, at the eastern edge of Lake Leman, makes it mainly a stop on the road from Martigny to Lausanne [63, 65, 66]).

The countryside was equally intensively remodeled by the Roman presence. Several villas, such as the villas of Champ d'Asile, Crissier, Lussery, Ecublens, the villa of Le Buy or the villa of Saint-Prex [22, 75] to name just a few presenting the main characteristics of the *villae suburbanae* around Lousonna, were well connected to each other and to the main *vici* through the established Roman roads. They often preserved a clear division into *pars urbana* and *pars rustica* and sometime have conserved rich ornaments like mosaics, wall paintings, pottery and bronze tableware [64]. The biggest and most spectacular are those in Pully, Commugny or Baugy, which show a clear preference of the owners for elevated sites, close to the shores and with a wide view on the lake and the Alps [29, 64, 101].

4.8.2 Archaeological Dataset

The archaeological dataset containing the inventoried evidences for all the periods identified and discovered within the administrative limits of the Canton of Vaud was provided by the local Service Archéologique.[10] The original data was extracted from

[10] Etat de Vaud, Direction générale des immeubles et du patrimoine, Division archéologie.

Table 4.2 Extract of the original database structure of the Canton of Vaud

Field name
OBJECTID
SITE_ID
REGION_ID
SITE_NO
SITE_LIEU_DIT
SITE_ALTITUDE_MIN
SITE_CENTROIDE_X
SITE_CENTROIDE_Y
SITE_FOUILLE
SITE_SONDAGE
SITE_TEXT_NAT
SITE_TEXT_PER
FONCTION_PRINCIPALE
PERIODE_PRINCIPALE_txt
PERIODE_PRINCIPALE
QUALITE_PERIODE
QUALITE_FONCTION
QUALITE_LOCALISATION
SITE_COMMENTAIRE

the internal GIS database in the Access format. It contains 3623 rows of different sites and typologies, from which 1181 belong to the Roman epoch. The database is organized according to the fields shown in the Table 4.2.

The fields and corresponding rows retained for the creation of the new database are the following:

- SITE_CENTROIDE_X (Site centroid X)
- SITE_CENTROIDE_Y (Site centroid Y)
- FONCTION_PRINCIPALE (Principal Function)
- PERIODE_PRINCIPALE (Principal Period)
- QUALITE_PERIODE (Quality of the Period)
- QUALITE_FONCTION (Quality of the Function)
- QUALITE_LOCALISATION (Quality of the localization).

The field FONCTION_PRINCIPAL contains the following categories:

- Indeterminée
- Défensive
- Religieuse
- Communication
- Funéraire

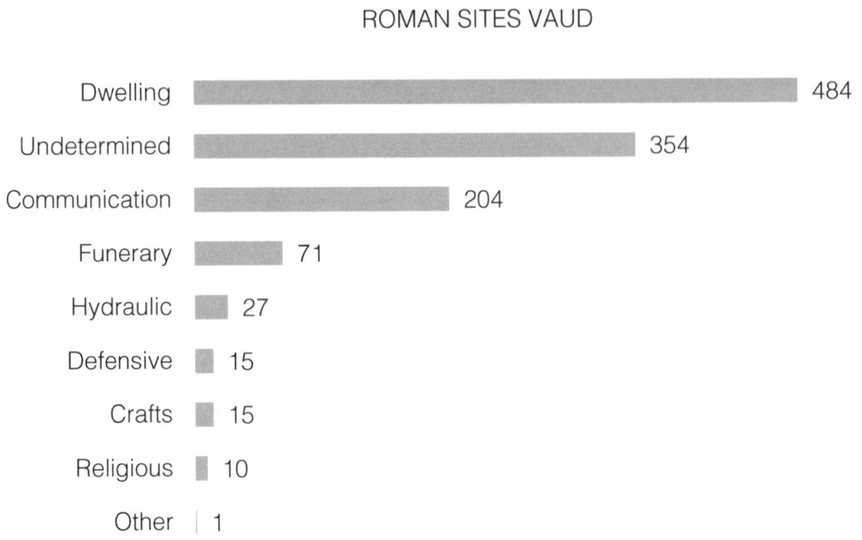

Fig. 4.8 Site typological classes for the Roman sites in the original database of the Canton of Vaud

- Hydraulique
- Artisanat
- Autre
- Mégalithes.

Figure 4.8 shows the quantity of settlements and other types of evidences for the Roman period in the original database. As for the other datasets, many rows contain empty cells and no geographic coordinates. Thus, in the data pre-processing phase, the dataset was rebuilt to be fitted to the modeling needs (Table A.5).

4.9 Canton of Geneva

4.9.1 General Framework of the Region

Bordering the Canton of Vaud to north, the Canton of Geneva, with an area of 242 km^2, is located at the westernmost tip of Switzerland at the foothills of the Savoyan Alps. It is part of the Swiss plateau and is located in a vast topographic depression limited to the north by the Jura range (reaching altitudes of over 1700 m), and to the south by the mountain of Salève (maximum altitude of 1379 m). The Geneva depression is limited to the southwest by the Vuache mountain (reaching an altitude of 1105 m), which was uplifted by important geological movements linking the Jura to the Salève and subalpine mountain ranges in the area of Annecy [103]. Situated on the shores of Lake Leman, the Canton of Geneva is mainly watered by the Rhone and Arve rivers. The present-day water level of lake Leman is situated at 372 m a.s.l., while in the past the level of the lake lay at more than 374.6 m, over 2 m above the current average level [41]. Such a water level rise cannot be only the consequence of a climate crisis (increase of rainfall and temperature drop), but rather of a natural dam occurred at the confluence of the Arve and the Rhône. This dam may also have been caused by other events like the collapse of the cliffs in the Saint-Jean district, facing the hill of the Bois-de-la-Bâtie [103]. The edges of the Geneva Basin, at the foothills of the Jura and Salève mountain ranges, are located at some 500 m a.s.l. The region's lands are mostly flat and very suitable for agriculture [103, 118].

The region was already occupied by the *Helvetii* and *Allobroges* Celtic tribes when the Romans established the first contacts with the populations on the other side of the Alps [10, 68]. For almost a century, from about 120 to 58 BC, Lake Leman was a link between the province of Transalpine Gaul and the Swiss Plateau. Moreover, no element of everyday life reflects a difference between *Helvetii* and *Allobroges* [121]. The literature and archaeological evidences for this period, however, are not very homogeneous and are mostly focused on Geneva.

Geneva itself, at the foot of Saint-Pierre hill, has been occupied since the middle of the second century BC. The territory around the Celtic oppidum of *Genua* (Geneva) was incorporated in the Gallia Transalpina province in 121 BC then named *Narbonensis* (27 BC) [8, 9]. At the end of the first century BC, while the left bank of the Rhone constituted part of the *Colonia Transalpina* or *Colonia Iulia Vienna,* the right bank of the river was part of the Colonia Iulia Equestris established in Nyon. It's only during the first century AD that the romanization process start to be tangible, confirmed by the monumental buildings and several rural villas occupying the countryside [31].

Genava became a *vicus* and then a *civitas romana*, assuming the topographic organization of a classical Roman city: the *decumanus* for example, crosses the upper city in an east–west direction and is still today recognizable in the urban fabric. As early as around 100 BC, the city was equipped with an omega-shaped harbor basin, made of oak piles logged in 122–121 BC. [8, 65].

Across Lake Leman, it was possible to reach the Saône via Lausanne, Orbe and Besançon, or the Rhine from Vevey, via Yverdon or Moudon, Avenches, Solothurn and Augst. Navigation on the lake was organized by the corporation of *nautes* of the Leman. From an epigraph, we also know about another corporation: *retiarii superiores*, which was based in Geneva and ensured the navigation from Lyon to Geneva, while the *nautes* of the Leman ensured that from Geneva to Messongex [68, 78].

Historically attested, a Roman bridge crossing the Rhone River, reconstructed in place of a more ancient one, is another evidence of the coherent development of a comprehensive road network in this area. The islands located in the middle of the Rhône bed are composed of clay and sandy blocks attributed to the Last Ice Age upper basal moraine. This material was certainly particularly favorable to the installation, allowing the sinking of piles for the construction of bridges [41, 103].

The ancient settlement of Geneva covered nearly 20 ha between the shore of the lake and the hill of Saint-Pierre, probably fortified by a network of trenches, defining a promontory. At the highest point of the hill, a trench delimited a privileged area, either a citadel (arx) or a religious area [9, 119].

The reconstruction of the ancient plan of the *vicus* remains incomplete. It is nevertheless the most important agglomeration of the Lake Leman area, extending over areas of nearly 35 ha on the left bank and 10 ha on the right bank of the Rhone river during the second century of our era. While the left bank seems to accommodate the majority of commercial activities, as shown by the construction of a new harbor and the presence of storehouses and thermal baths along the shores of the lake [8, 91, 116, 120], on the right bank, only the remains of a sanctuary whose origin dates back to the middle of the first century BC are well identified. The *vicus Genevensis* flourished till the second century AD, when a violent fire probably destroyed extensive parts of the *vicus* [121].

Few villas have been completely excavated on the shores of the Lake Leman. However, aerial surveys give a density of about one villa/km^2 for the Canton of Geneva [64, 116].

The best known of these villas is that of the Parc de la Grange. Established in a 400 m $\times$ 200 m enclosure, in a place occupied since the final Bronze Age, the *pars urbana* included a network of trenches and a dump dating back to the end of the 2nd or the beginning of the 1st century BC. It was repeatedly expanded until the fall of the Roman Empire and partly occupied until the 12th century. The buildings of the *pars rustica* were used until the 4th century, before being rebuilt during the 5th or 6th century [45].

4.9.2 Archaeological Dataset

The data concerning the archaeological evidences discovered in the administrative limits of the Canton of Geneva were provided by the local Service Archéologique[11] as an extract, in digital format, of the local GIS database. The original dataset contains 865 rows corresponding to the evidences covering all periods, from which 185 belong to the Roman period. The original database is structured as follows:

- ID_SITE (Site identifier)
- COMMUNE (Municipality of discovery)
- ADDRESSE (Address)
- TYPE (Type of site—settlement/single find)
- NATURE_SITE (Site typology—interpretation)
- LOCALISATION (Location)
- CIRCONSTANCES_DECOUVERTE (Discovery conditions)
- DATE_DECOUVERTE (Date of discovery)
- DATE_FOUILLE (Date of the excavation)
- ETAT_ANCIEN (Ancient status)

[11] République et canton de Genève, Office du patrimoine et des sites, Service cantonal d'archéologie.

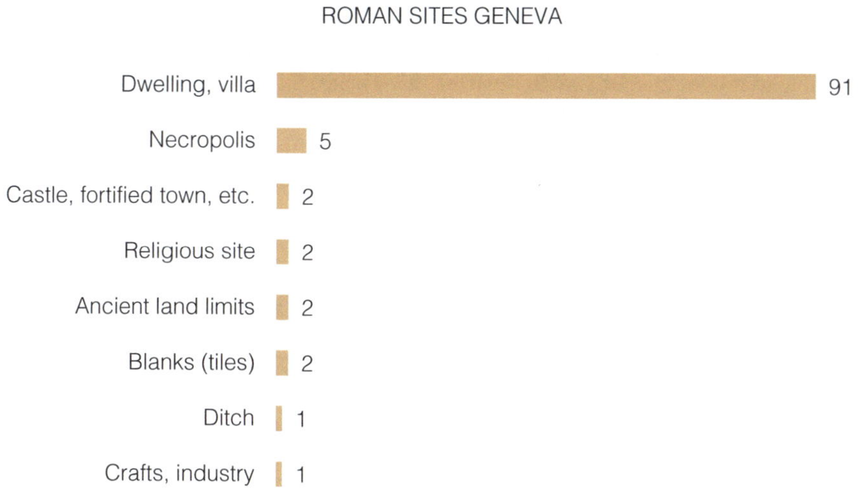

Fig. 4.9 Site typological classes for the Roman sites in the original database of the Canton of Geneva (excluding rows with incorrect geographic coordinates or missing values)

- ETAT_ACTUEL (Current status)
- REMARQUES (Comments)
- BIBLIO (Literature).

Unlike the other databases, the geographic coordinates are not embedded in the table and the field incorporating the epoch information is missing. While the other datasets consist of points as a shape of representation on the map, this database contains georeferenced polygons. In the pre-processing phase, the information regarding the epoch of reference was extracted and embedded within the new database for each row (Table A.3). The classes contained in the original dataset are shown in Fig. 4.9.

4.10 Geo-Environmental Predictors

The geo-environmental features assumed to have somehow influenced the location of human settlements in the regions analyzed here were carefully studied and then selected in order to be quantified and used in the modeling procedure. As the approach chosen in this Thesis is a data driven one, no prior assumption about which specific class of soil or land may have influenced site location choices have been formulated for the selection of variables. Moreover, only sites of the Roman period were considered for the model. Hence, agriculture, breeding and fishing are assumed to be the main livelihoods. Site location preferences must have taken into account specific soil characteristics and a certain distance from rivers and lakes [94] in order to carry on such activities. It is possible to assume that a part of the livelihood derived also by the trade established along the well structured terrestrial and water routes, but no sufficiently exhaustive information are given about its capacity that could made it possible to model this information within this Thesis.

Thus, if we rely on classical Roman texts and on the recommendations passed on by ancient authors, they would describe the best suitable terrains for agriculture as shown in Table 4.3, for example. Whilst it is probable that we cannot trust the sources for reconstructing a 'true' picture of productivity, an important factor in selecting geographic criteria is the availability of agricultural suitability data. It can be considered that the environment has not been affected by substantial variations in the lowlands since the Roman age [25] and that during the last 2,000 years, environmental and climatic conditions have remained relatively stable. With this in mind, several modern geographic data regarding the topography and the underlying geology, agricultural suitability and soil types were derived and processed in a GIS environment and complemented by other computed geographical data (e.g. distance from rivers and lakes).

Table 4.3 Recommended characteristics for agriculture according to ancient authors (in: [42])

Author	Recommendations (Characteristics)
Vitruvius (I, 4–5; VI, 1)	Raised terrain East or west exposition Distant from swamps Fertile soil Good traffic infrastructure
Cato (I, 2)	Protected from weather Fertile soil South exposition At the foot of a mountain Available manpower Source of water Near market or traffic infrastructure
Columella (I, 2–5)	Healthy climate Fertile soil level and inclined ground East or south exposition Traffic infrastructure Main house near the level part Quarry and water source nearby Protected from weather On hillside on slight elevation (safe from floods) Available water source Near, but not directly on road Exposition "towards sunrise at equinox"
Varro (I, 2; I, 4; I, 6–7)	Ground resources Near street Healthy area (climate, winds, swamp) Slight inclination (for drainage)

The geo-environmental data and the archaeological data are expressed in a vector or raster format to store geospatial information in a GIS environment and both are used within this Thesis. While in the first instance most of the variables come in a vector format, for the modeling procedure itself all of them are turned into raster in order to be compatible with the 100 m grid constituting the model (see Chap. 5 for further details).

Vector data (archaeological sites) exists as x and y coordinate pairs representing points, as lines or as polygons, with an associated attribute table. Raster data (variables) are usually grid data such as images, categorical data such as land use classification or continuous surface data, such as digital elevation models. A raster file consists of x, y and z data. X and y are the two-dimensional location of the cell, and z is a value for the elevation.

The selected geo-environmental features prone to influence the site locations and acting as independent variables in the modeling procedure are as follow: The topography with the Digital elevation model (altitude), and the calculated derivatives: Slope, Aspect Northness and Eastness; the cantonal administrative borders; the Hydrology with calculated distance to water (using the main lakes and rivers); the soil map with reference to the agricultural suitability and finally the Geology.

4.10.1 Topography

Many physical processes in the environment are dependent on the properties of the terrain. A DEM (Digital Elevation Model) is a quantitative model of a topographic surface in raster format, consisting of an array of elevations for many ground positions at regularly spaced intervals. Also known as a 'digital terrain model' (DTM), it is often used in reference to a set of elevation values representing the elevations at points in a rectangular grid on the Earth's surface. Collins and Moon [110] already argued for the potential of DEM data in generating new types of terrain information and several computer cartographic techniques that may be applied to elevation data. The resolution, or the distance between adjacent grid points, is a critical parameter [95] however.

In this Thesis, a DEM at national scale was freely acquired from the Federal office of Topography Swisstopo, with a cell resolution of 100 m (pixel size $= 100 \times 100$ m).

Such DEM resolution was primarily chosen to match that of the other available raster variables.

The DEM provides details about the minimum and maximum altitude in the different regions analyzed. The altitude has an impact on the subsistence activities and hence on the settling location preferences. The altitude has been particularly important for farming activity because of the fact that the higher one cultivates crops, the longer that crop takes to grow [42].

From the national scale DEM, as previously said, is possible to derive in GIS a number of other variables describing important terrain characteristics such as the slope and the aspect (Northness and Easteness in particular will be further discussed in the Chap. 5).

In order to analyze each region independently, the administrative Cantonal boundaries were used to mask (cut) the DEM shape. The vector SWISSBOUNDARIES3D provided a linear vector with an accuracy of 0.5 m for areas below 2000 m.

Slope is a direct function of the topography of a region and is a very important aspect within subsistence activities and land exploitation by means of agriculture, which becomes more difficult with steeper gradients. However, according to the study carried out by Goodchild [42] one should also consider a different scenario, where the modern terrain could not necessarily reflect the topography of the Roman period. In this context, only the introduction of machinery in agricultural activities in modern times would have strongly facilitated the land exploitation activities. In recent times, along with the intensive use of machinery flat lands in close proximity of water bodies are exploited for agriculture, which might have been avoided in Roman times. This is not only due to the restriction of machines to flatter surfaces, but also to the fact that the flatter areas are also mostly located in the floodplain of the river valleys or in former marshy areas. These flat floodplains are characterized by heavy alluvial soils that, although fertile, are difficult to work manually or with ox-drawn ploughs but are easily cultivated using modern machines. As suggested by their role in the hydrologic systems, floodplains have been prone to flooding before large canalization works. Alluvial deposits in the valleys can be several meters in depth, therefore obscuring possible sites, which would exaggerate eventual differences between ancient and modern practice [42].

Although, it is not possible to reconstruct the Roman terrain with exactitude, it may be possible to draw attention to the areas that were subject to the most significant changes and discuss how the model handles them.

Areas at low altitudes and with low slope have always been favored for cultivation activities, but not many exist in Switzerland, particularly in certain regions [122]. Steeper slopes sometimes had thus to be cultivated. As early agronomists had understood, the use of steeper slopes could also bring some advantages. For example, drainage can be more effective on a slope and problematic on flat terrain (Cato, *de Agricultura.* 155). One possibility to make steeper slopes suitable for agriculture to build terraces, similar to those present in the region of Lake Leman [118]. However, these kind of construction work is extremely difficult to identify in the other regions and even harder to date. Even if we assume that that the cultivation was also taking place on steeper slopes, specific instruments like oxen and ploughs were necessary (Pliny the Elder *NH* 18.178). Accordingly, these instruments were prevalent in the more mountainous regions [42].

Aspect is also an important factor for settlements location preferences, as it provides information about the orientation (between 0° and 360°), which might be particularly important for the cultivation of arable lands, where many crops depend on a certain level of sunlight to grow successfully [88]. In ancient times, the solar exposition of a site had surely an important role in the allocation choice. Zones with a longer solar exposition took advantage of more hours of daylight, heating and prolonged visibility [42]. Moreover, ancient agronomists argued that south-facing slopes yielded better harvests than north-facing slopes and were thus preferred for productive agriculture and farming activities (see for example Varro *Rust.* 1.39.1, Cato *de Agr.* 1.2–4).

Climatic features such as wind direction, rainfall, and frost can all affect the cultivation of crops and would ideally also be considered. However, they are rarely documented in ancient literature [100] and very difficult to derive and quantify for past times on a local scale. No reliable information is available, as regular measurements of climatic phenomena like these have only started to be recorded recently. It seems however evident that the Romans were aware of their importance. Pliny the Elder (*NH* 18.24) for example discusses the importance of wind direction [100].

4.10.2 Hydrology

Topography is shaped by hydrological processes and the water presence is another important factor in settlement location analysis. For this study, the information related to the main rivers and lakes were used and treated within the model as a source of water supply and communication routes.

The GIS data for the rivers and lakes system covering the entire territory of Switzerland was retrieved from the Federal Office of Topography Swisstopo as a vector layer named 'Vector200' and containing both major and minor rivers.

The "Topic Hydrography" in the Vector200 describes the different elements of the water bodies (Table 4.4). Within this precise Topic, five (GIS) feature classes were contained.

Only the polyline vector of "FlowingWater" (VEC200_FlowingWater) (Table 4.5), representing the river system and the polygonal vector "Lake" (Table 4.6) corresponding to the lakes, were considered in the modeling procedure.

Topic Hydrography

See Table 4.4

Feature Class Flowing Water—Specific Attributes

See Table 4.5

Table 4.4 Topic hydrography contained in the geodata set Vector 200

Feature class	Geometrie	Beschreibung
Flowing water	Polylinie	Fliessgewässer
Stagnant water	Polylinie	Stehende Gewässer
Dam	Polylinie	Staudämme
GWK_FW_Node	Punkt	Gewässerknoten der Fliessgewässer
Lake	Polygon	Seen

Table 4.5 Extract of the table of attributes for the feature class "Flowing Water" from the original geodata set Vector 200 with the associated metadata description accompanied by a short translation

Attribute	Kurze Beschreibung/short description
GEWISSNR[a]	Gewässernummer/water body number
Name[a]	Name des Gewässers/name of the water body
LaufNr[a]/serial Nu	Nummer des Gewässerlaufs/number of the watercourse
Breite[a]/width	"Kartografische" Breite des Abschnitts/"cartographic" width of the section
Klasse[a]/class	Breite, einheitlich über die ganze Länge / width, uniform over the whole length
Linst[a]	Strukturinstanz des Gewässerlaufs/structural instance of the watercourse
GWLNR[a]	Eindeutiger Identifikator des Gewässerlaufs/unique identifier of the watercourse

[a]Gewisse attribute/certain attributes

Table 4.6 Extract of the table of attributes for the feature class "Lake" from the original dataset Vector 200 with the associated metadata description accompanied by a short translation

Attribute	Kurze Beschreibung/short description
GEWISSNR[a]	Gewässernummer/water body number
LaufNr[a]	Nummer des Gewässerlaufs/number of the watercourse
Linst[a]	Strukturinstanz des Gewässerlaufs/structural instance of the watercourse
TopOrt[a]	Topologie des Referenzorts zu den Gewässern/topology of the reference site to the waters
GWLNR[a]	Eindeutiger Identifikator des Gewässerlaufs/unique identifier of the watercourse
Measure[a]	Adresse (Gewässermeter)/address (water meter)
GWK_FW_Node_OID[a]	*GTDBOID* von GWK_FW_Node/*GTDBOID* from GWK_FW_Node

[a]Gewisse attribute/certain attributes

Feature Class Lake—Specific Attributes

See Table 4.6

4.10.3 Soil and Agriculture Suitability Map

A soil suitability map, the *Digitale Bodeneignungskarte der Schweiz/Carte des aptitudes des sols de la Suisse/Carta digitale delle attitudini dei suoli della Svizzera* was used to provide with the most valuable information about the terrain suitability for ancient subsistence activities. This map was originally produced by the Federal office for Agriculture and made available after a series of improvements by Swisstopo in its latest version in 2012. The soil suitability map of Switzerland consists of 144 different cartographic units distributed over approximately 11'000 polygons. It includes the suitability of terrains for vegetation cover and cultivation. Generic soil properties and agricultural suitability are summarized in 7 classes. The cartographic units were constructed on the basis of six soil properties: 'Soil depth', 'Soil skeleton', 'Water storage capacity', 'Nutrient storage capacity', 'Permeability' and 'Water saturation'. Each cartographic unit corresponds to a combination of these properties.

Moreover, this map also contained information about the Soil suitability as arable land, which is a special assessment of the soil aptitude for agricultural management. For obvious reasons, agriculture has been a fundamental aspect of human societies for centuries. As Duchaufour [23] pointed out, people have therefore looked out for soils with ideal properties for cultivation throughout the ages and have logically chosen spaces where: (i) the depth is sufficient for the development of the roots, (ii) the grain size is balanced (optimum clay content) (iii) the structure ensures aeration and natural drainage, (iv) low coarse skeletal content is present, and (v) adequate humidity content is found [23]. Soil properties and soil suitability are very important factors especially in determining agricultural productivity, which, in turn, can shape the Roman site distribution patterns [86, 99]. The intensive land use changes and deforestation occurred during the Roman period in Europe, probably related to the introduction of agriculture and to the mass movements of human population, is discernible from the soil properties and confirmed also by recent studies on pollen-based land-cover reconstructions, focused on northern and central Europe [76, 102], hence, the decision of using this map as a variable in the modeling procedure [58].

4.10.4 Geology

Finally, the Geological map of Switzerland 1:500,000 (ed. 2005) is part of the variable set and provides information about the distribution of the uppermost rock strata. Most of the geological maps in Switzerland are recorded and published by the Swiss Geological Survey (incorporated in Swisstopo) in collaboration with universities, commissions and private agencies. This map ("GK500", edition 2014) was revised and made available by Swisstopo[12] as a vector dataset (containing lines and polygons) describing geological data in an exhaustive way.

A geological map classically gives a 2D modeling of a complex 3D environment. It combines a great variety of information (lithological, chronological, structural, and morphological) corresponding to various spatial data types (point, line, surface). Each feature of a geological map holds a high semantic content [79].

[12] https://shop.swisstopo.admin.ch/en/products/maps/geology/GK500/GK500_DIGITAL.

The Swiss geological formations are recorded in GK500 vector with different graphical elements (symbols, shapes and colors) on a topographical foundation. The polygonal vector layer contains surface objects of geological formations lithostratigraphically distinguishable. This variable helps to identify where arable lands and Roman sites were situated related to geological formations. Goodchild [42] conducted a comprehensive study on the geological properties of the most suitable terrains for agriculture. For example, if considering the geological properties related to agriculture, limestone will tend to produce a fairly poor soil, sands and conglomerates, as well as clay, tend to produce soils prone to severe erosion and as such are not particularly conducive to farming, and by consequence for settlements location. On the other side, clay, unlike what is thought, represents the finest element of the soil. It has high colloidal properties, i.e. it is made up of very small minerals that cement each other, creating a very compact and hardly permeable layer [21]. The tendency of agricultural soils that are too clayey is to cause water stagnation and radical asphyxia. This creates problems of physical fertility of the soils. On the other hand, clay has the ability to retain nutrients and therefore improves biological fertility. The presence of sand makes the agricultural land easily permeable and workable. It must be said, however, that an excessive amount of sand causes fertility problems. The sand, in fact, does not retain water and mineral salts, necessary elements for the nutrition of plants [21].

Although some argue that the character of modern soils, such as their fertility, is unlikely to reflect the situation in the Roman period, as the climate has changed and soil has been lost to erosion and other factors [85], some studies [87, 124] assert that the environment as well as the agriculture production (especially on the Swiss Plateau) has not substantially changed and represented a relatively stable resource from the Late bronze age to the Middle Age. Accordingly, they support the use of modern data for modeling procedure [122].

References

1. Anderson JT, Augustoni C, Duvauchelle VS, Castella D (2003) Des Artisans à la campagne. Carrière de meules, forge et voie gallo-romaines à Chablais (FR). Archéologie Fribourgeoise, vol 19
2. Anderegg JP (2002) Une Histoire du Paysage Fribourgeois. Espace. Territoire et habitat. Service Cantonal des biens culturels Fribourg
3. Archaologie Schweiz (2001) Avenches. Hauptstadt Der Helvetier. as. 24(2001):2
4. Bayard D, Collart J-L (éd) (1996) De la ferme indigène à la villa romaine. Actes du 2e colloque de l'association AGER tenu à Amiens (Somme) du 23 au 25 septembre 1993, Amiens
5. Berti Rossi S, Castella MC, Pierre A (2005) La fouille de Vidy "Chavannes 11", 1989–1990 : trois siècles d'histoire à Lousonna : archéologie, architecture et urbanisme Cahiers d'archéologie Romande102
6. Binford LR (1981) Behavioral archaeology and the "Pompeii premise." J Anthropol Res 37(3):195–208
7. Binford LR (1992) Seeing the present and interpreting the past—and keeping things straight. In: Rossignol J, Wandsnider L (eds) Space, time, and archaeological landscapes. Plenum Press, New York, NY, pp 43–64

8. Blondel L (1934) Fortification préhistorique et marche romain au Bourg-de-Four. Geneva 12:38–63
9. Blondel L (1941) De la citadelle gauloise au forum romain .Geneva 19:98–118
10. Bögli H (1962) La Suisse à l'époque romaine. Ed. Société Anonyme Chocolate Tobler
11. Bögli H (1996) Aventicum: la ville romaine et le musée. Guides archéologiques de la Suisse n. 19
12. Bolliger S (2003) Untersuchungen zum römischen Strassennetz in der Schweiz. Bonner Jahrbücher 202(203):237–266
13. Braillard L, Mauvilly M (2008) Morphogenesis of the Sarine canyon in the Plateau Molasse, Switzerland: new data from an archaeological site. Geographica Helvetica 63:181–187
14. Bronner FX (1844) Der Kanton Aargau, historisch, geographisch, statistisch geschildert, 2 voll. St. Gallen; Bern: bei Huber und Compagnie
15. Buchsenschutz O, Curdy P (1991) L'Habitat Helvète sur le Plateau Suisse. AS 14/1, pp 89–97
16. Camilli EL, Ebert JI (1992) Artifact reuse and recycling in continuous surface distributions and implications for interpreting land use patterns'. In: Rossignol J, Wandsnider L (eds) Space, time, and archaeological landscapes. Springer US, Boston, MA, pp 113–136. https://doi.org/10.1007/978-1-4899-2450-6_6
17. Canning S (2003) Site unseen: archaeology, cultural resource management, planning and predictive modelling in the Melbourne metropolitan area. PhD thesis, La Trobe University, Australia
18. Coulon G (1990) Les Gallo-Romains. I. Les villes, les campagnes et les échanges; II. Métiers, vie quotidienne et religion. Paris
19. Cramatte C (2012) Turicum (Zürich). In: Bagnall RS, Brodersen K, Champion CB, Erskine A, Huebner SR (eds) The encyclopedia of ancient history.https://doi.org/10.1002/9781444338386.wbeah16202
20. Curdy P (1997) Les Grisons. in: Kolloquium ARS 1997:53–54
21. Derungs N (2018) La gestion durable des sols agricoles : sécuriser les démarches ou légitimer les controverses ? L'exemple des politiques agroenvironnementales autour de l'érosion hydrique des sols arables en Suisse. Université de Neuchâtel
22. Dubois Y, Paratte C-A (2001) La pars urbana de la villa gallo-romaine d'Yvonand VD-Mordagne. SSPA 84:43–57
23. Duchaufour P (2000) Introduction à la science du sol: Sol, végétation, environnement. Dunod, Paris
24. Dunnell RC (1992) "The notion site." In: Rossignol J, Wandsnider L (eds) Space, time, and archaeological landscapes, pp 21–41. Interdisciplinary Contributions to Archaeology. Springer, Boston, MA:
25. Ebersbach R (2016) Paleoecological recontruction and calculation of calorie requirements at Lake Zurich. Forschungen Zur Archäologie Im Land Brandenburg 8:69–88
26. Ebnöther C (1995) Der römische Gutshof in Dietikon, Zürich/Egg (Monographien der Kantonsarchäologie Zürich 25)
27. Ebnöther C, Monnier J (2002) Ländliche Besiedlung und Landwirtschaft. In: Laurent Flutsch, Urs Niffeler und Frédéric Rossi (Hrsg.), Die Schweiz vom Paläolithikum bis zum Mittelalter (SPM) V: Römische Zeit. Basel, 135–178
28. Egloff M, Farjon K (1983) Aux origines de Lausanne. Les vestiges préhistorique et gallo-romains de la Cite. Cahier d'Archéologie Romande
29. Felka H, Loï ZF (1982) La villa gallo-romaine de Cuarnens, Études de Lettres. Lausanne 1:49–75
30. Fellmann R (1992) La Suisse gallo-romaine: cinq siècles d'histoire. Payot, Lausanne
31. Flutsch L, Niffeler U, Rossi F (2002) Die Schweiz vom Paläolithikum bis zum frühen Mittelalter: vom Neandertaler bis zu Karl dem Grossen. Römische Zeit. Basel: Verlag Schweizerische Gesellschaft für Ur- und Frühgeschichte, 2002, p 432
32. Flutsch L (2010) L'époque romaine, ou, La Méditerranée au nord des Alpes. Le savoir suisse Hist. 26

33. Frei Stolba R, Marting Benedetti I (1991) La Svizzera in epoca romana. Revue Suisse D'histoire 41:111–125
34. Frei Stolba R (1995) Die Helvetier im römischen Reich: Überlegungen zu ihrer Integration und Gesellschaftsstruktur, In: Herzig et HR, Frei-Stolba R (eds) La politique édilitaire dans l'empire romain IIe - IVe siècle après J.-C..Actes du IIe Colloque Roumano-Suisse 12–19 Sept 1993, Berne, pp 167–186
35. Frei Stolba R (1998) Nyon, une colonie romaine sur les bords du Lac Léman. Dossiers d'Archéologie n. 232, avril, (Ed.) Frédéric Rossi (contributions épigraphiques), pp 14–35
36. Frei-Stolba R, Bielman A, Lieb H (2009) Recherches sur les institutions de Nyon, Augst et Avenches. In: Dondin-Payre M (dir.), Raepsaet-Charlier MT (dir.). Cités, municipes, colonies: Les processus de municipalisation en Gaule et en Germanie sous le Haut Empire romain. Nouvelle édition [en ligne]. Paris: Éditions de la Sorbonne. https://doi.org/10.4000/books.psorbonne.28144
37. Fuchs M (1992) Ravalement à Vallon – Les peintures de la ville romaine. AS 15.1, pp 86–93
38. Furger A, Isler-Kerenyi C, Jacomet S, Russenberger C, Schibler J (eds) (2001) Die Schweiz zur Zeit der Römer. Archäologie und Kulturgeschichte der Schweiz, 3. Verlag Neue Zürcher Zeitung, Zürich, p 352
39. Gaffney VL, Stančič Z (1991) Gis approaches to regional analysis: a case study of the Island of Hvar. Znanstveni inštitut, Filozofske fakultete
40. Gallant TW (1986) 'Background noise' and site definition: a contribution to survey methodology. JFA 13:403–418
41. Gallay A (ed) (2008) Des Alpes au Léman: images de la préhistoire, 2nd edn. Infolio, Gollion
42. Goodchild H (2007) Modelling roman agricultural production in the middle tiber valley, central Italy. PhD Thesis, University of Birmingham
43. Goudineau C (1992) Cesar et la Gaule. Paris
44. Guyan WU, Zürcher A, Schneider Jürg E (1985) Turicum – Vitudurum – Iuliomagus. Drei Vici der Ostschweiz (Zürich)
45. Haldimann M-A, André P, Broillet-Ramjoué E, Poux M (2001) Entre résidence indigène et domus gallo-romaine. Le Domaine Antique Du Parc De La Grange, Archéologie Suisse 24:2–15
46. Haldimann M-A, Rossi F, Berti S (1997) Le bassin lémanique. Une charnière entre archéologie et histoire, D'Orgétorix à Tibère, Actes du colloque de PARS 1995 à Porrentruy. Lausanne 1997:65–76
47. Hanel N, Schucany C (eds) (1999) Colonia, Municipium, Vicus. Struktur und Entwicklung städtischer Siedlungen in Noricum, Tätien und Obergermanien. Beiträge der Arbeitsgemeinschaft « Römische Archäologie » bei der Tagung des West- und Süddeutschen Verbandes der Altertumsforschung, Wien. Oxford, Archeopress, pp 93 (BAR International Series, 783)
48. Hartmann M (2018) Epoca Romana. Argovia, In: Dizionario Storico della Svizzera, Versione del. https://hls-dhs-dss.ch/it/articles/007392/2018-02-06/
49. Hitz F (2018) Grigioni, In: Dizionario Storico della Svizzera. Versione del 11.01.2018. consulted 06.02.2018 https://hls-dhs-dss.ch/it/articles/007391/2018-01-11/
50. Hochuli–Gysel A (1986) Chur in romischer Zeit. Bd.1 Areal Dosch. Basel
51. Horisberger B (2017) Zurigo, Cantone - Epoca romana. Historische Lexicon der Schweiz. https://hls-dhs-dss.ch/it/articles/007381/2017-08-24/
52. Kaenel G, Klausener M, Fehlmann S (1980) Nouvelles recherches sur le vicus gallo-romain de Lousanna (Vidy/Lausanne). Cahier d'Archeologie Romande, Lousonna 2
53. Kaenel G, Paunier D (1991) Qu'est-il arrivé après Bibracte? Les Helvètes et leurs voisins, 153–168
54. Kilcher MS, Zaugg M (1983) L'Helvétie au temps des Romains. La Suisse antique 3, Lausanne. Les voies romaines. Guide romain de voyage. 1992. Office national du tourisme Berne
55. Martin-Kilcher S (2018) La Rezia in Epoca Romana. Caratteristiche Generali. In: Grigioni. Dizionario Storico della Svizzera. Versione del 11.01.2018. https://hls-dhs-dss.ch/it/articles/007391/2018-01-11/

56. McCoy MD (2020) The site problem: a critical review of the site concept in archaeology in the Digital Age. J Field Archaeol 45(sup1):S18–S26. https://doi.org/10.1080/00934690.2020.1713283
57. Morel J, Amstad S (1990) Un quartier romain de Nyon: De l'epoque Augusteenne au III siecle, Noviodunum II. Lausanne
58. Nussbaum M, Spiess K, Baltensweiler A, Grob U, Keller A, Greiner L, Schaepman ME, Papritz A (2018) Evaluation of digital soil mapping approaches with large sets of environmental covariates. Soil 4:1–22. https://doi.org/10.5194/soil-4-1-2018
59. Paratte CA (1994) Rapport préliminaires sur la campagne des fouilles d'Orbe VD-Bosceaz 1993. JbSGUF 77(1994):148–152
60. Patrik LE (1985) Is There an archaeological record? Adv Archeol Method Theory 8:27–62
61. Pauli L (1991) I passi alpini e le migrazioni celtiche. In : I Celti. Ausstellungskatalog Venedig. Milano, pp 215–219
62. Paunier D (1982a) La présence de Rome. In: Histoire de Lausanne. Toulouse-Lausanne, 44–79
63. Paunieret al (1989) Le vicus gallo-romain de Lousonna-Vidy: le quartier occidental; le sanctuaire indigène; rapport préliminaire sur la campagne de fouilles 1985. Bibliothèque Historique Vaudoise
64. Paunier D (1996) La romanisation des campagnes; un état des recherches en Suisse. Revue Archéologique De Picardie, No Spécial 11:261–270
65. Paunier D (1998a) Le Leman, de l'epoque gallo-romaine au Moyen Age. Arch Sci Geneve 51(1):91–102
66. Paunier D (2000) La villa gallo-romaine d'Orbe-Boscéaz. Rapport sur les campagnes de fouille 1996–1997. Le mithraeum. Le château d'eau. Institut d'archéologie et d'histoire ancienne de l'Université de Lausanne
67. Paunier D (2006) La romanisation et la question de l'héritage celtique. In: Celtes et Gaulois l'Archéologie Face à l'Histoire, Actes de La Table Ronde de Lausanne 17–18 Juin 2005. CAE Européen Mont-Beuvray
68. Paunier D (2010) Celtes et Gaulois, l'Archéologie face à l'Histoire. In: Goudineau C, Guichard V, Kaenel G (eds) Colloque de synthèse, Collège de France, 3–7 juillet 2006, Centre archéologique européen, pp 105–127
69. Pichard Sardet N (1993) Lousonna. La ville gallo-romaine et le musee., AFS 27, Lausanne
70. Raget J (1995) Il percorso attraverso i valichi dello Julier, del Settimo e dello Spluga in epoca romana, In: L'Antica Via Regina. Tra gli itinerari stradali e le vie d'acqua del Comasco. Como, 363–389
71. Raget J (1998) Chur – Welschdorfli, Schutzbau Areal Ackermann, AFS 29
72. Raget J (2000) Graubunden in romischer Zeit. Archäologie Der Schweiz 23:46–56
73. Rychner J (1997) Die Nordostschweiz. Kolloqium ARS 1997:95–99
74. Rychner J (1999) Der Romischer Gustof von Neftenbach, Zurich/Egg
75. Reymond S, Eschbach F, Perret S (2009) La villa romaine du Buy et sa forge : dernières découvertes a Cheseaux, Morrens et Etagnieres (Canton de Vaud, Suisse). Cahiers d'Archéologie Romande 115. Lausanne
76. Roberts N, Fyfe RM, Woodbridge J, Gaillard MJ, Davis BAS, Kaplan JO, Leydet M (2018) Europe's lost forests: a pollen-based synthesis for the last 11,000 years. Sci Rep 8(1). https://doi.org/10.1038/s41598-017-18646-7
77. Ruoff, E. 1993 Chur in romischer Zeit, In: Churer Stadtgeschichte. Bd.1 Von den Anfangen bis zur Mitte des 17. Jh., Chur 1993, 136–179
78. Sauter MR (1976) Chronique des découvertes archéologiques dans le canton de Genève en 1974 et 1975. Genava n.s. 24, 259–279
79. Sartori M, Ornstein P, Metraux C, Schreiber L, Kuehni A (2006) From geological cartography to digital maps: spatial data model and GIS tool. In: Proceedings of the 5th European congress on regional geoscientific cartography and information system, pp 189–191
80. Schiffer MB (1987) Formation processes of the archaeological record. University of New Mexico Press, Albuquerque, NM

81. Schucany C, Krause MMF (2002) Das tägliche Leben. In: Laurent Flutsch, Urs Niffeler und Frédéric Rossi (Hrsg.), Die Schweiz vom Paläolithikum bis zum Mittelalter (SPM) V: Römische Zeit. Basel 2002, 217–266
82. Schucany C (2006) Die römische Villa von Biberist- Spitalhof/SO (Grabungen 1982, 1983, 1986–89). Ausgrabungen und Forschungen 4. Remshalden
83. Schucany C (2011) The villa landscape of the Middle Aare valley and its spatial and chronological development. In: Derks T, Roymans N (eds) (2011) Villa landscapes in the Roman North. Amsterdam archaeological studies
84. Schwab H (1981) Les débuts de l'homme. In : Histoire du Canton de Fribourg, vol I
85. Shiel RS (1999) Reconstructing Past Soil Environments in the Mediterranean Region". In: Leveau P, Walsh K, Trément F, Barker G (eds) The archaeology of Mediterranean landscapes 2: environmental reconstruction in Mediterranean landscape archaeology. Oxbow Books, Oxford, pp 67–79
86. Simpson I, Adderley WP, Guðmundsson G, Hallsdóttir M, Sigurgeirsson M, Snæsdóttir M (2002) Soil limitations to Agrarian Land production in Premodern Iceland. Hum Ecol 30:423–443. https://doi.org/10.1023/a:1021161006022
87. SPM V (2002) Epoque romaine/Età Romana, Société suisse de Préhistoire et d'Archéologie, Bâle 2002 (La Suisse du Paléolithique à l'aube du Moyen- Age, vol V)
88. Spurr MS (1986) Arable cultivation in Roman Italy c. 200 B.C–c. A.D. 100 (J Roman Stud Monogr 3). Society for the Promotion of Roman Studies, London
89. Steigmeier A (2018) Argovia. In: Dizionario Storico della Svizzera. Versione del 06.02.2018. https://hls-dhs-dss.ch/it/articles/007392/2018-02-06/
90. Tarpin M (2015) I Romani in Montagna: tra immaginario e razionalità. In: Il capitale culturale, Studies on the value of cultural heritage XII, pp 803–822. ISSN 2039–2362
91. Terrier J (2002) Découvertes archéologiques dans le canton de Genève en 2000 et 2001. Genava, N. 50:355–388
92. Thomas DH (1975) Nonsite sampling in archaeology: up the creek without a site? In: Mueller JW (ed) Sampling in Archaeology. University of Arizona Press, Tucson, pp 61–81
93. Tomasevic Buck T (1993) Romische Siedlungstrukturen im Gebiet der Scweiz. Actes 1993:39–60
94. Van Leusen PM (2002) Pattern to process: methodological investigations into the formation and interpretation of spatial patterns in archaeological landscapes. PhD thesis, Faculty of Arts. http://dissertations.ub.rug.nl/faculties/arts/2002/
95. Verhagen P (2007a) Testing archaeological predictive models: a rough guide in layers of perception. In: Proceedings of the 35th computer applications and quantitative methods in archaeology conference, Berlin, Germany, 2–6 April, Bonn, pp 285–291
96. Vion E (1989) L'analyse archéologique des réseaux routiers. Paysages Découverts 1:67–99
97. Vogt E (1948) Der Lindenhof in Zürich (Zürich 1948)
98. Wallenfeldt J (2008) Aargau. In: Ecycloapedia Britannica (eds) https://www.britannica.com/place/Aargau-canton-Switzerland
99. Wescott K, Brandon R (eds) (2000) Practical applications of GIS for archaeologists: a predictive modelling kit. Taylor & Francis, London
100. White KD (1988) Farming and animal husbandry. In: Grant M, Kitzinger R (eds) Civilization of the ancient Mediterranean. Scribner's, New York, pp 211–245
101. Weidmann D (1982) Villa romaine de Cologny et du Buy (comm. de Morrens), Revue Historique Vaudoise, Lausanne, p 1 76
102. Wickham C (2006) Framing the early middle ages. Europe and the mediterranean, 400–800 –Chris Wickham. The Econ Hist Rev 59(2):417–419. https://doi.org/10.1111/j.1468-0289.2006.00351_17.x
103. Wildi W, Corboud P, Girardclos S, Gorin G (2017) Guide: géologie et archéologie de Genève = Guidebook: geology and archaeology of Geneva. 2e éd. Genève: Section des sciences de la Terre et de l'environnement, 93 pp
104. Wolf C, Burri E, Hering P, Kurz M, Mautewolf M, Quinn DS, Winiger A (1999) Les sites lacustres du Néoiithique et de l'àge du Bronze à Concisesous-Colachoz (Canton de Vaud)

au bord du lac de Neuchàtel : premiers résultats concernant en particulier le Bronze ancien. Annuaire De La Société Suisse De Préhistoire Et D'archéologie 82:7–38
105. Zürcher A (1985) Vitudurum. Geschichte einer romischern Siedlung in der Ostschweiz, In: Turicum – Vitudurum – Iuliomagus. Drei Vici in der Ostscheweiz, Festscrhrift O. Coninx, Zurich, pp 165–233
106. Plog S, Plog F, Wait W (1978) Decision making in modern survey. In: Schiffer MB (ed) Advances in archaeological method and theory. Academic Press, New York, pp 383–420
107. Vitali D, Kaenel G (2000) Un Helvète chez les Etrusques vers 300 av. J.-C. Archäologie Schweiz 23(3):115–122
108. Ebnöther C, Schucany C (1998) Vindonissa und sein Umland. Die Vici und die ländliche Besiedlung, Jahresbericht der Gesellschaft Pro Vindonissa pp 67–97
109. Vautey PA (1985) Riaz/Tronche-Bélon Le sanctuaire Gallo-Roman Archéologie fribourgeoise/Freiburger Archäologie 2. Fribourg Suisse: Ed. universitaires
110. Collins SH, Moon GC (1981) Algorithms for dense digital terrain models. Photogramm Eng Remote Sens 47:71–76
111. Kaenel G, Crotti P (eds) (1992) Celtes et Romaines en Pays d'en Vaud. Musée cantonal d'archéologie et d'histoire Lausanne
112. Tarpin M (2000a) Colonia, Municipium, Vicus : Institutionen und Stadtformen. In: Hanel N, Schucany C (eds) Colonia, municipium, vicus. Struktur und Entwicklung städtischer Siedlungen in Noricum, Rätien und Obergermanien, Colloquium, Wien, 21-23 May 1997, BAR International Series, 783, Oxford, 1999, pp 1–10
113. Tarpin M (2000b) Urbs et oppidum le concept urbain dans l'Antiquité romaine. In: Guichard V, Sievers S, Urban OH (dir), Les processus d'urbanisation à l'âge du Fer. Eisenzeitliche Urbanisationsprozesse. Actes du colloque de Bibracte (8-11 juin)
114. Motschi A, Wild W (2011) Städtische Siedlungen — Überblick zu Siedlungs - entwicklung und Siedlungs topografie: Zürich, Winterthur, Weesen In: Siedlungsbefunde und Fundkomplexe der Zeit zwischen 800 und 1350. Akten des Kolloquiums zur Mittelalterarchäologie in der Schweiz. Archäologie Schweiz AS, Schweizerische Arbeitsgemeinschaft für die Archäologie des Mittelalters und der Neuzeit SAM, Schweizerischer Burgenverein SBV (eds) Verlag Archäologie Schweiz, Basel, 2011. 483
115. Habermehl D (2011) Exploring villa development in the northern provinces of the Roman empire. In: Roymans N, Derks T (eds) 2011 Villa landscapes in the Roman North: economy, culture and lifestyles (vol 17). Amsterdam University Press
116. Paunier D (1982b) L'archéologie gallo-romaine en Suisse romande: bilan et perspectives. In: Etudes de lettres. Revue de la Faculté des Lettres de l'Université de Lausanne Ser.4, 1, pp 5 28
117. Paunier D (1998b) Dix ans d'archéologie gallo-romaine en Suisse: espuisse d'un bila. In: Revue du Nord, 80:328 235–252
118. Studer J, David-Elbiali M, Besse M (dir) (2011) Paysage... Landschaft... Paesaggio. L'Impact des activités humaines sur l'environnement du Paléolithique a la période romaine. Actes du colloque du Groupe de travail pour les recherches préhistoriques en Suisse (GPS/AGUS), Museum d'histoire naturelle, Genève, 15–16 mars 2007
119. Bonnet C (1984) Les premiers edifices du groupe episcopal de Geneve In: Actes du Xe Congrès International d'Archéologie Chrétienne Tl. 2 S. pp 21–32
120. Bonnet C (1989) Les premiers ports de Genève. In: Archäologie der Schweiz, vol 12, pp 2–24
121. Haldimann MA, Ramjoue EC, Simon C (1991) Les fouilles de la cour de l'ancienne prison de Saint-Antoine: une vision renouvelée de la Genève antique. Archäologie Schweiz 14(2):194–204
122. Ebersbach R (2015) Eine Potentialkarte Archäologie für den Kanton Bern. Archäologie Bern/Archéologie Bernoise 2015:212–233
123. Flutsch L (2005) L'époque Romaine, Ou La Méditerranée Au Nord Des Alpes. Lausanne: Presses polytechniques et universitaires romandes (Le savoir suisse Histoire)
124. Müller F, Kaenel G, Lüscher G (eds) (1999) SPM IV Eisenzeit/Âge du Fer. Die Schweiz vom Paläo- lithikum bis zum frühen Mittelalter – La Suisse du Paléolithique à l'aube du Moyen-Âge 4. Basel 1999

Chapter 5
Modeling Approach

We don't understand human experts and yet we trust them,
so why should we not extend the same degree of trust to an expert computer?
Pearson, 2016

The present chapter will introduce the methodology applied in this Thesis and the steps followed for designing the database and the model for each case study. A conceptual model has been developed to construct a new, uniform database for modeling purposes and a fuzzy logic structure has been introduced in order to manage and embed the uncertainty related to the archaeological data. Particular attention is paid to structure a clear methodology from the data preprocessing to the final application of the Machine Learning algorithm.

The first part of this chapter is mostly dedicated to an essential exploration of the data, the definition of the archaeological database architecture and the solution experimented to deal with data uncertainty. An exhaustive locational preference analysis (also named in this text as Exploratory Spatial Data Analysis, ESDA) is performed on the data in order to assess which fundamental criteria where taken into consideration by ancient populations when assessing the best location to settle. This analysis further allows examining whether the site distribution over the different geomorphological regions considered followed similar criteria to those underlined in the literature and by other studies of this kind (if applicable).

In the final part, several models have been constructed and run using machine learning techniques, one or more in each region, plus a final model at a national scale. While dissecting the defined model structure, managing uncertainty in archaeological information will be explored on the cases of Aargau and Geneva, in the optic of improving model accuracy.

Although finding the best modeling approach for such different case studies was a challenging task with respect to the quality and quantity of archaeological records available, a range of regional predictive maps have been produced.

As seen in Chap. 3, prediction methods in archaeology vary widely based on the outcome required and the nature of data available. The current trends although rely heavily on mathematical and statistical toolsets. Computational models involving

M. E. Castiello, *Computational and Machine Learning Tools for Archaeological Site Modeling*, Springer Theses, https://doi.org/10.1007/978-3-030-88567-0_5

Machine Learning (ML) techniques and geo-referenced databases enjoyed increasing popularity in archaeological fields in recent years. As discussed, numerous studies have been carried out by various authors, attesting the great potential of applying ML approaches to explain socio-natural questions influencing human behavior and interactions between individuals or groups and the environment [1].

While designing an APM strategy, the basic sources of information commonly used, besides the archaeological data, are the environmental data (described in more details in Chap. 4). Once the available data have been gathered, they must be synthesized and evaluated in terms of their applicability for predicting site location, in other words they must be quantified.

Before to perform computations of any kind, the modeling procedure should generally start with a first exploratory analysis of the collected archaeological information, "acknowledging that a collection of data never starts without a question in mind", to cite Verhagen and Withley [2].

That said, the data structure was carefully examined and in a first stage of the research, the regional GIS databases were restructured (conceptual modeling) in order to match the model needs.

This restructuring phase looks at what common components all the regional archaeological databases posses. By using a simplified structure with respect to the original one, overclassification is then corrected from the original dataset, allowing for later cross-referencing and the identification of common patterns to understand to what extent sites were selected according to similar criteria for settlement locations.

Although there is probably no procedure capable of correcting all biases and certainly it is not possible to make good data out of bad data, several techniques (to cite a few: [3–5]) exist to remove, or at least reduce, apparent biases from existing databases. Such techniques have been experimented in this Thesis with the aim of obtaining better-quality analysis datasets for model development or testing purposes.

As stated by Maguire and Dengermond [6], the main interest from a statistical point of view lies in how well a model works in application, how accurately it performs on future cases. To reduce the effects of this problem, modern quantitative methods can attempt to limit biases and contribute by improving the replicability of the entire modeling workflow for the use in other locations. The authors also mention the difficulty of producing robust archaeological predictive models that can be reproducible, so that the results obtained can also be attained by other researchers using the same methods, as well as interoperable and transferable, in order to be applicable to other contexts, situations and times, by means of integrating data from different sources [6].

Thus, this initial conceptual modeling procedure for archaeological data plays an important role in the definition of the methodology. Basically, human concepts of space and evidence have to be formalized, as computer systems work by using sets of formal rules. Conceptual models are an abstraction of the real world and incorporate only relevant data [6]. Hence, identifying the fields in the original datasets that can help to answer the research questions is an essential step on which successful model outcome will depend.

Table 5.1 RF final models created and retained for each region

RF classification	RF regression
1. ZURICH	–
1. AARGAU (settlements)	1. AARGAU (single finds)
2. GRISONS	–
1. FRIBOURG	–
1. VAUD	–
1. GENEVA	1. GENEVA

On the other hand, predictor variables (proxies/futures) are rigorously analyzed as happened for the archaeological known sites, as they represent the criteria under which the presence of unknown sites locations will be verified. In this context, the choice of a location settlement is dependent upon the presence of a set of favorable criteria (e.g. close to water, flat terrain, dry ground, etc.). The absence of any single one of those characteristics negates the probability of a 'presence' site.

While these features are in principal extracted at national level, they are rescaled at regional level; transformed from vector to raster format and combined together with the archaeological data into a matrix to facilitate the database query. The physical limitation in easily displaying the high number of variables and archaeological sites is bypassed by using this specific technique.

Only when this preliminary work is fully accomplished, it is finally possible to set up the APM model. The basic structure of an APM using ML techniques requires the use of a comprehensive input dataset grouping information from independent and dependent variables (extracted from the matrix abovementioned) and the subsequent division of the dataset into a training and a testing subset in order to obtain a final prediction.

As every single region analyzed in this Thesis presents distinctive specifications concerning the archaeological data, different models have been run to suit local specific needs. Finally, the ensemble of the results obtained are compared and discussed. The total number of models created is presented in Table 5.1.

As shown in Table 5.1, the RF regression has been applied to compute uncertainty in the archaeological dataset of the Cantons of Aargau and Geneva.

In the following paragraphs, the approach used for creating an adequate database architecture and modeling procedures are outlined (Fig. 5.1).

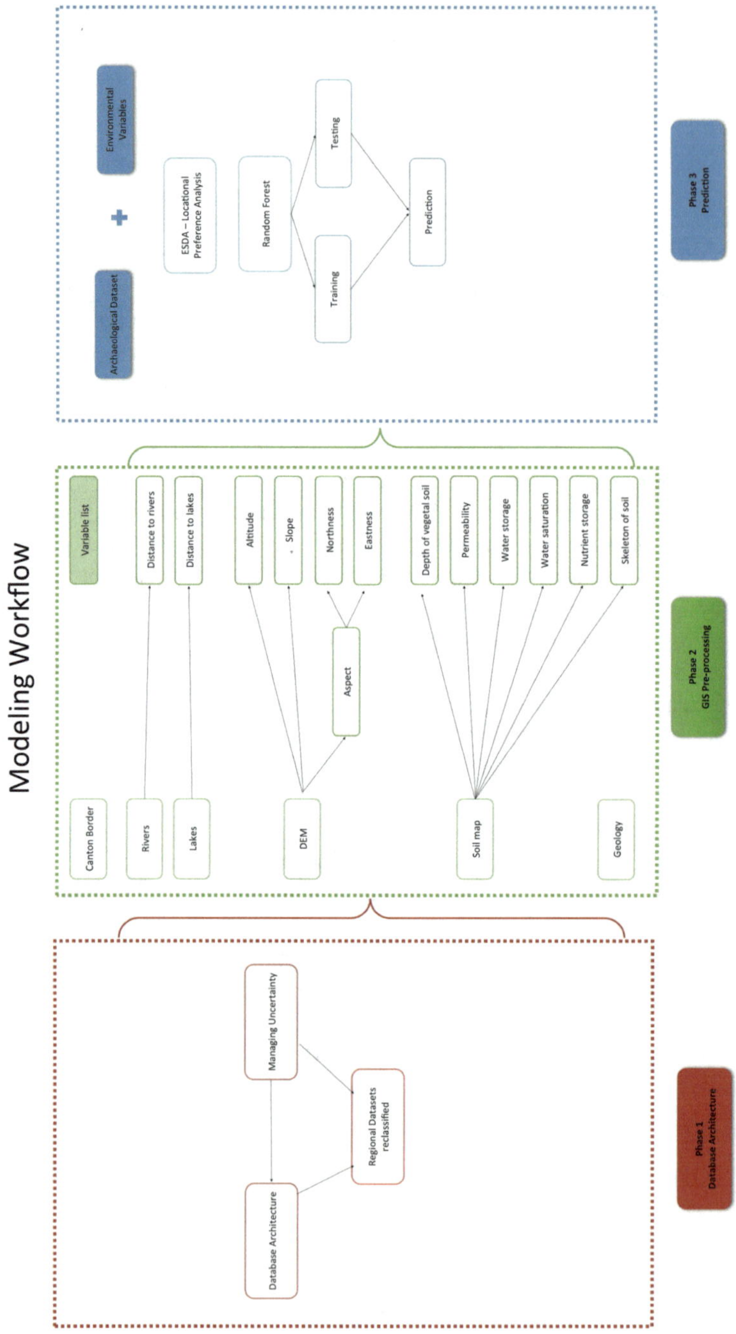

Fig. 5.1 Predictive modeling workflow

5.1 GIS Preprocessing

5.1.1 Conceptual Modeling for the Archaeological Database

This section presents the most relevant steps undertaken to create a georeferenced, relational database[1] (DB) for each of the regional case studies, from the establishment of data requirements to the database design, data modeling and implementation. The database design is one of the most crucial components of the project, not only for model building purposes, but also for the reusability in future research. The following pages describe the conceptual data model and the underlying theory of how and why the model was developed.

The description of the procedure followed is explained through a fluid narration of the steps undertaken, in order to facilitate the knowledge transmission. Moreover, a more detailed description of these steps is rather referred to as the software development life cycle (SDLC) [7], that would veer far away from the purpose of this research, which instead wish to focus on the presentation of an APM methodology and structure. Nevertheless, all the relevant steps allowing for the APM setup are summarized and described.

Likewise, the most basic GIS manipulations (as for example the import of a dataset or the transformation of a set of geographic coordinates) performed within this study will be skipped to pay more attention to those procedures considered relevant to assist in the predictive modeling process performed (extensive manuals and literature on GIS processing exist to address these most basic operations).

Generally speaking, establishing a database structure and defining its data fields at the beginning of a research project is a common practice, but it also carries methodological problems. The risk always persists that once the process of data collection is underway, it appears that the structure of the database should be adapted in order to remove unnecessary fields or incorporate new ones [8]. It is therefore of crucial importance to store the data in a well thought manner from the beginning. Moreover, as Ossa and Simon [9] pointed out, a "database design should aim to minimize potential errors during data entry and subsequent data modification, and to maximize analytical flexibility—including the potential for other, later researchers to use the data."

Bearing in mind that the requirements for this research are primarily spatial, the data formatting (and associated attributes) was designed to match such standards. Thus, from the original regional datasets, a selection of the most consistent and significant fields for each of the dataset entries was kept relatively straightforward for creating a new GIS DB.

Within ArcGIS (10.6 and 10.7) environment, the original archaeological datasets where imported from excel files based on their geographic location (the longitude

[1] Relational databases have multiple tables of data that are designed to interact with each other for efficiency of data organization and minimum of repetition of data entry and data error.

Fig. 5.2 Left: polygonal features for Roman sites. Right: point features for Roman sites

and latitude generally stored within the fields X and Y). Since no information was provided about the size of the settlements, once imported, they were converted into point shapefiles[2] to allow editing the information.

Almost all the regional DBs consisted of point features classes, which means that to each site corresponds a point on the map with its associated geographic coordinates. The unique exception is the DB of Geneva, where sites are represented by polygonal features (Fig. 5.2). In this case, each site is defined by a surface area corresponding to the approximate areal extent of the archaeological excavations. Since for the modeling procedure all sites had to be turned into point features, the polygon features had to be transformed. In order to maintain the information about the spatial extent of the sites, the site data from the polygons was extracted on a grid of points with 100 m spacing, corresponding to the resolution of the independent variables, by performing a geostatistical analysis (*Extract value to point* tool[3] of ArcGIS 10.6). This specific analysis results in an output feature class[4] containing the new site point features of Geneva (Fig. 5.2).

Finally, the field attributes retained from the original datasets in the new DBs were the same for each region, with few exceptions due to the specific dataset structures. The fields retained from the dataset are the following:

Zurich

- FS Fundstellenart
- FS Epoche

[2] For more details about Esri Shapefile refer to Esri documentation: https://desktop.arcgis.com/en/arcmap/10.3/manage-data/shapefiles/what-is-a-shapefile.htm.

[3] For more details on this tool see https://pro.arcgis.com/en/pro-app/tool-reference/spatial-analyst/extract-values-to-points.htm.

[4] For more details on Feature classes: https://desktop.arcgis.com/en/arcmap/10.3/manage-data/geodatabases/feature-class-basics.htm.

- FS Nord-Koordinate
- FS Ost-Koordinate

Aargau

- Befund/Qualität
- Datierung/Qualität
- X Koordinate
- Y Koordinate

Graubünden

- FS Fundstellenart
- FS Epoche
- FS Nord-Koordinate
- FS Ost-Koordinate

Fribourg

- Fonction
- Y
- X
- 'other epochs'
- Nature du site
- Epoque

Vaud

- Site_centroide_x
- Site_centroide_y
- Fonction_principale
- Periode_principale
- Qualite_periode

Geneva

- Type
- X
- Y
- Nature Site
- Remarques

After this first operation, further manipulations were necessary. When clearly absent, the epoch assignment (as in the case of Geneva DB) was manually entered according to the information provided in the description field ('Remarques'). The same procedure was performed and applied to the other cantonal DBs.

The fields containing the description (Remarques, Beschreibung, etc.) generally provided valuable information about the site discovery, sometime supplying the missing information of blank cells in other fields (e.g. coordinates and epoch). When it was not possible to recover information from other fields, empty rows were deleted, as they could jeopardize the proper functioning of the model. For example, since the final DBs had to consist in correctly geo-localized punctual entities, the geographic coordinates were an indispensable attribute for further consideration.

Once all the DBs had the same format and the same common set of attribute columns, a further operation was performed in order to isolate only the sites belonging to the Roman period. As seen in Chap. 4, the various definitions used in the DBs for naming a same epoch would have made this a twisty process, moreover because the epoch assignment is sometimes not contained in the 'Epoch' field, but has to be deduced from corresponding information in the 'Description' field. Thus, the new spatial DB system allows performing SQL (Structure Query Language) syntax queries by using Python scripting tools. This allowed to finalize this step, using manually implemented query expressions. The queries examined in each DB whether the term 'Roman', in all its variations, was contained in one or more field. The entries matching with such requirements were verified and, if corresponding, selected and exported into a new feature class. This feature class thus contains all sites assigned to the Roman epoch in the corresponding 'Epoch' field as well as all sites that had not been assigned to a period, but could be attributed to the Roman epoch thanks to the information given in other fields (such as 'Description'). Within this new feature class, a field named *Presence* defined the rows as presence when corresponding to a Roman settlement.

The set of procedures illustrated above was manually repeated likewise for the ensemble of the regional DBs. Considerable efforts were made in order to check and correct all the information and to elaborate the data for obtaining a consistent new format. In relational database technology all of these standardization issues are referred to as *normalization*, which means that examples of modification anomalies can be handled by creating a new and efficient data structure [10].

Furthermore, the sites were assigned different types of geographic coordinates system according to when they were registered (CH1903+/LV95 or CH1903/LV03). In order to permit a spatial association, they had to be projected into the same system.

All information contained in the 'Description' field of the selected sites was verified and if considered correct (by means of a literature comparison), the sites were retained. This procedure was necessary not only to confirm or deny the classification into the Roman epoch, but also for better defining the class of evidence. Indeed, an in-depth screening of the field describing the type of evidence was performed. Since providing consistent naming conventions for data fields and the information they contain is an important aspect of database design, a data field like "type" should refer to the same kind of information wherever it is used [9].

The regional DBs however often use different definitions to refer to the same kind of information. Furthermore, as data in different language were considered (German and French), the crucial 'type' field had to be re-classified. For the purposes of interpretation and discussion throughout this Thesis, and in order to create a standardized DB allowing supra-regional comparisons, the following 11 generic site type categories were defined as a common nominator for all regional DBs:

- Settlements
- Single finds
- Religious sites
- Fortifications
- Graves
- Roads
- Bridges
- Water infrastructures
- Quarries
- Others
- Unknown.

As shown by the example of Aargau (Table 5.2), the re-classification tables for all Cantons are listed in the appendices A.1, A.2, A.3, A.4, A.5, A.6, these generic categories allowed to considerably reduce the number of site type classes and to group all similar cases into single categories. After the geo-spatial normalization and the epoch selection outlined above, this procedure represents a further step towards a standardized DB.

For the further modeling procedure, two categories (settlements and single finds) were retained (see Chap. 4 for selection criteria).

The above outlined restructuration of the datasets into a shared database architecture not only serves the modeling needs, but also allow to combine the six different dataset into one single DB to perform a comparative analysis. Figure 5.3 shows the number of occurrence of Roman sites on the 11 site type categories, differentiated by Canton. The settlements and single finds appear to be by far the most occurring categories. This further justifies the choice of these categories for the modeling procedure, as more data means more confident results. As experienced by Casarotto et al. [11] this is by no means a definitive set of archaeological evidence classifications, and the literature is exhaustive regarding typological and functional classifications. Nevertheless, the scheme adopted in this context was chosen for its simplicity and replicability, rather than any analytical power.

5.2 Mapping Uncertainty

Defining what is meant by quality of information can be very difficult, given the diversity of dimensions that this concept takes on [12]. According to Goodchild [13], "Quality [...] is a measure of the difference between the data and the reality that they represent, and becomes poorer as the data and the corresponding reality diverges. Thus, if data are of poor quality, and tell us little about the geographic world, then they have little value".

In the context of uncertainty underlying the archaeological data, as described previously, the need to produce a clearer and more comprehensive picture of the information collected became essential in order to visualize variations in the amount of data and the levels of quality and reliability of the information regarding the

Table 5.2 Re-classification of the Roman site categories for Aargau. The first column shows the new generic categories, the two right columns the original DB structure (with English translation)

DATABASE AARGAU—Roman categories		
Class	English classification	Original classification
Bridges	River crossing	Flussübergang
Fortifications	(Military) camp	(Militär) Lager
	Castella	Burg, Schloss, Kastell
	Other fortifications	Andere Befestigung
	Refuge, earthwork	Refugium, Erdwerk
	Tower, Watchtower	Wachturm, Hochwacht
Graves	Cemetery (group or field of graves)	Gräbergruppe, -feld
	Flat grave	Flachgrab
	Flat grave (1–2)	Flachgrab (1–2)
	Flat grave group	Flachgräbergruppe
	Grave	Grab
	Grave (1–2)	Grab (1–2)
	Grave hill	Grabhügel
Others	Ditch, Wall	Graben, Wall
	Form: others	Form: andere
	Hint from local name	Hinweis aus Flurname
	Others	Andere
	Pit, shaft, holes	Grube, Schacht, Mulde
	Various	Diverses
	Wall, Border ditch	Weidmauern, Grenzgraben
Quarry	Quarry	Abbau, Steinbruch
Religious sites	Religious building—Temple, Church, Chapel	Kultbau - Tempel, Kirche, Kapelle
	Religious site—Offertory	Kultplatz, Opferstelle
Roads	Road	Strasse
	Road network (supposedly)	Strassennetz (angeblich)
	Traffic	Verkehr
Settlements	(Single) Building, House, Dwelling Tower, Basement	(Einzel-)Gebäude, Haus, Wohnturm, Keller
	Camp, Village with walls	Lagerplatz, Dorf, mit Umwehrung
	Camp, Village, Vicus	Lagerplatz, Dorf, Vicus
	City (unfortified)	Stadt (unbefestigt)
	Construction stones, ceramic, wood	Bausteine, -keramik, -hölzer

(continued)

Table 5.2 (continued)

DATABASE AARGAU—Roman categories		
	Construction structures (wood and stone)	Baustrukturen (Holz und Stein)
	Court, Farm, Installation (Villa)	Hof, Gutshof, Anlage
	Crafts (oven)	Verarbeitung, Raffinerie
	Occupation layer, Stone concentration	Kulturschicht, Steinkonzentration
	Other installations	Sonstige Anlagen
	Production, Industry, Crafts	Produktionsplatz, Industrie, Gewerbe
	Settlement	Ansiedlung
	Settlement	Siedlungsstelle
	Structures and observations (unidentified)	Festgestellte Strukturen und Beobachtungen
Single finds	Artifact	Artefakt
	Coin	Münze
	Coin deposit	Münz-Depot
	Coins	Münzen
	Concentration of single finds (ceramic)	Konzentration von Einzelfunden
	Material, deposit	Material, Depot
	Other single finds	Anderer Einzelfund
	Single finds, scattered finds	Einzelfund, Lesefund, Streufund
	Treasure	Schatz
Unknown	Unknown	Unbekannt
Water infrastructure	Canalization	Kanalisation
	Water distribution	Wasserversorgung
	Well, Water canalization	Sodbrunnen, Wasserleitungen

time periods and typologies reported. Moreover, the modeling procedure consists of many parameter values that can influence the model outcome. Such assigned parameter values are based on a large number of assumptions, inevitably leading to further uncertainties.

While it is not possible to estimate all uncertainties in the data used within this modeling procedure and to undertake a full uncertainty assessment, the main sources of uncertainty with regard to the archaeological databases and the potential implications for the obtained results were addressed trough a fuzzy logic approach. This theory and its methodology have been described in the Chap. 3, but as useful recall stands Gacôgne's [14] definition: "Fuzzy logic, or more generally the treatment of uncertainties, is to study the representation of imprecise knowledge and reasoning

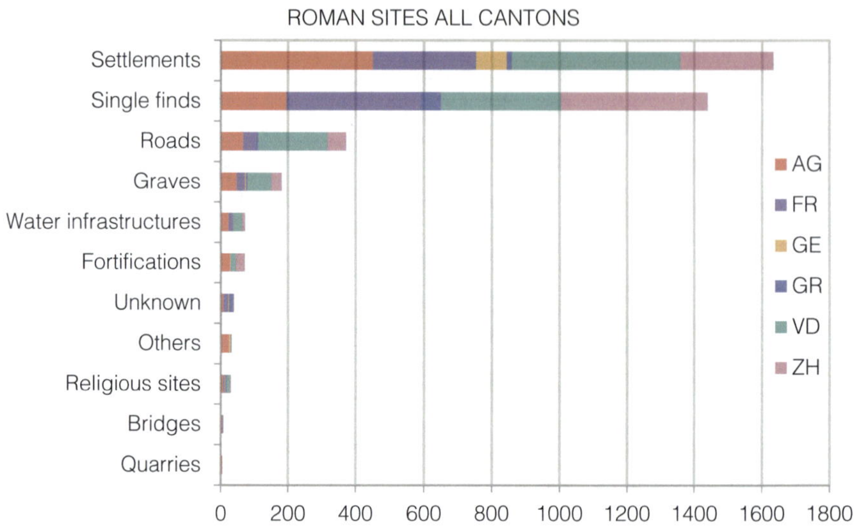

Fig. 5.3 Cumulative occurrences of Roman site categories for all study areas according to the new database architecture

approached". Although this approach has not yet been widely adopted in archaeology [12], its theoretical and methodological framework is particularly well suited for the kind of issues addressed by this Thesis, particularly when modeling unknown sites based on uncertain knowledge.

As pointed out by Fusco [12], this approach also has the advantage of keeping all the available data, rather than considering only those estimated as "reliable" and eliminating the "unreliable" from the databases. This selection in fact could even be seen as counterproductive: in addition to loosing data, it could lead the analysis to overestimate the quality and certainty of the data considered to be reliable. The quantification of the data quality and its direct integration into analyses and modeling procedures would represent a much more significative and performant approach. In this context, the reproducibility of the methodological protocol is an important aspect of the process. This approach can be applied to all type of issues, geographical and temporal scale [12].

Two different datasets out of six incorporated information regarding uncertainty and were thus computed taking it into consideration. In the case of the Canton of Aargau, the uncertainty is related to the time period. The degree of uncertainty of the information was originally and explicitly stored in the column DATIERUNG_QUALITÄT (Quality of the dating), with qualitative expressions such as 'romisch-unsicher' (roman-unsure), 'romisch-sicher' (roman-sure), 'unbestimmt-unsicher' (unknown-unsure).

The Canton of Geneva presented uncertainty with regard to the site typology. The degree of uncertainty of the information was stored in the column TYPE (type) of the original dataset, with qualitative predications such as ‘site presumé’ (Suspected site) and ‘zone potentielle’ (Potential area).

The solution experimented in this Thesis consisted in assigning a numerical degree of membership, which expresses the subjective level of ’confidence’ in the assignment under consideration. Although this coefficient of membership is assigned “subjectively”, the procedure roots in the tradition of a subjective approach to the uncertainty that has seen among its major exponents De Finetti [15] for probability theory and Savage [16] for statistics. Those numeric values are the expression in numerical terms of a series of elements evaluated subjectively by the researcher, in which the experience and scientific correctness converge [17, 18], aiming at giving the scientific status of measurability and verifiability to the reliability problem.

Table 5.3 shows the example of uncertainty for the linguistic variable “Type” in the database of Aargau, which applies to the single finds in particular. The uncertainty quality is expressed with 4 linguistic terms: “sure”, “unsure”, “unknown”, “impossible” and a coefficient of membership or a degree of reliability is assigned in percentage. Thus, the numeric value for the uncertainty can be considered as follows: “impossible” equals 0, “unknown” equals 1, “unsure” equals 2 and “certain” or “sure” equals 3.

Table 5.4 shows the example of uncertainty for the linguistic variable “Type” in the database of Geneva, which applies to the settlements. The uncertainty quality is expressed with 5 linguistic terms: “sure”, “potential”, “presumed”, “possible”, “impossible” and a coefficient of membership or a degree of reliability is assigned in percentage. Thus, the numeric value for the uncertainty can be considered as follows: “impossible” equals 0, “possible”, “presumed” or “potential” equals 1, “certain” or

Table 5.3 Uncertainty quantification for single finds in Aargau

Uncertainty	Period	Type	Percentage (%)	Numeric
Certain	Roman	Sure	100	3
Unsure	Roman	Unsure	66.6	2
Unknown	Roman	Unknown	33.3	1
Unknown	Roman	Blank	33.3	1
Impossible	Neolithic	Sure	0	0

Table 5.4 Uncertainty quantification for settlements in Geneva

Uncertainty	Period	Type	Percentage (%)	Numeric
Certain	Roman	Sure	100	2
Unsure	Roman	Potential	50	1
Unsure	Roman	Presumed	50	1
Unsure	Roman	Possible	50	1
Impossible	Neolithic	Sure	0	0

"sure" equals 2. Only the numerical values were taken into account while modeling within ML algorithm.

The quantification protocol developed allows assigning quantitative uncertainty values to qualitative linguistic labels (such as "unsure" or "unknown", etc.) in a systematic manner. Once obtained, quantitative values can be algorithmically processed as part of the predictive model [3].

The process followed at this stage of the research consisted of both proposing a global, clear and at the same time, synthetic image of archaeological databases concerning vast geographical areas very heterogeneous in their levels of data quality, as well as to propose an original point of view. The problem of uncertainty was approached with a fuzzy logic structure application and the results embedded in the Machine Learning modeling procedure.

The approach as developed in this section is part of an exploratory line of reasoning where one first tries to identify the different levels of uncertainty through which the data is expressed, going from the general to the specific, before to model some of the uncertain sites and single finds as certain. In fact, according to Farinetti et al. [19], uncertainty should ideally not be considered only a posteriori, once modeling results have been obtained, as it is very often the case in archaeology today, but should be integrated at the time of data collection not only as an attribute of the artifact or site in question (such as its epoch or typology for example) but as the point of view, or even the theoretical basis from which the data are considered. In this regard, Farinetti et al. [19] also stated that "we therefore should deal with the degree of uncertainty since the beginning of the processing of our collected datasets, and therefore at the shared database level, as we saw, in order to get less biased results at the end of the process."

5.3 Preparing the Environmental Variables

Following the predictive modeling workflow, the geo-environmental variables were modeled within ArcGIS 10.6 and 10.7. This section will illustrate the specific procedure for preparing each variable for its later use in R environment.

The independent variables are represented by three main categories: topography, water resources and soils, with a particular attention on arable soils. Chapter 4 addressed why each variable was used in creating the predictive model. For example, the topography played a role in site location through altitude, slope, etc. Soil properties influenced the suitability for agriculture, which played a crucial role for supporting communities. Water bodies were important in that they provided fresh water but also served as communication and commercial routes. This chapter outlines the preprocessing performed on the independent variables prior the predictive modeling procedure.

5.3.1 Topography

The first GIS operation made, as briefly sketched in Chap. 4, was to mask (*cut*) the shape of the DEM acquired at the national scale on the administrative borders of each Cantons. Thus, the original vector SWISSBOUNDARIES3D was split in 6 independent polygonal shapefiles named Borders_* (plus the initials of each Canton).

Since a certain number of archaeological sites fell outside or on the exact line representing the cantonal borders, a 1 km buffer was added to this vectors in order to avoid any archaeological site omission. The extent of each of the cantonal borders with a buffer was extracted from the DEM of the entire territory of Switzerland, resulting in the basemaps on which the other environmental variables (also acquired at national scale) could be further cut out and processed.

Previous to any spatial operation the national geographic reference system was assigned to the variables, as performed already for the archaeological DBs.

Thus, the individual DEMs now contained the highest and lowest altitude values for each of the case studies and provided a base on which to calculate and derive the slope and aspect.

Slope is easily calculated in ArcGIS and produced directly in raster format. It has been computed for each region and in degrees. No further reclassifications were therefore performed on this feature to highlight possible threshold.

Likewise, the Aspect[5] or the terrain orientation was calculated and derived from the DEM for each region. It also consists in a raster format, which expresses the terrain orientation with regard to the compass direction. Further computations had to be performed on this variable prior to its use in the model. As explained in the Chap. 4, Aspect is generally expressed with values from 0 to 360 degrees. In order to avoid the use of a circular variable that can introduce bias, as very distinct values (0° and 360°) represent the same situation in reality (north orientation), it was preferred here to transform the aspect variable into north and east orientation by trigonometric functions [20].

Two features were thus obtained through a Map Algebra expression using Python syntax[6]: Northness and Eastness. They correspond respectively to the cosine and to minus sine of the aspect angle. Northness will take values ranging from 1 if the aspect is orientated northwards, to -1 if the aspect is orientated southward, and towards 0 if the aspect is either east or west . Eastness behaves similarly, except that values

[5] "Aspect identifies the downslope direction of the maximum rate of change in value from each cell to its neighbors. It can be thought of as the slope direction. The values of each cell in the output raster indicate the compass direction that the surface faces at that location. It is measured clockwise in degrees from 0 (due north) to 360 (again due north), coming full circle. Flat areas having no downslope direction are given a value of -1" from the definition provided by the ArcGIS documentation.

https://desktop.arcgis.com/en/arcmap/10.3/tools/spatial-analyst-toolbox/how-aspect-works.htm

[6] https://desktop.arcgis.com/en/arcmap/10.3/tools/spatial-analyst-toolbox/raster-calculator.htm.

close to 1 represent east-facing aspect, -1 west-facing aspect, and values close to 0 either north- or south-facing aspect.

5.3.2 Hydrology—Distance to Water

In general, to quantify the concept of 'attraction' concerning Roman settlements and single finds, it is necessary to connect it to the idea of 'distance': the closer site x is to place y, the greater the attractiveness of y on x.

A site's proximity to water bodies such as rivers and lakes, considered as source of water supply and particularly as communication and commercial routes was quantified and computed based on the linear and polygonal features representing these landscape entities in Vec200 (see Chap. 4).

The Swiss map of the rivers and lakes consisted of both major and minor rivers, as well as of stagnant waters and dams currently visible in the topographical maps. Since today many lakes in Switzerland are dammed lakes, which did not exist during the Roman period, a first selection of the lakes was performed, eliminating all modern artificial lakes. In order to allow for a pertinent spatial differentiation, and considering lakes primarily as communication routes and fishing grounds, a further selection was applied, considering only lakes with a surface above 1 km^2 (main lakes) in the further procedure. Furthermore, it was considered that these larger lakes have probably existed more consistently over time, whereas very small lakes might have appeared more recently.

In a similar way, rivers with a certain water flow and width were taken into account and interpreted as possible communication routes in use during ancient times. The main watercourses (classified as 4–6 in the field KLASSE of the original vector, which include the rivers defined as main rivers and showing a certain width) were thus selected for further modeling purposes.

At this point, through a further effort of abstraction, the Earth's surface was considered flat and smooth, eliminating all geomorphological roughness, as if it were a two-dimension Cartesian plane (Euclidean space). Using the GIS-based 'Euclidean distance' analysis, the distance of every point in space (pixel) to the nearest lake and river was computed, resulting in two features, distance to main lakes, and distance to main rivers. These computed features allow a better understanding of the distribution of the archaeological evidences in various intervals of linear distances, in relation to the attractors (rivers and lakes), in order to numerically verify the different levels of proximity [21]. The outputs of this analysis, two raster features, were resized into 100-m cell grids. As for the Slope, no prior reclassification or thresholds class was computed. Instead, a continuous distance calculation was preferred.

5.3.3 Soil Map—Agricultural Suitability

The soil map *Digitale Bodeneignungskarte der Schweiz* (*Carte des aptitudes des sols de la Suisse/Carta digitale delle attitudini dei suoli della Svizzera*, 2012) was also acquired at national scale, but as performed with the other variables, it was necessary to extract the extent of the regional DEMs.

This original polygonal map contains information concerning the current soil qualities and the suitability of soils for cultivation. The following soil properties, such as surface stoniness or permeability etc., relevant for agricultural activities were considered:

The agricultural suitability
The depth of vegetal soil
The soil skeleton
The water saturation capacity
The water storage capacity
The soil permeability capacity
The nutrient storage capacity

Derungs [22] explains that the development of soils is based on a number of characteristics. Depending on the biogeographic context and its "age", the soil can be as deep as a few centimeters to several meters. Divided in horizons, its upper part is rich in more or less decomposed organic matter, and rich in living organisms, called humus. Soils are also defined by scientists according to their properties: the texture, structure, porosity, water regime, temperature, clay-humic complex, Ion exchange, cation exchange capacity, pH, redox potential, soil fertility/quality [23–25]. These properties have a direct relationship with their propensity to be cultivated, e.g. sufficient soil depth, adequate but not excessive water drainage, etc. [22].

To perform the extraction procedure, a selection based on each specific attribute or property was necessary. Through *Polygon to raster* tool, which converts polygonal features into rasters, the soil map was split into 7 raster submaps, each one focusing on one specific attribute per time. Each of the soil properties, by product, is classified in up to 7 classes based on the different soil aptitude levels (Table 5.5).

Each simplified soil attribute was resized into a raster map with 100 m resolution, in order to comply with the other input data. Thus, for each Canton a set of 7 soil maps was provided.

5.3.4 Geology

Finally, the digitized Geological map of Switzerland provided valuable information about the terrain of each Canton. The map contained a list of 30 categories of geological properties (see Table 5.6), accompanied by very detailed information about the chronological epoch, the description of the specific type of substrate and

Table 5.5 List of the soil properties and their classes

Class	1	2	3	4	5	6
Soil properties						
Agricultural suitability	Unknown	Hindered	Good	Very good	Sufficient	–
Deep soil	Unknown	Very shallow	Shallow	Medium	Deep	Very deep
Skeleton	Unknown	Not stony	Slightly stony	Stony	Very stony	Extremely stony
Water saturation	Unknown	Absent	Humid	Slightly wet	Wet	–
Water storage	Unknown	Very poor	Poor	Medium	Good	Very good
Permeability	Unknown	Very slow	Slower	Slightly slower	Normal	Extreme
Nutrients storage	Unknown	Very poor	Poor	Medium	Good	Very good

its formation, the description of lithology/petrology, etc. The original map covering all Switzerland was extracted and masked on the DEMs extension of each Canton of interest and transformed into a raster through the same process followed for the soil map. The extraction procedure was based on the specific field "Geology", thus 6 different cantonal geological maps were created.

The ensemble of the spatial layers was processed in order to correct and eliminate construction errors (e.g. no-data and resampled to match with the same spatial resolution of 100 m). Another aspect to be considered while dealing with the archaeological datasets and geo-environmental variables is the constant of missing values. However, missing data is usually a result of disrupted sampling, or the repurposing and combination of previously collected datasets. In this study, missing values derive principally from the soil map, as it did not always contain the same number of classes for each region analyzed, which means, that not for all cantons the same quantity of information was available. The no-data or missing values problem is endemic in the real-world datasets. In this case, the missing values hold no particular significance in the overall variable's meaning.

Last but not least, the ensemble of feature class layers was computed for the overall extent of Switzerland following the same procedures as described above. These countrywide features were then used as independent variables to perform a predictive model covering the entire Swiss territory.

The input variables were finally imported in R environment (free software environment for statistical computing and graphics), where further computations were carried out. In addition to the preprocessing described in this Chapter, a post-processing stage was performed after the modeling procedure to visualize the results of the analyses in intuitive graphical manner. The results of this post-processing will be presented in Chap. 6.

Table 5.6 Classes of the Swiss geological map

Geology	
Category	
1	Lakes
2	Glaciers
3	Silt, Clay, Loess, Ground Moraine, Frontal Moraine
4	Clayey silts
5	Gravel and Sand (Glacial Deposit)
6	Gravel and Sand (Modern Deposit)
7	Gravel Grit, blocs, Talus slopes
8	Marls with weakly solidified Sandstone inclusions
9	Marls and Clayey Schists
10	Non calcareous red sandstone, sandy clay schists
11	Clay thin to thick, ferrous clays
12	Limestone with medium-solidified Sandstone inclusions
13	Conglomerates, from weakly to mildly solidified
14	Conglomerates, from weakly to mildly solidified
15	Conglomerates, breccia
16	Clayey shists, calcareous phyllite
17	Clayey shists, calcareous phyllite, sandstone
18	Limestone phyllites, calcareous gneiss and limestone schists
19	Solid Limestone
20	Limestone, sandy, marly, shisty limestone
21	Schist Deposit
22	Dolomites and cornieules
23	Granites, quartz diorites, quartz syenites, diorites
24	Quartzite porphyres, porphyrites, porphyritic tufs
25	Quartzite
26	Gneiss with two micas or biotite
27	Schisty Conglomerates and breccia
28	Sericite Gneiss
29	Green Schists
30	Serpentines

5.4 Locational Preference Analysis

According to Antoni et al. [26], we should represent the data in such a way that the structure and the underlying phenomena not immediately visible can emerge “il ne s’agit plus simplement de représenter le monde mais également de mieux faire émerger des phénomènes non directement visibles ou mal perçus et de mieux réfléchir aux structures spatiales qui s'y organisent" that is, by means of quantitative solutions, we should let the data speak for itself and observe if any spatial structure emerge before proceeding with further manipulations.

The roots of the exploratory approach at the core of archaeological locational preference analysis [27] may be seen in the philosophy of Grounded Theory, which consists of researching original viewpoints by exploring the research subject without any assumption behind the analysis or any preconceived idea [28–30]. This approach can be very appropriate to help answering the archaeological questions this Thesis is dealing with and can constitute an original response to the 'fixism' of models, and to the conditioning of analyses by strong and preconceived hypotheses [12]. As stated by Banos [31], it is important that the researcher explores what the data reflects with an openness to what is not expected to find what he/her is not looking for.

According to Fusco [12], the exploratory approach can moreover be understood as the ability of the scientist to be guided by the data in the search for the unexpected [32]. Exploratory Spatial Data Analysis (ESDA) as defined by Anselin [33] is: "the collection of techniques to describe and visualize spatial distributions, identify atypical locations (spatial outliers), discover patterns of spatial association (spatial clusters), and suggest different spatial regimes and other forms of spatial instability or spatial non-stationarity". This approach can be particularly useful in the manipulation of archaeological databases, taking into account the complexity of a large amount of data. Generally speaking, it represents the crucial process of performing initial investigations on data in order to discover patterns, to spot anomalies, to test hypothesis and to check assumptions with the help of summary statistics and graphical representations [34]. In archaeology, this set of procedures constitutes the core of locational preference analysis.

In the present context, the locational preference analysis performed by the combined use of two softwares (GIS & R), allowed to overlay the archaeological data in association with the environmental features and to visualize statistical trends in their spatial distribution. To summarize, by considering different kind of analytical and visualization techniques, ESDA can provide valuable information about the nature of the data before applying any specific model. Finally, it aims at a better processing and understanding of the results [26, 35–37].

5.4.1 All Cantons

By combining the regional data, it is possible to extract and compare information at a supra-regional scale. Although some of the regions considered for the analysis are very different from each other, most Roman settlements (85% of 1488 sites) endorse a preference for lowlands (between 350 and 650 m a.s.l.). As shown in Fig. 5.4, Roman settlements in AG, GE, VD and ZH are highly concentrated around altitudes of 400–450 m a.s.l. The location of settlements at higher altitudes in the Cantons of Fribourg and Grisons can be explained by the natural geomorphology of

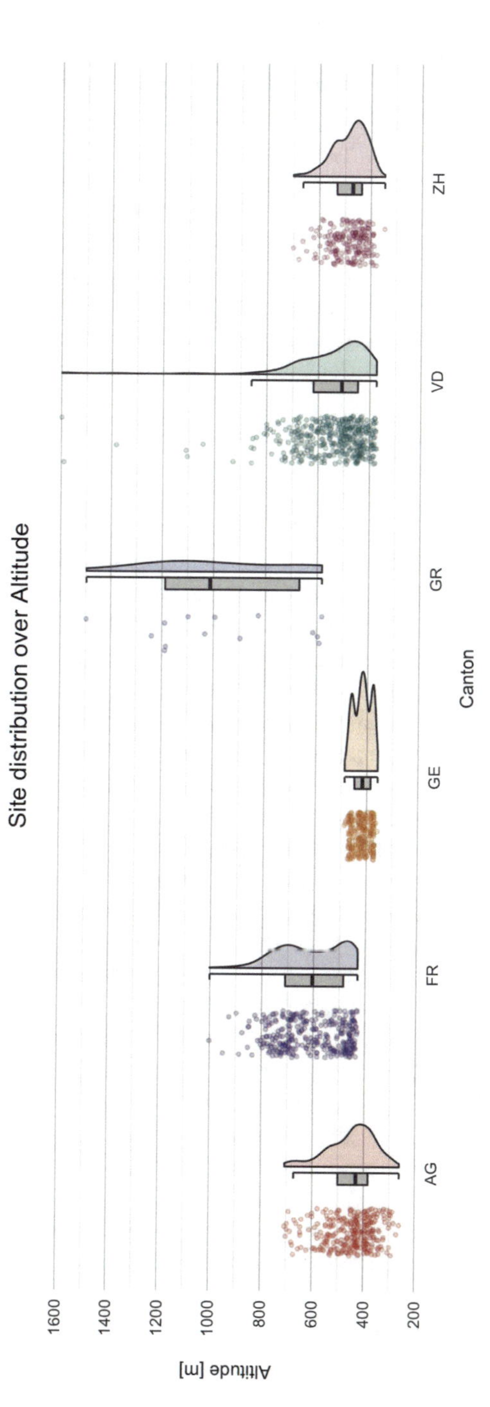

Fig. 5.4 Distribution of all Roman settlements over Altitude per Canton. The point jitters show all occurrences. The boxplot indicate the median value and the interquartile range IQR (25th (Q1) to 75th (Q3) percentile (half of all occurrences lie within the gray box), the whiskers show the "minimum" (Q1-1.5*IQR) and "maximum" (Q3 + 1.5*IQR) values. The colored area indicates the concentration of occurrences (Kernel probability density)

these regions. The Canton of Vaud shows a few outliers represented by settlements discovered in the Jura Mountains.

Similarly to the distribution over altitude, the Roman settlements are located on gentle slopes (85% of the sites located at less than 5°). Only very few extreme outliers in the Cantons of Fribourg (1) Grisons (1) and Vaud (2) lie on slopes above 20° (Fig. 5.5).

As can been observed in the Figs. 5.6 and 5.7, the settlements generally respect a northwest orientation (28% north and 25.4% west). The Canton of Vaud shows an opposite trend with settlements mainly oriented in southeastern direction. However, no strong preference for a particular orientation can be discerned. Furthermore, the fact that the very large majority of settlements lie on slopes under 5° corroborates the relevance of theirs northwestern orientation. On flat ground, orientation has less importance.

Figure 5.8 shows the distance from the main lakes. In 4 out of 6 cantons (FR, GE, VD and ZH) more than half of the settlements are located within 10 km of a lake. In Aargau and Grisons, some settlements show values above 20 km. This can be explained by the low number of main lakes in these regions.

In five Cantons out of six, Roman settlements are mainly located within less than 5 km from the main rivers, with exception of Vaud, where some sites have been found at a distance above 15 km. As can be seen in Fig. 5.9, in AG, FR and GE a majority of settlements are even located within 2500 m from the next main river. As can be observed from the kernel density representation (color-filled areas), the density of settlements decreases again in immediate proximity to the rivers. This may be explained by the risks associated with floodplains in flat areas. In the Canton of Vaud, greater distances from main rivers can be observed due to the absence of such rivers for large parts of the canton.

With regard to the agricultural suitability map, the large majority of the settlements (53%) lay on soils classified as very good for agricultural production, followed by the unknown and unsuitable classes (Fig. 5.10). The two latter classes often refer to the urbanized areas. The classes of the other soil properties variables holding the largest percentages of sites are those that describe the soils as 'slightly stony' (37%) or 'stony' (35%); with a 'good' water storage capacity (45%); with a 'slightly slow' permeability (33%) and normal permeability (32%); with an 'absent' water saturation (48%); with a 'good' nutrients storage capacity (56%) and deep vegetal soil (59%). All these soil characteristics point at lands suitable for agriculture. The class 'unknown' accounts for 20% of all settlements in every variable, as it represents the urbanized areas. The exhaustive results of this analysis are shown in Appendix A.7.2.

Also regarding the geology, a clear preference appears, where more than 50% of the Roman settlements are located on terrains classified as silt, clay, ground and frontal moraine (Fig. A.119).

By looking regionally at each single Canton, it is possible to isolate and extract the following patterns with regard to the sites position over the landscape and with respect to the environmental features considered:

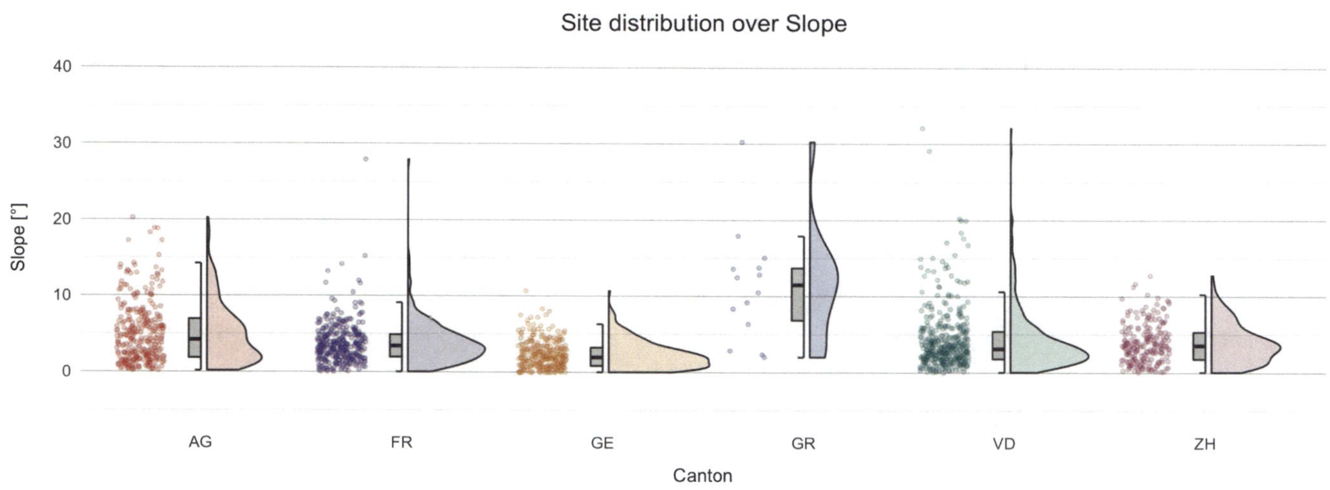

Fig. 5.5 Distribution of all Roman settlements over Slope per Canton. The point jitters show all occurrences. The boxplot indicate the median value and the interquartile range IQR (25th (Q1) to 75th (Q3) percentile (half of all occurrences lie within the gray box), the whiskers show the "minimum" (Q1-1.5*IQR) and "maximum" (Q3 + 1.5*IQR) values. The colored area indicates the concentration of occurrences (Kernel probability density)

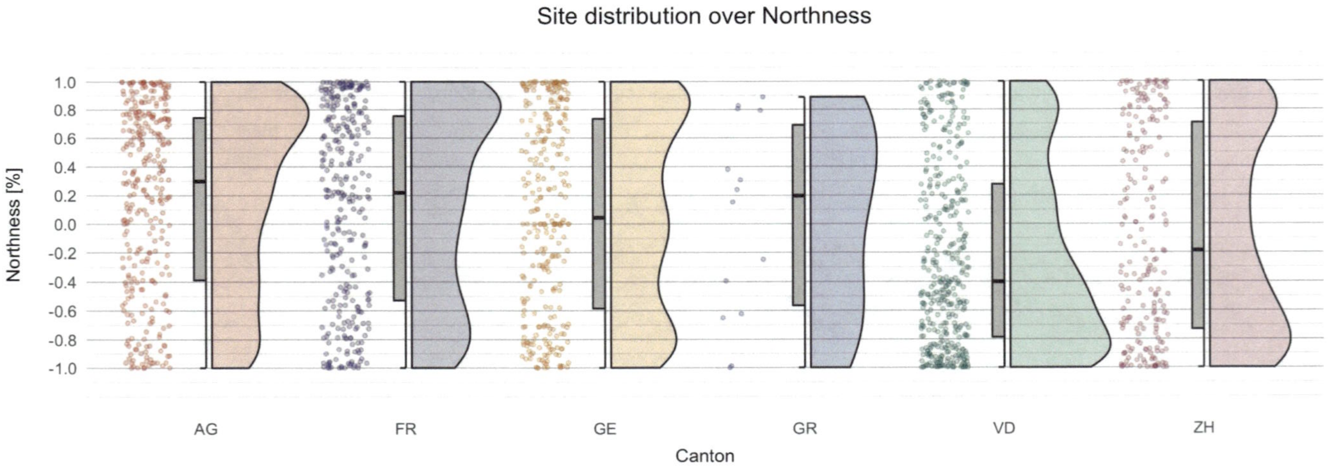

Fig. 5.6 Distribution of all Roman settlements over Northness per Canton. The point jitters show all occurrences. The boxplot indicate the median value and the interquartile range IQR (25th (Q1) to 75th (Q3) percentile (half of all occurrences lie within the gray box), the whiskers show the "minimum" (Q1-1.5*IQR) and "maximum" (Q3 + 1.5*IQR) values. The colored area indicates the concentration of occurrences (Kernel probability density)

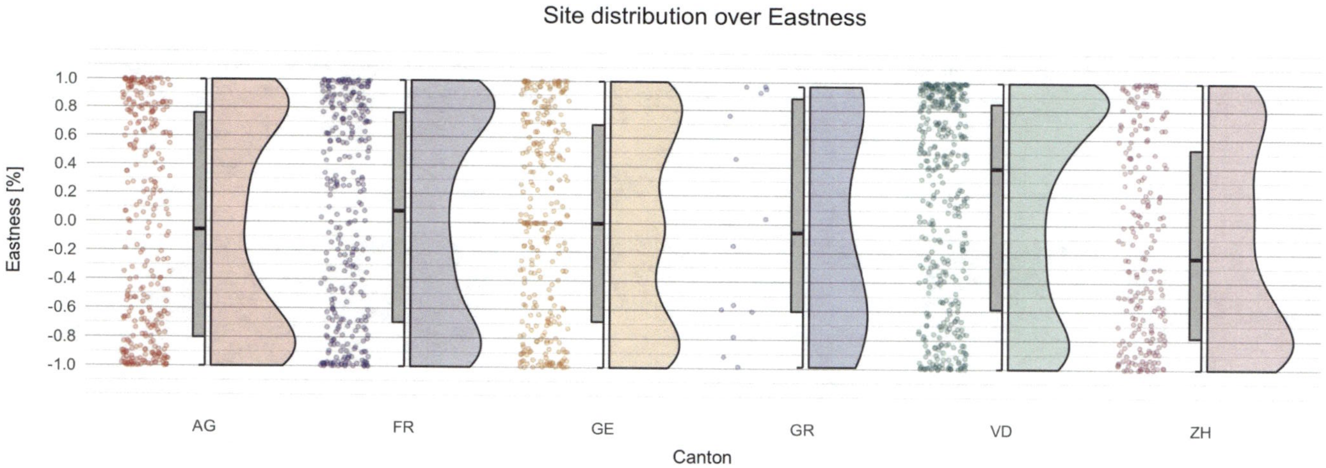

Fig. 5.7 Distribution of all Roman settlements over Eastness per Canton. The point jitters show all occurrences. The boxplot indicate the median value and the interquartile range IQR (25th (Q1) to 75th (Q3) percentile (half of all occurrences lie within the gray box), the whiskers show the "minimum" (Q1-1.5*IQR) and "maximum" (Q3 + 1.5*IQR) values. The colored area indicates the concentration of occurrences (Kernel probability density)

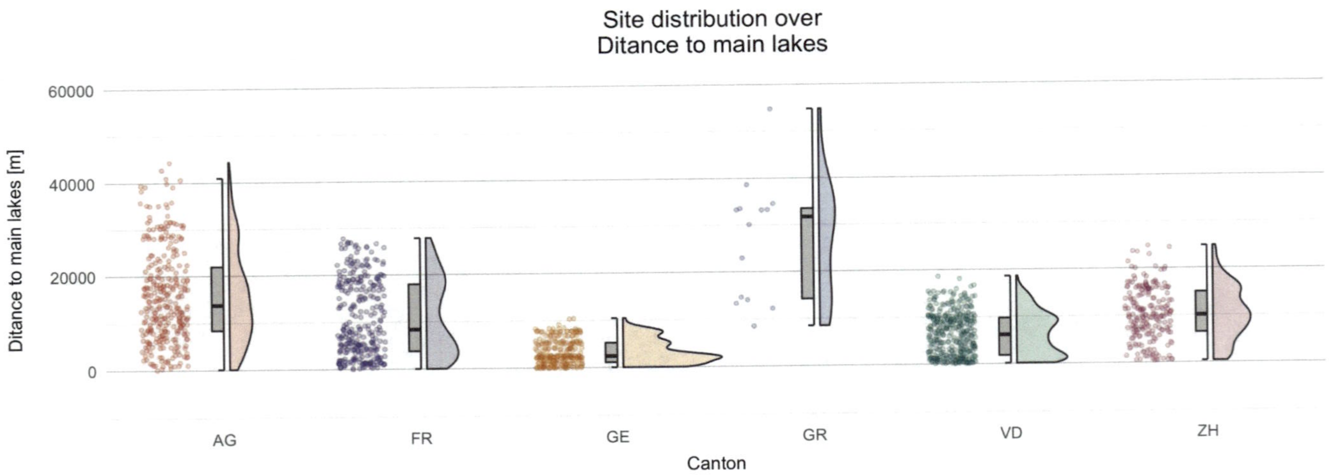

Fig. 5.8 Distribution of all Roman settlements over the Distance to main lakes per Canton. The point jitters show all occurrences. The boxplot indicate the median value and the interquartile range IQR (25th (Q1) to 75th (Q3) percentile (half of all occurrences lie within the gray box), the whiskers show the "minimum" (Q1-1.5*IQR) and "maximum" (Q3 + 1.5*IQR) values. The colored area indicates the concentration of occurrences (Kernel probability density)

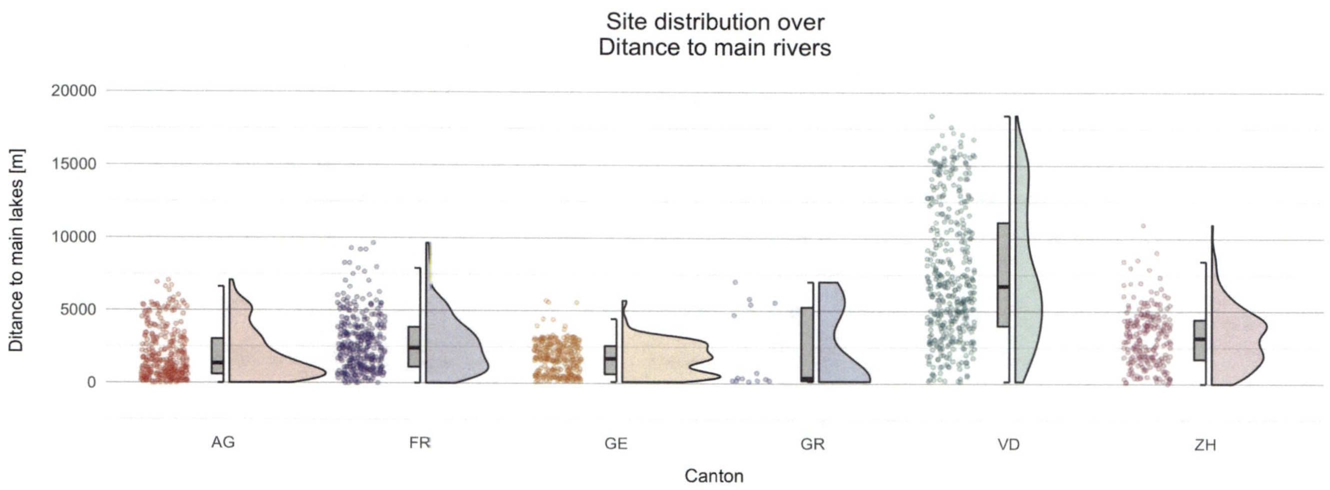

Fig. 5.9 Distribution of all Roman settlements over the Distance to main rivers per Canton. The point jitters show all occurrences. The boxplot indicate the median value and the interquartile range IQR (25th (Q1) to 75th (Q3) percentile (half of all occurrences lie within the gray box), the whiskers show the "minimum" (Q1-1.5*IQR) and "maximum" (Q3 + 1.5*IQR) values. The colored area indicates the concentration of occurrences (Kernel probability density)

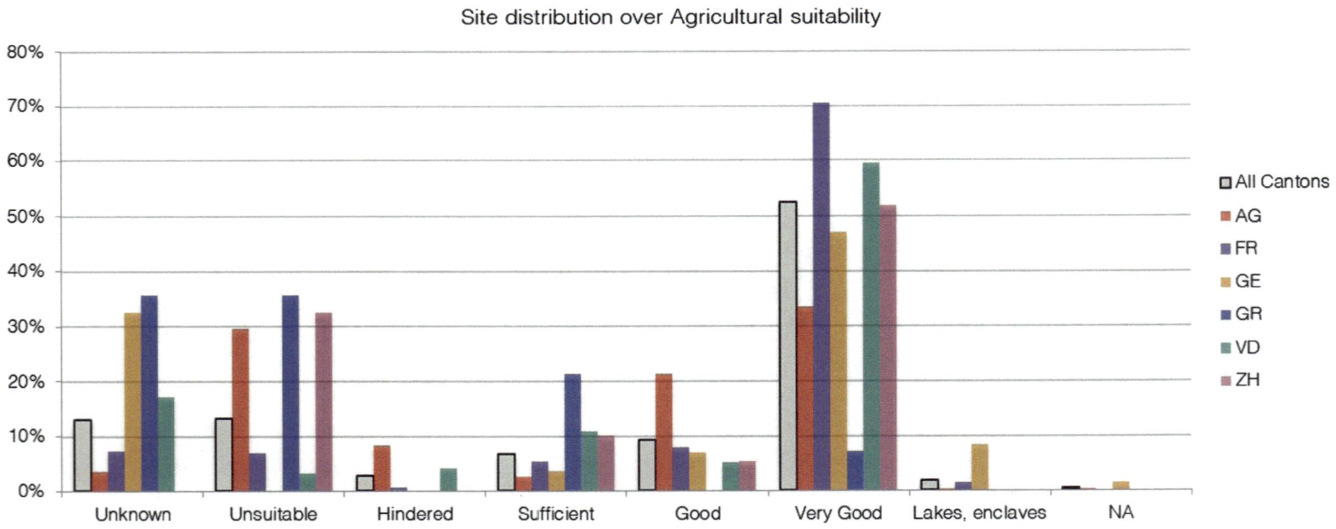

Fig. 5.10 Distribution of all Roman settlements over the clases of Agricultural suitability. The first column (All Cantons) shows the average values for the 6 computed Cantons

5.4.2 Canton of Zurich

The settlements (tot. number of 227 considered for the analysis) are mostly distributed below 500 m of altitude (58% of the total); 28% of them are oriented west while no significative preference in north or south orientation is observable: respectively 23% and 25%. The majority of the sites in this region (73%) is located on gentle slopes (under 5°), within 3 km from the main rivers but more than 3 km away from the main lakes. With regard to the agricultural suitability, it seems that the soils allowing very good production are favored, while the slightly stony soils with a good nutrient storage capacity and slightly lower permeability seem to prevail, grouping 57% of the sites. As defined by the geological map, the silt, clay, ground and frontal moraine are those types that accommodate 47% of the Roman settlements in the region (Appendix A.6.3).

5.4.3 Canton of Aargau

From the total number of settlements considered for the analysis (303), 74.5% are situated at less than 500 m of altitude and 56% preferred gentle slopes (with less then 3° of terrain inclination). Most of the settlements are oriented northwest within 3 km from the main rivers, but more than 5 km away from the main lakes; they lay on 'not stony' soils, with good nutrient storage capacity and very deep vegetal soils. There is great variety on the type of geological classes emerging from the analysis: only 37% of the settlements are located on gravel pits and sand (Appendix A.1.3).

5.4.4 Canton of Fribourg

Most of the sites (70.5% on the total 302) are located between 500 and 1000 m of altitude, favoring gentle slopes (below 5° of terrain inclination). A strong preference for northeast orientation and a close proximity to the main lakes (less than 1 km away) and rivers (less than 3 km) can be discerned. The 70.5% of the sites are based on areas where the modern soils are classified as very good for agriculture activity and production. Half of the settlements lie on silt, clay, ground and frontal moraine (Appendix A.2.3).

5.4.5 Canton of Geneva

The Roman settlements of this region (276) are equally distributed over the low heights and gentle slopes. While 59% of all settlements are located very close (less than 3 km) to the main lake, 88% lie within 3 km from the main rivers; preferably oriented north, they favored those soils where a 'very good' agricultural activity can be carried out. Additionally, the grounds on which most of the sites are located are classified as stony, with a slight permeability, and correspond to the silt, clay, ground and frontal moraine geological class (Appendix A.3.3).

5.4.6 Canton of Vaud

The sites (385) are equally distributed over the different heights: 36.4% lie between 371 and 500 m a.s.l., while 39.4% are located above this range and up to 1000 m. Only few sites lie above this range. A majority (56.5%) favored gentle slopes (with less that 5° of inclination). Only 25% of them are within 3 km from Lake Leman and the main rivers, with preference for southwest orientation. The advantaged soils are those that can support a very good agricultural production. The other soil qualities preferred are: deep vegetal soil, normal permeability, slightly stony soil, absent water saturation, good water storage and good nutrient storage. The geology confirms that the same type of grounds as in the other cantons are selected: silt, clay, ground and frontal moraine (Appendix A.5.3).

5.4.7 Canton of Grisons

The few Roman settlements (14 in total) considered for the analysis lie above 1000 m of altitude with a slope within 7°. Only one site is located on a 30° slope, while 8 sites are oriented northeast. All the settlements are located above 8 km distance from the main lakes and within 500 m from the main rivers. The agriculture suitability of the soils is defined both as 'unsuitable' both as 'very good production'. Again, the geology of the interested areas corresponds to silt, clay, ground and frontal moraine (Appendix A.4.3).

5.5 Random Forest Based Approach

This section will proceed by describing the model rules, the parameterization and calibration by illustrating the approaches experimented step by step as a narrative computing guide. It aims at providing the reader with a more in depth understanding of how the models were built. The model construction is kept intentionally separated from the results obtained and their interpretation, in order to provide with a first and exhaustive guideline attempt that could serve as a methodological support for future investigations, both in Cultural Heritage Management and in academic research.

Several independent models have been constructed: RF classification-based models have been used to predict the probability to discover unknown Roman settlements in the study areas and RF regression-based models to predict the probability to discover Roman single finds in the Canton of Aargau and Roman settlements in the Canton of Geneva. This expedient was explored due to the uncertainty contained in the original databases and to finally embed it in the machine learning-based modeling approach.

Basically, all types of RF models share the same settings, while specific differences concerning the number of trees to be grown can vary based on the spatial regional extension. Another important aspect is the number of variables retained at each modeling process. Mainly, the use of the agriculture suitability map combined with a geological map, topographic indexes (Digital Elevation Model (DEM), Slope, the orientation to North and East) and the distance to rivers and lakes, provide such excellent results.

According to the protocol followed, after the preprocessing in GIS, a series of steps have to be performed before it is possible to run the RF algorithm (as implemented by Breiman [38]) in R studio[7] (IDE for R).

At this stage, the ensemble data are named as dependent and independent variables. The dependent variables correspond to what is going to be modeled, the Roman settlements and the single finds for each region. In other words, dependent variables are the labels of the classification model, meaning that if in a region (cell-pixel) there is an archaeological site, it is labeled as a presence and labeled as absent otherwise [39]. As far as site location is concerned, there can be a variety of potential dependent variables. It is possible to predict site presence or absence, site density, site types, site functions, or various combinations, but whether the dependent variable changes, this would require internal revisions of model settings [40].

The labels for each canton are imported in R Studio as text file. The files contain only the columns considered essential for the computing procedure, i.e. the geographic coordinates X and Y, and the Presence/Absence indication. Presence/Absence value is expressed with 0 or 1. Thus, the dependent variables (Roman settlements) for each case study are listed as follow:

Zurich: 227 settlements
Aargau: 303 settlements; 113 single finds
Graubünden: 14 settlements; 113 settlements (until Middle Ages)
Fribourg: 302 settlements
Geneva: 276 settlements
Vaud: 385 settlements.

However, in the case of Geneva and Aargau, the procedure undertaken for managing the uncertainty has led to process it with a different RF type (Regression, Chapter 3.3.5). The original datasets contained information about uncertainty with regard both to the nature of the finds and to the epoch assignment. This additional information made it possible to incorporate uncertainty in the ML based models and

[7] https://rstudio.com/

motivated the experimenting of a new approach (the approach is better detailed later in this section).

On the other hand, the independent variables correspond to the list of geo-environmental factors as prepared in the preprocessing stage, imported as a list of 14 raster (TIF) files:

Digital Elevation Model (DEM)
Slope
Northness
Eastness
Distance to main Rivers
Distance to main Lakes
Agriculture Suitability
Permeability
Depth of Vegetal Soil
Water Saturation Capacity
Water Storage Capacity
Nutrient Storage Capacity
Skeleton of Soil
Geology.

The first part of the computing procedure is standardized and equally valid for all the case studies and models constructed in this Thesis. It proceeds by combining the different TIFF files into a single, multi-layer raster object (raster brick) that, having a predefined extension, can encompass all the variable's spatial extent. This is a fundamental step while computing several spatial analysis and spatial data manipulations.

The next steps consist in the generation of pseudo-absences [41]. As previously seen (Chap. 3), RF works by combining presences and absences datasets. Unfortunately, the archaeological datasets used for the regional models have little or no information on site absence. Thus, the predictions can only be made using 'pseudo-absence' data, by assuming that the prevalence of non-sites is approximately equal to a random or uniform distribution over the whole study region. This is justified by the argument that the proportion of sites compared to non-sites is very small, and thus the prevalence of non-sites will be very similar to such a random or uniform distribution [42, 43]. The pseudo-absence set consists in a sample of pixels selected randomly over the study areas, where no Roman sites are located. To ensure a good generalization of the model and to avoid the overestimation of lower classes, a balanced number of absences i.e. in a number equal to the observed presences, is specified.

The presences and pseudo-absences are merged and then combined with the independent variables to produce a spatial matrix or grid of 100 m equidistant cells, where each cell contains the information about the presence or absence of a Roman site and the information concerning the environmental features (quality of soil, degree of slope, altitude values, distance to lakes, etc.).

When dealing with spatial data and very few presences are available, as experienced in this research, especially when modeling each canton separately, the observations (or cells) close to each other hold similar characteristics. This phenomenon also known as "spatial autocorrelation" [44] can lead to an overestimation of the predictive performance of the model (i.e. fitting the model with input points close to each other).

One way to avoid the bias introduced by spatial autocorrelation consists in selecting the training and testing data far enough apart in the geographic space. This can be achieved by adopting, for example, a statistical technique called spatial *k*-fold cross validation [45, 46]. This technique consists in splitting the original dataset into a number *k* of non-overlapping groups called folds, training the model on k-1 sets and then testing it on the hold out sets. The process is repeated *k*-times and the *k*-error estimates are finally averaged to yield the overall error rate. Through *k*-fold cross validation, the input data for the model are randomly selected across the study areas. The size and the number of *k* folds are manually defined (using the *Range Explorer* tool to determine the spatial range of autocorrelation) as suggested by [47].

In order to train the RF algorithm on the training dataset, hundreds of decision trees are created, where a random archaeological data subset is combined with a reduced number of independent variables, randomly sampled as candidates at each split, by measuring the node impurity. Not all the input data is used to train the model, as 1/3, the so-called out-of-bag (OOB), is kept out for testing operations and to assess the generalization capacity of the algorithm. The trees grow and eventually stop when each terminal node contains less than a pre-fixed amount of presences. The 'out-of-bag' (OOB) are used to optimize the parameters of RF (the number of trees and the reduced number of variables) trained on independent data (the training dataset), while the testing dataset is normally used to evaluate the error rate of the final optimized model and to assess its ability to make good prediction on unused data (the testing dataset) [48–50].

The prediction of new data is computed taking the majority or the soft voting. This consists in converting the results of a binary classification, such as the prediction of presence ("1") or absence ("0") of a Roman site/single find, by counting how many times each cell is classified as "positive" or "negative" and normalizing the result over the total number of predictions. Alternatively, the proportion of how many times a cell is classified as "positive" or "negative" can be computed to asses the probability that the cell is a presence or an absence. Once the algorithm is trained and its performance has been assessed on the testing data set, it can be applied on the spatial matrix representing the entire study area to predict the presence or absence for every cell of the study area. The probabilistic output can be used to elaborate the final predictive map: a visual support that can help in identifying those areas susceptible to unveil unknown Roman sites/single finds, ranked into low to high probability (0–1) [49, 51].

Model accuracy and error rate is computed for each observation using the out-of-bag predictions, and then averaged over all observations (*Confusion matrix*). Many statistical methods related to regression and classification implement an indirect measurement of variable importance by computing criteria such statistical significance or Akaike's Information Criterion [52] in order to select the model variables in the first instance. RF however implements a very different approach.

Variable importance is assessed by evaluating the mean decrease accuracy computed by looking at how much the tree nodes, which use that precise variable, reduce the mean square errors estimated with the out-of-bag across all the trees in the forest. Additionally, the partial dependence plots give a graphical depiction of the marginal effect of each variable on the class probability [49, 51, 53].

Finally, a Receiver Operating Characteristics curve (ROC) measures the quality of the model prediction. This is a graphical technique based on the plotting of the true positives rate (TPR) against the false positives rate (FPR), both expressed as a percentage of the total number, where true positives (TP) (as for the true negatives, TN) are the correct classifications, and the false positives (FP) occurs when outcomes are incorrectly predicted, as "yes" when it is actually "no" (and vice-versa for false negatives, FN) [54]. According to Hosmer and Lemeshow [55], when the value of the "Area Under the ROC Curve (AUC)" lies close to 0.5 denotes a bad classifier while is equal or closer to 1 denotes an excellent classifier. Both the ROC curve and the corresponding AUC were estimated to evaluate the performance of the different models constructed.

For the Canton of Zurich, two models were compared: a first one, including all the geo-environmental features and a second one considering only the first six most important variables, as resulting from the previous model. The final probability map was elaborated based on the results of the second model. The number of trees was set at 700 with 3 variables combined at each split.

Based on the experience acquired while modeling the Canton of Zurich, several settings with regard to the number of trees and number of variables combined at each split were tested. A parameterization using 1000 decision trees and 4 variables at each node produced the optimal results and was thus applied for all models presented here. For all models, 5 k-folds were used to implement the k-folds cross validation described above. As mentioned before, the size of the *k*-folds was determined based on a range of autocorrelation analysis and was set as follows after several tests: 5000 m for Aargau, 9639 m for Fribourg, 4500 m for Geneva, 12000 m for Grisons, 13,000 m for Vaud and 6000 m for Zurich.

With respect to the Cantons of Aargau, Fribourg, Geneva and Vaud, RF classification models were also run for each study area considering first the full list of variables then a reduced number of it. However, no substantial variations were observed in the results, and finally good performing models were obtained using the entire set of variables.

The final dataset of Grisons, despite the wider geographic study area, did not contain enough settlements belonging to the Roman period to permit the construction of a reliable model, which prompt for modeling all the data available and belonging to the Roman period at once, with no particular distinction for the type of evidence. Here, no significant difference in the model performance accuracy was observed while using the full list of independent variables or a reduced number, so that the results presented in the next chapter were computed using all environmental features.

For the Cantons of Aargau and Geneva, besides running the RF classification on the Roman settlements dataset with similar settings to the ones for Zurich, RF regression was run on the settlements (GE) and single finds (AG) dataset. The regressor predicts continuous values lying between the minimum and maximum value of the input dependent variable. As for the RF classification, a coefficient is also assigned to each independent variable expressing whether a variation in that variable makes a single find or a settlement more or less likely to occur. In short, RF regression was used for Aargau and Geneva to predict a quantitative response, where the output is a continuous value [0–3 for AG and 0–2 for GE] that expresses the probability of finding an 'absence' or an unknown single find or settlement, simply holding similar characteristics to those already known and rated by the archaeologists as 'sure', 'unsure' or 'unknown' (AG) or 'potential/suspected' (GE).

In the final step, by merging the regional datasets, a more consistent database containing 1488 Roman settlement sites was generated and used as dependent variable for a predictive model designed at a national scale. 'CH' model aimed at predicting the probability to find new archaeological evidence of Roman settlements based on the knowledge on the already known sites analyzed in the six study areas. This procedure finally also allowed to extend the prediction to the other Swiss regions not analyzed in this study. Setting up this kind of model required more computational power as well with regard to the GIS spatial analysis as for the R data manipulations.

The protocol defined for this large scale model, does not differ much from the one illustrated before, up to the implementation of the RF function. The variables were all taken into account for the computational procedure and the excellent results allowed to discard the option of considering just a reduced list of variables for improving model accuracy. Only the *k*-folds size (50 km) was changed.

References

1. Brughmans T (2019) Formal modelling approaches to complexity science in Roman studies: a manifesto. 2(1):1–19
2. Verhagen P, Whitley TG (2012) Integrating archaeological theory and predictive modeling: a live report from the scene. J Archaeol Theory Method 19/1:49–100. https://doi.org/10.1007/s10816-011-9102-7
3. Gonzalez-Perez C (2018) Information modelling for archaeology and anthropology. Inf Model Archaeol Anthropol. https://doi.org/10.1007/978-3-319-72652-6
4. Martin-Rodilla P, Pereira-Fariña M, Gonzalez-Perez C (2019) Qualifying and quantifying uncertainty in digital humanities: a fuzzy-logic approach. In: ACM international conference proceeding series, pp 788–794. https://doi.org/10.1145/3362789.3362833
5. Niccolucci F (2006) Managing uncertainty in archaeological GIS applications. Reading historical spatial information from around the world studies of culture and civilization based on geographic information systems data
6. Maguire DJ, Dangermond J (1991) The functionality of GIS. In: Maguire DJ, Goodchild M, Rhind D (eds) Geographic information systems: principles and applications, vol 1. Longman Scientific and Technical, New York, pp 319–335

7. Watt A, Eng N (2014) Database design, 2nd edn. BC campus, Victoria, BC. https://opentextbc.ca/dbdesign01/
8. Sewell JP, Witcher R (2015) Urbanism in ANCIENT Peninsular Italy: developing a methodology for a database analysis of higher order settlements (350 BCE to 300 CE). Internet Archaeol 40. https://doi.org/10.11141/ia.40.2
9. Ossa A, Simon A (2010) Basic archaeological database design. Arizona State University, Archaeological Research Institute
10. McMurdo G (1982) Database file normalization as an information science related activity. J Inf Sci 4(1):9–17. https://doi.org/10.1177/016555158200400103
11. Casarotto A, de Guio A, Ferrarese F, Leonardi G (2011) A GIS-based archaeological predictive model for the study of Protohistoric location-allocation strategies (Eastern Lessinia, VR/VI). Ipotesi di Preistoria, vol 4, n° 2, Bologna, pp 1–24
12. Fusco J (2016) Analyse des dynamiques spatio-temporelles des systèmes de peuplement dans un contexte d'incertitude. Application à l'archéologie spatiale. PhD thesis, Université Nice Sophia Antipolis
13. Goodchild H (2007) Modelling roman agricultural production in the middle tiber valley, central Italy. PhD thesis, University of Birmingham
14. Gacôgne L (2003) Logique floue et applications, Institut d'informatique d'entreprise d'Evry, p 128. http://www.ensiie.fr/~gacogne/polyflou.pdf
15. De Finetti B (1970) Teoria delle probabilità, Sintesi introduttiva con appendice critica, Torino, Einaudi
16. Savage L (1972) The foundation of statistics. Dover, New York
17. Hermon S, Niccolucci F (2002) Estimating subjectivity of typologists and typological classification with fuzzy logic. Archeologia e Calcolatori 13:217–232
18. Hermon S, Niccolucci F (2003) A fuzzy logic approach to typology in archaeological research. In: Doerr M, Sarris A (eds) The digital heritage in archaeology: computer applications and quantitative methods in archaeology. Archive of Monuments and Publications, Hellenic Ministry of Culture, Heraklion, pp 169–178
19. Farinetti E, Hermon S, Nicolucci F (2004) Fuzzy logic application to survey data in a GIS environment. In: Beyond the artefact. Computer applications in archaeology. ArcheoLingua, Budapest (Hungary)
20. Roberts DW (1986) Ordination on the basis of Fuzzy set theory. Vegetatio 66:123–131
21. Bertoldi S, Castiglia G, Castrorao Barba A (2019) A multi-scalar approach to long-term dynamics, spatial relations and economic networks of Roman secondary settlements in Italy and the Ombrone valley system (Southern Tuscany): towards a model? In: Verhagen P et al (eds) Finding the limits of the limes. Computational Social Sciences. Springer, Cham, pp 191–214
22. Derungs N (2018) La gestion durable des sols agricoles: sécuriser les démarches ou légitimer les controverses? L'exemple des politiques agroenvironnementales autour de l'érosion hydrique des sols arables en Suisse. Université de Neuchâtel
23. Duchaufour P (2000) Introduction à la science du sol: Sol, végétation, environnement. Dunod, Paris
24. Gobat J-M, Aragno M, Matthey W (2003) Le sol vivant. Presses Polytechniques et Universitaires Romandes, Lausanne
25. CABI (2004) Managing soil quality: challenges in modern agriculture. Schjønning P, Elmholt S, Christensen BT (éds) CABI Publishing, Tjele, Denmark
26. Antoni JP, Klein O, Moisy S (2004) Cartographie interactive et multimédia: vers une aide à la réflexion géographique, Cybergeo: Eur J Geogr [En ligne]. Systèmes, Modélisation, Géostatistiques, document 288. https://doi.org/10.4000/cybergeo.2621
27. Casarotto A (2018) Spatial patterns in landscape archaeology: a GIS procedure to study settlement organization in early Roman colonial territories. PhD Thesis. Leiden University Press
28. Glaser B (2012) No preconception: the dictum. Grounded Theory Rev 11(2). Sociology Press.

29. Voiron-Canicio C (2012) L'anticipation du changement en prospective et des changements spatiaux en géoprospective. L'Espace géographique, tome 41(2):99–110. https://doi.org/10.3917/eg.412.0099
30. Strauss AL, Corbin J (1998) Basics of qualitative research. Sage, Thousand Oaks (Californie), p 312
31. Banos A (2001) A propos de l'analyse exploratoire de données, Cybergeo Eur J Geogr (197)
32. Roncayolo M, Chesneau I (2011) L'abécédaire de Marcel Roncayolo, introduction à une lecture de la ville. Entretiens avec Marcel Roncayolo, InFolio, p 607
33. Anselin L (1996) Interactive techniques and exploratory spatial data analysis. Geographical information systems: principles, techniques, management and applications. Geoinformation International, Cambridge
34. Tukey JW (1977) Exploratory data analysis. Addison-Wesley, Reading, Mass.
35. Deluigi N (2018) Data-driven mapping of the potential mountain permafrost distribution. PhD thesis, University of Lausanne
36. Andrienko N, Andrienko G (2005) Exploratory analysis of spatial and temporal data: a systematic approach. Springer, 703 pp
37. Leuenberger M, Parente J, Tonini M, Pereira MG, Kanevski M (2017) Wildfire susceptibility mapping: deterministic vs. stochastic approaches. Environ Model Softw 101:194–203 (2018). https://doi.org/10.1016/j.envsoft.2017.12.019
38. Breiman L (2001) Random forests. Mach Learn 45:15–32
39. Lotfian M (2016) Urban climate modeling, case study of Milan city. Master thesis, Politecnico di Milano.
40. Altschul J (1988) Models and the modeling process. In: Judge W, Sebastian L (eds) Quantifying the past and predicting the past: theory, method, and application of archaeological predictive modeling. US Bureau of Land Management, Denver (CO), pp 61–96
41. Barbet-Massin M, Jiguet F, Albert CH, Thuiller W (2012) Selecting pseudo-absences for species distribution models: how, where and how many? Methods Ecol Evol 3:327–338
42. Verhagen P, Whitley TG (2020) Predictive spatial modelling. In: Gillings M, Hacıgüzeller P, Lock G (eds) Archaeological spatial analysis: a methodological guide, pp 231–246
43. Kvamme KL (1988) Development and testing of quantitative models. Quantifying the present and predicting the past: theory, method, and application of archaeological predictive modeling, pp 325–428
44. Getis A (2008) A history of the concept of spatial autocorrelation: a geographer's perspective. Geogr Anal 40(3):297–309
45. Hijmans RJ (2012) Cross-validation of species distribution models: removing spatial sorting bias and calibration with a null model. Ecology 93(3):679–688
46. Pohjankukka J, Pahikkala T, Nevalainen P, Heikkonen J (2017) Estimating the prediction performance of spatial models via spatial k-fold cross validation. Int J Geogr Inf Sci 31(10):2001–2019. https://doi.org/10.1080/13658816.2017.1346255
47. Valavi R, Elith J, Guillera-Arroita G (2019) blockCV: an r package for generating spatially or environmentally separated folds for k-fold cross-validation of species distribution models. Methods Ecol Evol 10(2):225–232. https://doi.org/10.1111/2041-210X.13107
48. Breiman L, Cutler A (2010) Random forests. http://www.stat.berkeley.edu/~breiman/RandomForests/
49. Breiman L (2001) Random forests. Mach Learn 45:5–32. https://doi.org/10.1023/A:1010933404324
50. Breiman L (1996) Bagging predictors. Mach Learn 24:123–140
51. Cutler DR, Edwards Jr TC, Beard KH, Cutler A, Hess KT, Gibson J, Lawler JJ (2007) Random forests for classification in ecology. Ecology 88(11):2783–2792. https://doi.org/10.1890/07-0539.1
52. Akaike H (1973) Information theory and an extension of the maximum likelihood principle. In: Petrov BN, Csáki F (eds) 2nd international symposium on information theory, Tsahkadsor, Armenia, USSR, 2–8 Sept 1971. Akadémiai Kiadó, Budapest, pp 267–281. Republished in Kotz S, Johnson NL (eds) (1992) Breakthroughs in statistics, I. Springer, pp 610–624

53. Baudron P, Alono-Sarría F, García-Aróstegui JL, Cánovas-García F, Martínez-Vicente D, Moreno-Brotóns J (2013) Identifying the origin of groundwater samples in a multi-layer aquifer system with Random Forest classification. J Hydrol 499(2013):303–315. https://doi.org/10.1016/j.jhydrol.2013.07.009
54. Fawcett T (2006) An introduction to ROC analysis. Pattern Recogn Lett 27(2006):861–874
55. Hosmer DW, Lemeshow S (2000) Applied logistic regression. Wiley, New York. https://doi.org/10.1002/0471722146

Section IV

Chapter 6
Results and Discussion

In the following section, the results obtained from the machine learning modeling procedure will be compared and discussed with those obtained from the Locational preference analysis (Chap. 5.4).

Furthermore, a comparison of the results obtained in each of the regional models is offered to discern if patterns existed in the data and if any evident criteria in the sites distribution can be recognized at supra regional scale.

To address this goal the regional results have been further outlined and compared with those obtained from the model performed on the entire Country.

6.1 Zurich

The first model run on the Canton of Zurich returned a ranking of variables importance that establishes the DEM, slope, geology, water saturation, distance to lakes and skeleton of soil, as the most influencing factors in determining the areas holding the highest probability to hide Roman settlements. These variables were then input as the exclusive environmental variables in a second model (Fig. A.95). The map in Fig. 6.1 shows its final prediction.

The partial dependence plots evaluated how much each specific class for every single variable have influenced positively or negatively the prediction of Roman settlements location, for example: the high probability is mainly located at a maximum altitude of 650 m a.s.l. with a slope of less than 10°. The highest probability is located at more than 7 away from the lakes and between 6 and 15 km from the main rivers (Fig. A.96, Fig. A.97).

With respect to the agricultural suitability partial plot the soils described as unsuitable influenced at most the prediction (Fig. A.97, Fig. A.98). The *unsuitable* class recurs often over the different environmental variables as one of the most influencing class, for example for skeleton soil and water saturation. This class identifies the

M. E. Castiello, *Computational and Machine Learning Tools for Archaeological Site Modeling*, Springer Theses, https://doi.org/10.1007/978-3-030-88567-0_6

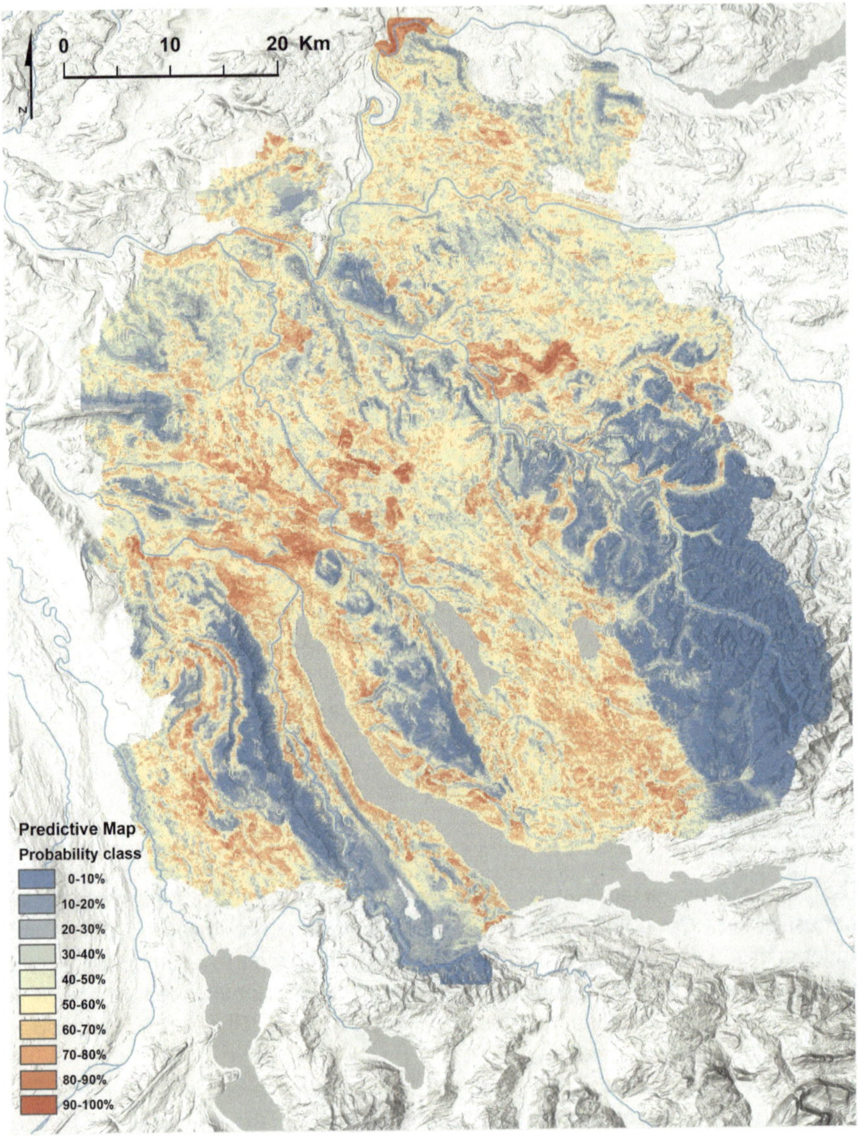

Fig. 6.1 Predictive map for the Canton of Zurich

non-productive areas (as the mountains), and the urbanized areas at once. The original variable used within the modeling procedure (the soil map; for more details see Chap. 4) presented no distinction between unsuitable areas and non-productive areas, hence it was not possible to separate them or considering the elements independently for performing further and specific analysis.

If it is true that the rescue archaeological excavations today are particularly intensive whereas new construction activity is planned, it is still possible to speculate about what is the cause and what the consequence of such situation. Modern urban centers, like the cities of Zurich or Winterthur, Lausanne or Geneva, were built upon the remaining of ancient settlements, in *continuum* with the old *vici*, as the historical sources can confirm (see Chap. 4). Thus, it should not surprise to find a high concentration of sites just in the neighboring of the modern agglomerations.

On the other hand, while interpreting the model results, one must keep in mind that the agriculture suitability map derives from a soil map (Bondenkarte; for more details see Chap. 4) produced with the primary goal to digitize the most suitable soils with respect to this kind of activity only. It therefore should come with no surprise that the digitization process may have produced less accurate information (or no data at all in some cases) with respect to those areas falling within modern urbanized agglomerations or high mountain areas.

Examining the partial dependence plot for the geological map (Fig. A.98), several types seem to have influenced the prediction in the Canton of Zurich: clays, gravels and glacial moraine deposits as wells as marls.

Since the model performance accuracy of the first model run was 70% it was decided to run a second model in order to improve the accuracy performance, which considered only the six most important variables but returned only a slightly better percentage of 72.3% (Fig. A.94).

6.2 Aargau

The classification model run for the Canton of Aargau and focused on the Roman settlements shows that the DEM, distance to main lakes, permeability, slope, skeleton soil and agriculture suitability are the 6 most important variables (see the Ranking of variable importance in the Fig. A.9).

The predictive map for the RF classification model is shown in Fig. 6.2. The highest probability areas are located below 600 m a.s.l. with a slope till 20° (Fig. A.10). Although, it is arguable that steep slope would reveal any settlement, because the lands are in general less favorable for arable cultivation, several assumptions may be formulated. It can be supposed that the existence of terracing or trenching in agricultural activities, attested in the Empire from the Early Republican time according to turned these areas in suitable settlement locations. Moreover, according to [1] steep slopes were often forested, and as such had other beneficial resources. They represented a source of raw materials and a source of animal product. Woodland may have been exploited for hunting and the wood was required for activities such as industry, heating and baths [1]. As suggested by the literature (Chap. 4.5) the Romans may have regarded these areas as a stable and reliable source of income.

The distance to main rivers and to main lakes as shown in Fig. A.10, seems not to be very significant for the final prediction.

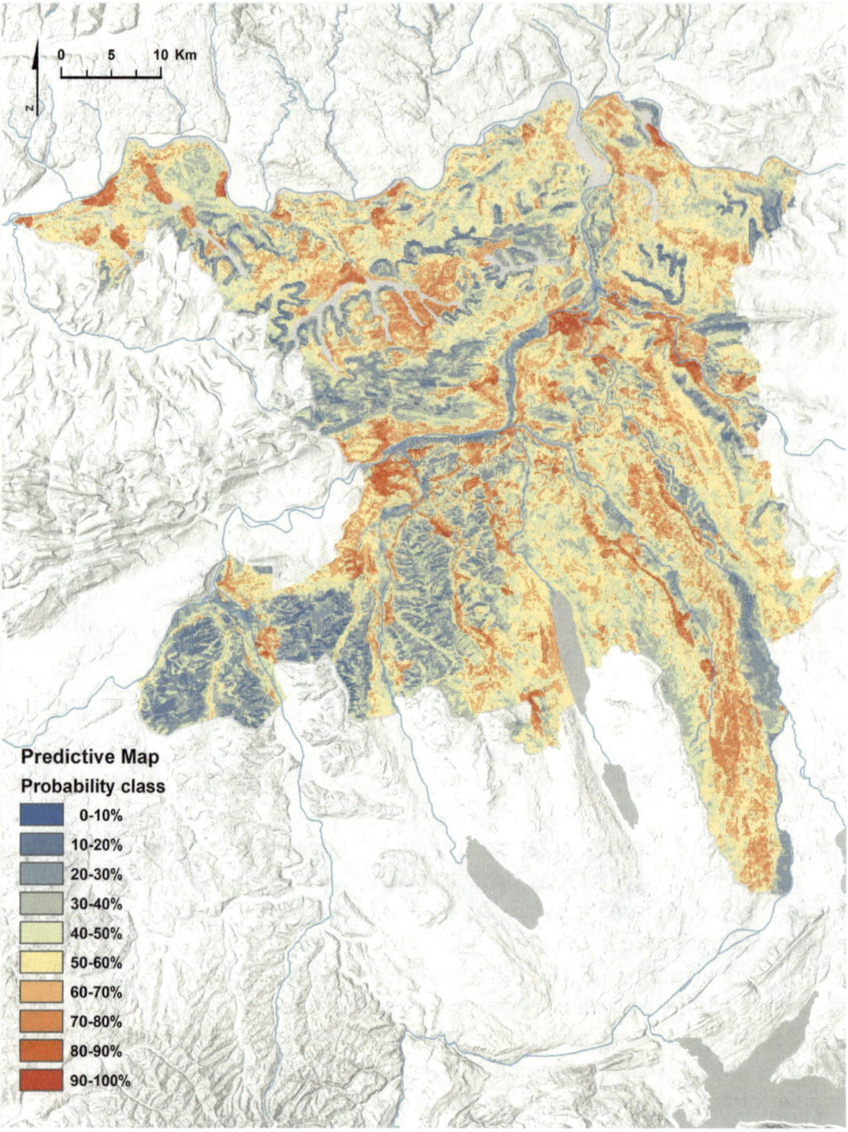

Fig. 6.2 Predictive map for the Canton of Aargau

Fig. A.11 shows that the types of soils classified as unknown, unsuitable and good for agricultural suitability have stronger influence in the location of high probability areas. Similar to the case of Zurich, the unknown and unsuitable classes in this region mainly correspond to the urbanized areas. With respect to the other soil properties, the unknown (urbanized areas) class shows the highest influence. Nevertheless, the Aargau soils are today considered the most productive of the Country. Likely it was as relevant in ancient times, as suggested by the model results. In fact, as shown in Fig. A.13, the model predicts high probability areas mainly on very good agricultural lands. The model performance accuracy is 68% (Fig. A.8) which identifies this model as good.

The result of the regression model for the single finds is visually expressed by the map in Fig. 6.3. As previously mentioned in Chap. 5, the regression returned a prediction on continuous values instead of a binary classification. The output of this map simply shows the probability of each pixel of the raster to contain a single find. The probability values of the region are expressed with a gradient color scale that goes from 0 to 3 and from brown to green, where in green is the probability to find a sure single find, in brown there is no probability at all. The scale of uncertainty is reproduced by values around 1 indicating unsure finds and values around 2 probable finds. The range absence-unsure-probable-sure is thus reproduced in the prediction output.

6.3 Grisons

For the canton of Grisons, two different datasets were used, depending on whether the Canton was modeled by itself or within the ensemble Swiss model. In fact, only 14 Roman settlements figured in the database. This was of a lesser issue for the Swiss model, as along with the other Cantons, enough settlement presences were available to run a model. When considering only the Canton of Grisons, the amount of data is not sufficient to sustain an accurate model. Hence, a dataset containing all the settlements belonging to the pre-medieval epochs (113 presences) was used for the regional model. This approach ensured coherence in the methodological procedure and output completeness. At the same time, it suggested a model of the settlement distribution across different periods for a defined region [2].

Two more solutions were explored with regard to the model parameters. A model was first run with the full list of variables and in a second stage it was run with the variables classified as the most relevant by the first one. This second model shows a ranking of variable importance (Fig. A.63) that classifies the DEM, slope, distance to main rivers, geology and distance to main lakes as the most relevant variables in determining the location of high probability areas. Hence, the probability to find settlements is higher between 500 and 1500 m a.s.l. on terrains with a slope below 17° (Fig. A.64).

With regard to the distance to lakes, distances below 26 km seem to have a certain importance for the final prediction which is probably justified by the scarce presence of natural lakes larger than 1km^2 in the region. An intense network of streams and rivers otherwise balances such wide distances. Figure A.64 shows that the highest probability to discover settlements is within 2 km from the main rivers. Moreover, the site presences are mostly predicted in the valley floors of the main rivers like the Rhine, the Inn or the Moesa, which were likely to be already navigable at the times of Romans. This would confirm their role as transit and commercial routes.

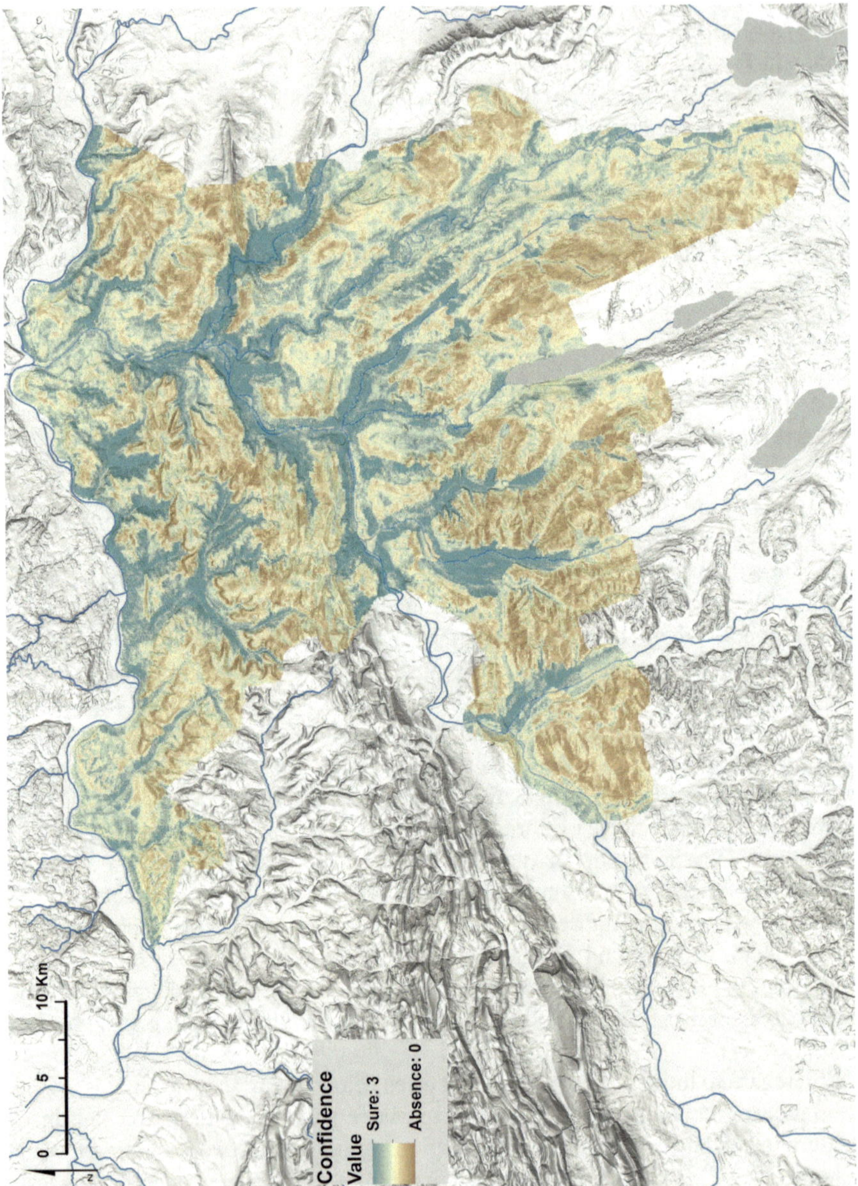

Fig. 6.3 Predictive map for Canton of Aargau (Single finds, RF Regression)

By observing the geology in Fig. A.66, it emerges a variety of influencing classes. This is probably due to the wide geographical extension of the study area and to the complexity of the landscape; nearly all the types of the geological classes are represented.

According to the AUC (Fig. A.62), the model performed very well in the Canton of Grisons, with 94% of accuracy. As expected, the highest probability areas are confined to the northern valleys of the Canton, while the south, mostly occupied by high mountain peaks, holds the lowest probability with few exceptions corresponding to the valleys (Fig. 6.4).

With respect to the agricultural suitability, the unknown class seems to be seen as of importance by the model (Fig. A.61). This could be due to the large area of this class (mostly the mountains). However, its importance in the variable ranking is clearly outweighed by the DEM and the distance to main rivers, so that it doesn't influence the predictive output significantly. Otherwise, typically agricultural soils are influencing the site presence prediction, such as good water and nutrient storage, slightly slower permeability and slightly stony soils.

6.4 Vaud

Figure 6.5 shows the final predictive map for the Canton of Vaud. As shown by Fig. A.78 the model run on the Canton of Vaud performed well with 80.9% of accuracy. The ranking of variables importance (Fig. A.79) corroborates the greater impact that the topographical indexes, such as the altitude and slope, the geology and the distance to the main water resources, have in defining high probability zones. The highest probabilities to discover new sites tend to cluster at 500 m a.s.l with a slope <8°, generally within 3 km from the main lakes and with a southeast orientation as seen in Fig. A.80.

The partial dependence plot for the distance to main rivers shows that the further away we are from the rivers (10 km and above), the higher the probability to find settlements. This is due to the high distance to main rivers on the fertile plateau between Lake Leman and Lake Neuchâtel. The optimal conditions (altitude, slope, soil properties) for high probabilities reunited in this area lead to a strong influence of distance from rivers (and to the second peak in distance to lakes around 10 km).

The partial dependence plot for geology (Fig. A.82) shows that clay and clayey terrains are the most favorable grounds. With regard to the agriculture suitability, terrains with very good agricultural suitability, deep vegetal soil and good to very good nutrient storage show important influence on model predictions, pointing at the importance of agricultural lands (Fig. A.81). In general, a soil layer of 10 cm is capable of easily retaining about 1 cm of water. The water storage is higher in clay-rich soils, as confirmed by the geology.

Finally, it is therefore not surprising that in the region with the bigger number of known farms and villas, the high probability to find settlements fall in the most favorable areas for agriculture, which is moreover confirmed by the initial ESDA.

Fig. 6.4 Predictive map for the Canton of Grisons

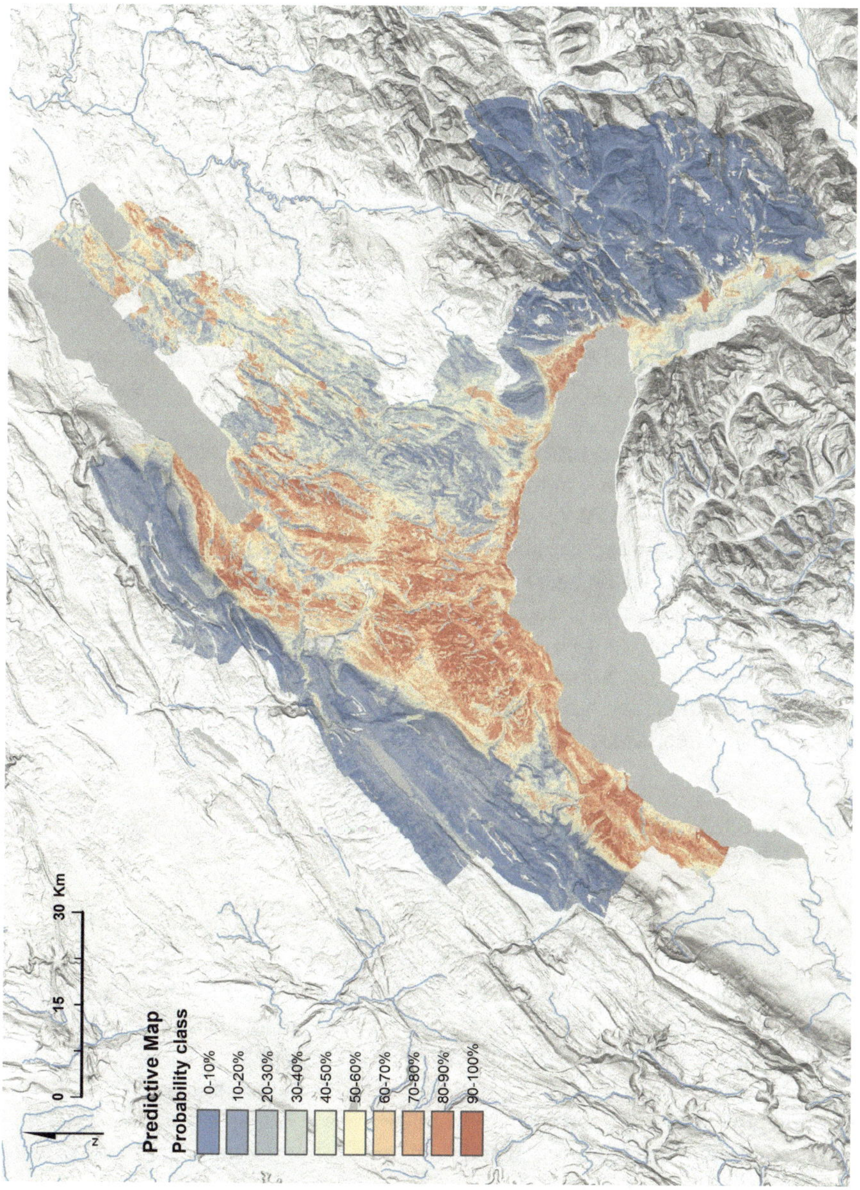

Fig. 6.5 Prediction map for the Canton of Vaud

6.5 Geneva

In this smallest Swiss region, two type of models were run: RF classification for Roman settlements (considering all the presences as sure) and RF regression that took into account the uncertainty in the archaeological dataset (78 settlements defined as "suspected" or "potential" and 198 as "sure").

According to the ranking of variable importance obtained from the RF classification (Fig. A.42), the distance to the main lakes and rivers, the DEM, the slope and the geology are the most relevant factors in locating high probability areas, followed by the agriculture suitability and the terrain orientation.

In particular, two peaks of distance to main rivers (500 m and 3 km) influence the model prediction, while the distance to main lakes has a positive influence until 3 km. These results only partly match with the ESDA analysis where the most suitable distance to rivers and lakes is equal to 3 km.

By observing the partial dependence plots in the Fig. A.43, the highest probabilities zones seem to be located within 380 and 450 m a.s.l., with 5° of slope, and a north east orientation.

The partial dependence plot for the geology (Fig. A.45) confirms the same geological types encountered in the other regions: clays and moraine deposits.

Figure A.41 shows the AUC with a model accuracy of 87% with all variables. The prediction is finally expressed through the predictive map in the Fig. 6.6

For the RF regression, the map in Fig. 6.7 shows the probability from 0 to 2 (white to green) to find a suspected site or a potential site.

Bearing in mind that the high density of modern constructions around the lake and the ground disturbances undoubtedly erased much of the human occupation traces, the results of the models run for Vaud and Geneva must be carefully handled as the variable human effect was not quantified.

6.6 Fribourg

The classification model run for this region and focused on the Roman settlements shows a performance accuracy stable around 73% (Fig. A.25). The prediction is expressed through the map in Fig. 6.8. The plot for the variable importance (Fig. A.26) confirms that the slope, the DEM, the distance to main lakes, the agricultural suitability and the nutrient storage remain the most relevant factors in assessing the highest probability areas. According to the results, these areas are more prone to be found around 500 m a.s.l., within 4 km from the main lakes, with a slope <10° (Fig. A.27) on terrains classified as good and very good for agricultural activities with a deep vegetal soil (Fig. A.28).

Examining the geological types in the partial dependence plot of the geology (Fig. A.29), the clayey silt, gravels and sands, conglomerates from weakly to mildly solidified are the most influencing in the final prediction.

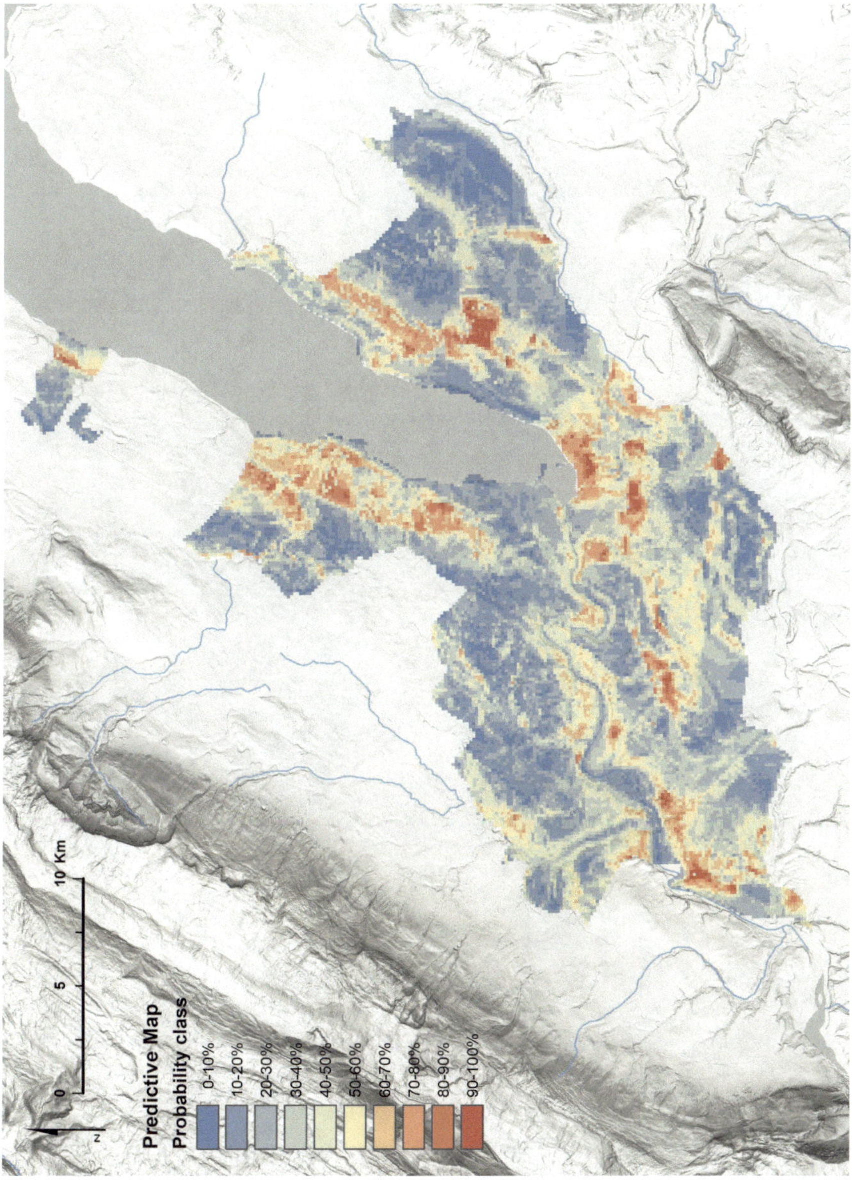

Fig. 6.6 Predictive map of the Canton of Geneva

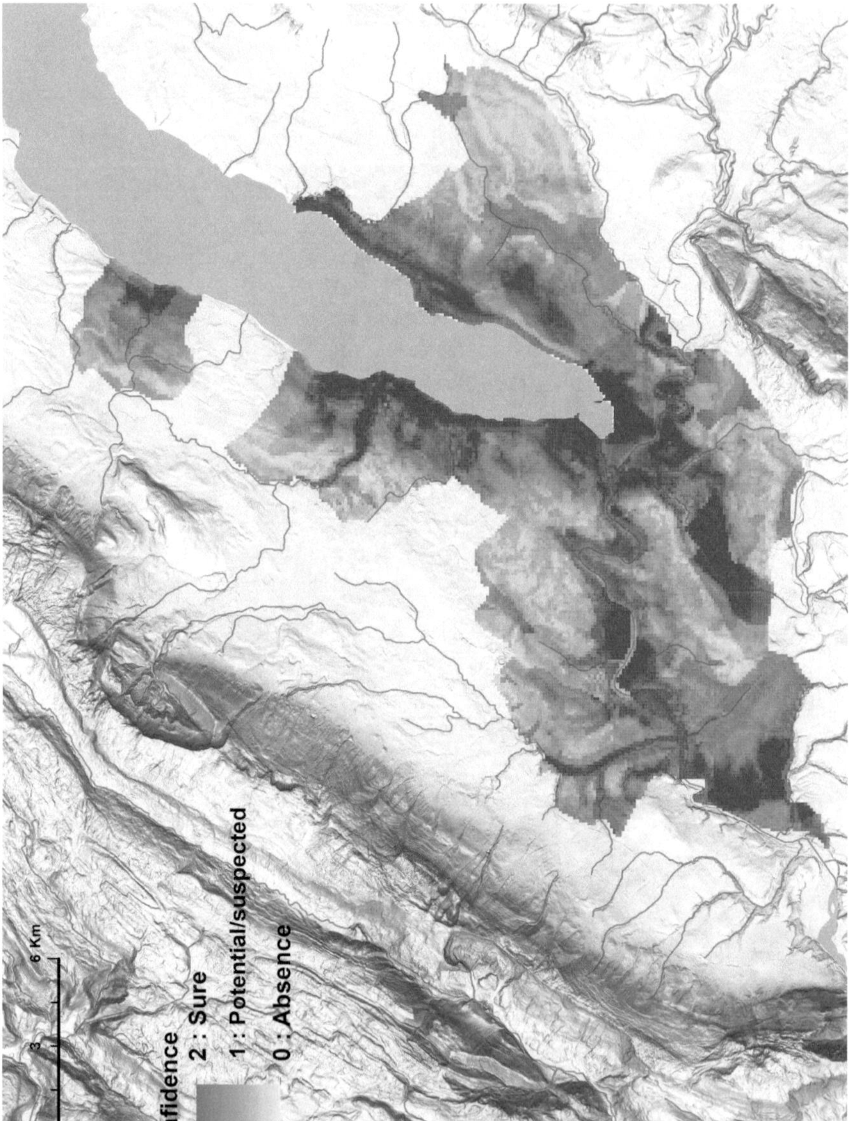

Fig. 6.7 Prediction map for the Canton of Geneva (RF Regression)

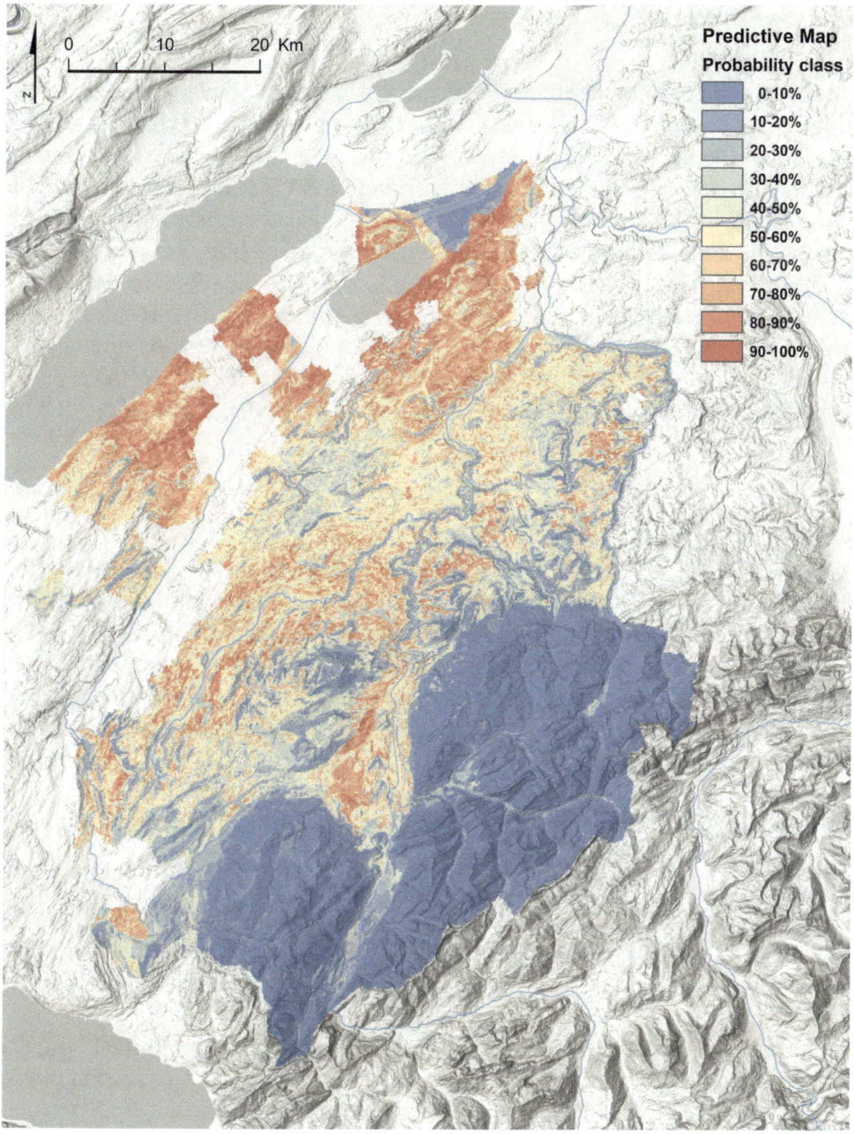

Fig. 6.8 Predictive map for the Canton of Fribourg

Next to the three lakes in the north of the canton, agricultural suitability seems to have played a major role in settlement location.

6.7 Switzerland

The model computed for the entire Country returned optimal results with regard to the performance accuracy of 89% (Fig. A.120). This was performed taking into account 1488 Roman settlements gathered from the six merged datasets and considering the full list of variables (see the predictive map in Fig. 6.9). The plot for the variable importance (Fig. A.121) shows that the DEM, the distance to main lakes, the slope, the geology, the distance to main rivers and the agricultural suitability are able to explain the majority of the areas that have a higher probability to hide Roman settlements, i.e. these variables are the most influencing factors in the prediction process. In more details, the highest probability areas lie on lowlands with a gentle slope, around 400–500 m a.s.l. within relative proximity to the main rivers (Fig. A.122) with a very good suitability for agricultural activity and deep vegetal soils (Fig. A.123, Fig. A.124).

6.8 Validity Assessment

The final step in every analysis is a common concern for any type of modelling procedure. This generally corresponds to the model validation process, which in this context refers to checking if the prediction obtained is a valid representation of the past landscape. However, in archaeological predictive modeling studies, the most common validation method is to compare the modeling results against the information and data derived from archaeological excavations and surveys. This approach is not always possible and concretely feasible in many cases.

Such kind of limitations led this research to explore different modeling approaches, until the Machine Learning technique (ML) that do not need of an external validation test. This method presents an internal validity and performance assessment by dividing the input dataset into training and testing sets, and by validating the prediction on a set intentionally kept aside. For these reasons, ML was considered an optimal solution to bypass more pragmatic issues related to the planning of special fieldworks for validation purposes.

The following pages present the different confusion matrix obtained with the ML modeling procedure for each regional model performed. The confusion matrix is generally used to visually express if the model is capable or not to distinguish between the real sites and absences (as explained in Chap. 5) thus offering a concrete validation tool. The class error for the presences prediction should be generally lower that the one estimated for the absences [3].

The confusion matrix helps in understanding if the model performed well and correctly generalized the prediction over the entire study area.

The confusion matrix for the Canton of Zurich (Table 6.1) shows that the model correctly predicted 141 site presences with 26.5% of error, which is better than the prediction of the 108 absences (with 38.6% of error).

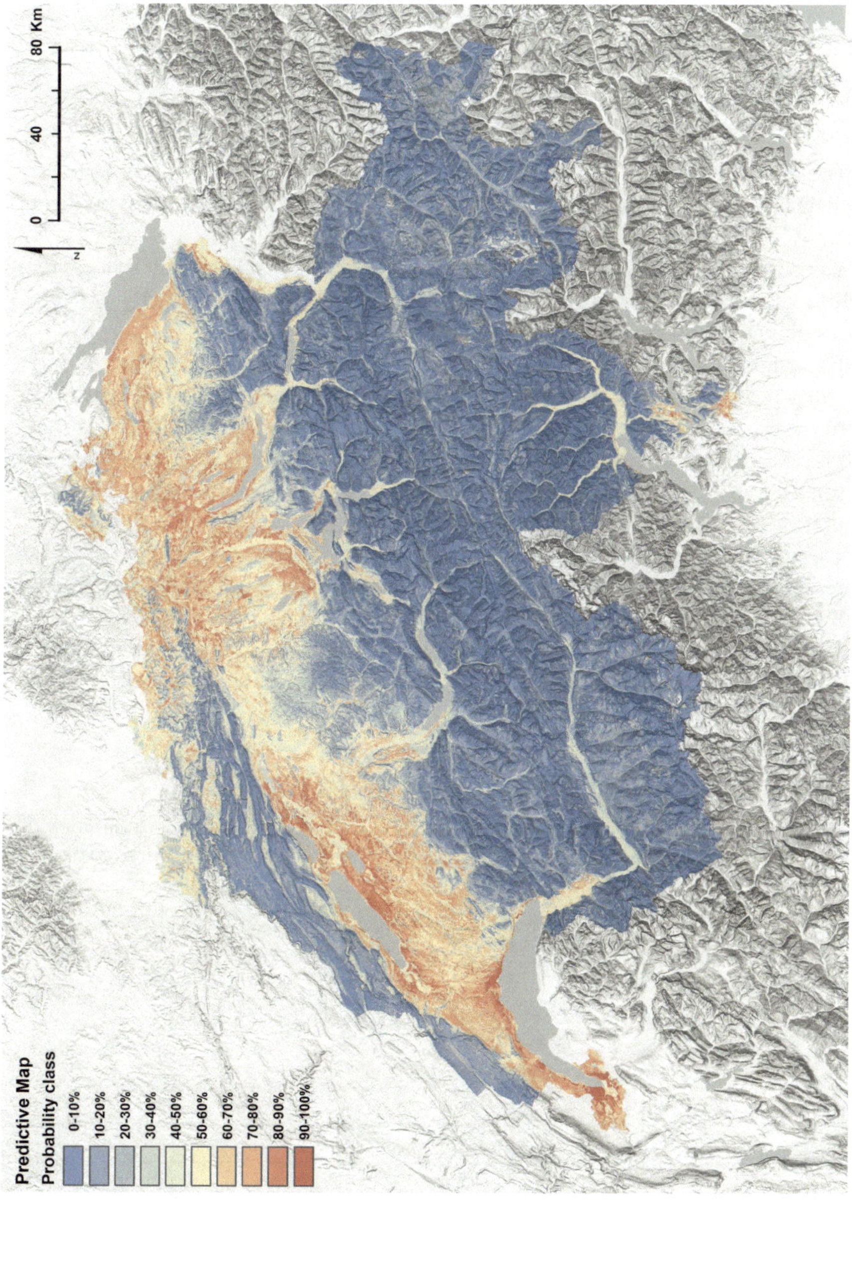

Fig. 6.9 Predictive map for Switzerland

Table 6.1 Confusion matrix for the Canton of Zurich

Confusion matrix Zurich				
		Real		
		Absences	Presences	Error %
Predicted	Absences	**108**	***68***	38.6
	Presences	***51***	**141**	26.5

The confusion matrix for the Canton of Aargau shows that the model correctly predicted 165 site presences with 30% of class error over the 147 absences predicted with 34% of class error (Table 6.2).

The confusion matrix for the Canton of Grisons shows that 88 site presences were well predicted with an error rate of 4% while 72 absences were correctly predicted with a class error of 15% (Table 6.3).

The confusion matrix for the Canton of Vaud shows that 271 presences were correctly predicted with a class error of 28% and 199 absences with 15% class error (Table 6.4).

The confusion matrix for the model run for the Canton of Geneva shows that 166 presences were correctly predicted with a class error of 12% while 89 absences were predicted correctly with a class error of 36% (Table 6.5).

The confusion matrix for the Canton of Fribourg shows that 202 presences are correctly predicted with a class error of 20% over 167 absences predicted correctly with a class error of 33% (Table 6.6).

The confusion matrix for Switzerland shows that 1129 presences are correctly predicted with a class error of 6.2% over 976 absences predicted correctly with a class error of 20.1% (Table 6.7).

Table 6.2 Confusion matrix for the Canton of Aargau

Confusion matrix Aargau				
		Real		
		Absences	Presences	Error %
Predicted	Absences	**147**	***76***	34%
	Presences	***74***	**165**	30.9%

Table 6.3 Confusion matrix for the Canton of Grisons

Confusion matrix Grisons				
		Real		
		Absences	Presences	Error %
Predicted	Absences	**72**	***13***	15.3%
	Presences	***4***	**88**	4.3%

Table 6.4 Confusion Matrix for the Canton of Vaud

Confusion Matrix Vaud

		Real		
		Absences	Presences	Error %
Predicted	Absences	**199**	*81*	28.9%
	Presences	*48*	**271**	15.0%

Table 6.5 Confusion matrix for the Canton of Geneva

Confusion matrix Geneva

		Real		
		Absences	Presences	Error %
Predicted	Absences	**89**	*52*	36.8%
	Presences	*23*	**166**	12.1%

Table 6.6 Confusion matrix for the Canton of Fribourg

Confusion matrix Fribourg

		Real		
		Absences	Presences	Error %
Predicted	Absences	**167**	*85*	33.7%
	Presences	*51*	**202**	20.1%

Table. 6.7. Confusion matrix for Switzerland

Confusion matrix Switzerland

		Real		
		Absences	Presences	Error %
Predicted	Absences	**976**	*247*	20.1%
	Presences	*74*	**1129**	6.2%

The descriptive Locational preference analysis performed by statistical and visualization means was an extremely important first phase of this study. Through this analysis, an overview of the Roman settlements distribution with regards to their geographic location and the environmental characteristics was provided.

6.9 Previous Knowledge—New Knowledge

The comparison between the Locational preference analysis and the ML model results allowed to cross-validate a newly produced information (obtained from the predictive maps) with the previous acquired knowledge (derived from the observation of sites and their relation with the environmental variables).

What can be clearly discerned from the Locational preference analysis results is that the good and very good soils for the agricultural activity had a great impact on the site location preferences, which is statistically evident not only at regional but also at supra-regional scale. The sites observe very similar criteria, with regard to the altitude, the slope, the distance to water resources, etc. no matter the region analyzed.

Now, by comparing these results with those of the ML procedure, common trends are again observable. In both procedures the altitude, the terrain inclination and the agricultural suitability are very important as with regard to the settlement location as for the predictive areas of high probability.

It is observable that the values for the altitude, slope and agricultural suitability match with those obtained by the locational preference analysis. The areas with the highest probability are those within 500–600 m a.s.l. with a slope of max 10° and with very good soils for arable lands.

Geology seems to have a determining influence on settlement location. Most settlements are found on fine-grained clayey soils and moraine deposits, and so are the predicted high probability areas. It appears that these types of soil correspond to good agricultural terrains most of the time.

Although all the variables used in this Study are based on modern maps, they played a relevant role in modeling Roman settlements, as in the case of the agricultural suitability. This map, despite its limitations (number of no data values, an unknown class combining lake areas with the urbanized and unsuitable areas without distinction), defined the boundaries of the settlement likelihood areas in most of the Cantons. When trying to model Roman settlements that based their principal livelihood on agriculture, the quality and the soil properties should not be overlooked.

The use of a historical map was initially considered. This was abandoned, as no digitized historical maps were available at the date of this research. Moreover, the oldest historical map in Switzerland with sufficient accuracy dates back to 1864 (Dufour map), and cannot ensure a realistic representation of the landscape in Roman time. As stated by Ebersbach [4], it is not possible to assume that historical maps represent ancient landscapes better than modern maps. In fact, even natural phenomena drawn on historical maps may originate no more than few hundreds years ago.

Finally, the results obtained from the locational preference analysis, representing here our a priori knowledge, are compared with the new results obtained from the ML modeling procedure. The model predicts the following share of areas with probabilities of more than 75% to contain a Roman settlement site:

Canton of Zurich: 4.3%
Canton of Aargau: 4.16%
Canton of Grisons: 8%
Canton of Geneva: 3.5%
Canton of Vaud: 11,7%
Canton of Fribourg: 12.9%

Such comparative analysis and its results offer an esteem of the high probability areas distribution at supra regional scale that can still offer today a possibility to (re)discover unknown Roman settlements.

6.10 Limitations and Advantages

The procedure applied for the present Archaeological Site Modeling study presented various limitations. In the context of this Thesis, the databases containing the information about the location and the type of archaeological sites have been collected and maintained by different Cantonal Archaeological Departments for over 30 years. The data inventoried follow different protocols and respect different infrastructure requirements. As seen in Chap. 4, extant archaeological data varied considerably in structure and quantity. Thus, an attempt was made to standardize the data by establishing a data structure that is flexible enough to be updated if necessary. The database architecture modeling was imposed as an essential step to be undertaken before the archaeological data could be further used. It provided a procedural methodology used in subsequent implementations.

It was not possible, within the limits of this geographically extensive research, to identify and integrate unpublished resources concerning the sites analyzed, neither to focus or deepen the analysis on more specific typologies of archaeological evidences and create more detailed thematic based maps.

Even though the management of uncertainty played a relevant role as introduced by missing data, incomplete information, errors and diverse sources of bias in the archaeological databases, the computational approach performed in this Thesis cannot create 'good' data where there is none, nor turn poor data into better data. It is difficult to determine whether the bias resulting from such diverse archiving techniques was transmitted to the modeling procedure. The same rules of empiricism and statistical inference apply to this data as to any other source of archaeological or geospatial data. The uncertainty was subjectively treated and the lack of a ready to use formula for incorporating it maybe be seen as a drawback.

Each regional model produced remarkably different outputs. The performance accuracy ranges between 68 and 94%. It must be pointed out at this stage that these results are extremely dependent on the quality of the data used. Such heterogeneity can be ascribed both to the very small number of sites used as dependent variable (the model performed better when combining the different datasets into the 1488 sites) and to the spatial autocorrelation of the environmental variables (most of the variables were derived from the digital elevation model and the soil map). Parametrizations and calibration decisions, such as the number of trees to grow, the number of variables to consider for the computation and the number of variables to include at each node split of Random Forest trees, may also have had a certain influence over the outputs.

Obviously, what is offered here is a methodology rather than an definitive answer. These maps cannot be taken as an absolutely exact representation of site location during the Roman period. Even if it is not possible to simply lift the models and use them 'as is' in a different region, the procedure designed in this Thesis offers an innovative and useful guideline for further applications and implementation in downstream research.

The GIS computing (detailed in Chap. 5) allowed to quantify modern topographic, geological and geographic phenomena and to correlate them with archaeological site location. While the correlation and statistical analysis performed through the locational preference analysis provided a strong basis for causality assumption, the archaeological phenomena were further interpreted and explained using stochastic methods, complemented by archaeological reasoning.

The ML modeling procedure revealed significant advantages with regard to the state of art in predictive modeling studies. The procedure defined produced a "good" alternative to the classical approaches because: (i) it is not affected by any kind of subjective assumption. The selection of variables as well as their weight in the prediction procedure is performed independently and without any previous assignment of reclassification or threshold values, (e.g. excluding areas above 700 m a.s.l from the model computations); (ii) no supplementary testing data is needed because the model quality assessment is part of the modeling procedure, as the input data are split in training and testing and the validation is performed by using the 'out of bag' (as defined in Chap. 5); (iii) random cross-validation is provided with the randomly selected training sets. This sampling strategy involving random and spatial fivefold cross-validation means that training observations were selected from polygons different than the ones chosen as testing examples. These sampling strategies avoided inferring a model from a training set that was very similar to the testing set, resulting in meaningless modeling performances. In other words, assuming that two neighboring pixels are (almost) identical, the main goal of spatial sampling is to avoid the use of an archaeological site evidence for training and an identical evidence (in terms of geo-spatial characteristics) for testing the model inferred during the training phase. The quality of the classification was thus more honest because it was computed on spatially different data; (iv) the model can be used to model any kind of archaeological evidences and epochs; (v) it provides graphical outputs easily interpretable by non-expert readers (predictive maps to visually identify the highest probability areas; the ranking of variable importance to discern which environmental factor influenced the prediction most and finally the partial dependence plots (to determine the influence of each class of the variables); (vi) the model can manage thousand of variables, both categorical and numerical, without internal conflict; (vi) it provides a first solution of this kind for quantifying and incorporating qualitative uncertainty and machine learning techniques.

References

1. Goodchild H (2007) Modelling roman agricultural production in the middle tiber valley, central Italy. Ph.D. Thesis, University of Birmingham
2. Brandt RW, Groenewoudt BJ, Kvamme KL (1992) An experiment in archaeological site location: modelling in the Netherlands using GIS techniques. World Archaeol 24:268–282
3. Breiman L (2001) Random forests. Mach Learn 45:15–32
4. Ebersbach R (2015) Eine Potentialkarte Archäologie für den Kanton Bern. Archäologie Bern/Archéologie Bernoise 2015:212–233

Chapter 7
Conclusions

7.1 Main Achievements and Conclusions

This Thesis developed and applied GIS and Machine Learning procedures on institutional data collections in archaeological site modeling analysis. Several case studies have been presented to show how the methodology works in practice (the Swiss Cantons of Zurich, Aargau, Grisons, Vaud, Geneva and Fribourg).

This work consisted of several procedural steps. One step regarded the database architecture and the assessment of uncertainty and vagueness that can affect the spatial patterns exhibited by the data collections. A second step aimed to shed light on the locational preferences and settlement strategies of ancient populations through an exploratory descriptive analysis (locational preference analysis). In a further step, this Thesis presented a Machine Learning-based approach to create predictive maps of the Roman settlements for each of the study area analyzed, as well as for the entire Swiss territory. The ensemble of the results obtained was examined in a comparative framework in order to investigate the settlement patterns at supra-regional scale and to assess the efficiency of the new methodology developed using the most sophisticated techniques for site detection.

The research carried out, especially with regard to the Machine Learning algorithm (Random Forest), represents a unique and innovative approach for modeling archaeological sites in Switzerland, at regional and supra-regional level. The machine learning techniques were furthermore combined with a descriptive site locational analysis complying with more common trends in archaeological research practices.

The work developed here tackled the traditional archaeological question of site detection and the issue of preservation and conservation of archaeological sites in the long-term perspective by combining cutting-edge technologies with analytical archaeological reasoning.

The methodological procedure, developed as a fluid and narrative protocol, was applied to six administrative units selected for their diversified urban and rural landscapes. The data modeled in this Thesis refer to; *i*) the archaeological sites discovered in those study areas, for which extensive regional databases exist and have

M. E. Castiello, *Computational and Machine Learning Tools for Archaeological Site Modeling*, Springer Theses, https://doi.org/10.1007/978-3-030-88567-0_7

been provided by the Cantonal Archaeological departments and *ii*) the environmental features identified and computed as influencing factors in the location of archaeological sites in the landscape.

In the history of archaeological research, the interest in enhancing our knowledge of past human behavior and in improving our skills to (post-)predict settlement location has always been a central focus of specialized archaeological landscape research and has remained unchanged until modern times. Furthermore, the extensive literature corroborates such focus. The discipline has been particularly committed to highlight the importance of studying societies in relation to the geography and environment (see Chap. 2), to produce results and more efficient tools that aim at facing current challenges, such as to avoid or mitigate anthropogenic and natural factors that could threaten or irreparably destroy our archaeological heritage.

This Study addresses the same questions as part of Landscape archaeology studies, for which a historical framework has been provided in Chap. 2. In particular, it provides an overview of the processes that have gradually driven the discipline towards a mature acceptance of computational and modeling approaches until recent years, as well as to the endorsement of archaeological predictive modeling (APM) as valuable approach and supportive decision making tool for addressing site preservation and conservation issues.

The archaeological discipline has benefited from many theoretical and technological advancements occurred in the last decades. The techniques and the supporting technology improved to prodigious size and so do theoretical concepts influencing methodological and interpretative frameworks. Undoubtedly, one of the greatest innovations for archaeological site modeling research has been the introduction of Geographical Information Systems (GIS) and the contribution of statistical analysis. Such spatial and quantitative methods, already largely employed in archaeological research since the 1960–1970s, experienced a boom at the end of the 1980s, when they served essentially to perform analyses that combined relational databases with spatial interpretations in order to better manage intra-site databases ([1–3], etc.) and to produce visual maps able to convey the research results in a most easy way. GIS are still used today to reconstruct ancient land use maps or migrations and movements of past societies over the landscape ([4, 5, 6] etc.).

Section III of this Thesis provided with several examples of Statistical and Quantitative applications for the development of APM, as well as with examples that make use of the most advanced Machine Learning techniques. According to the case studies presented, APM are reliable and highly performing tools in predicting the location of ancient settlements. From mountainous regions to alluvial plains, the role of the environmental variables in the modeling procedures is paramount and central. The only limitation that emerges form the list of examples cited and in general for performing predictive models is the availability of appropriate and reliable data upon which to construct the models.

The examples cited in Chap. 3 belong both to academia and to the milieu of cultural heritage management, highlighting the common purpose beyond such procedures. Regardless of the different domains in which they were originally developed, the models share a number of common attributes (e.g. archaeological data and environmental variables), and obtain significant results and accuracy.

Unfortunately, this kind of applications is still rare in the Swiss context, as well in archaeological research as in the cultural heritage management system. The case study presented in Chap. 3 is one of the few that explicitly addresses heritage management issues in Switzerland. It is primarily based on an exploratory data approach combined with subjectively weighted maps (altitude, slope, water distance etc.) to create site likelihood maps. The maps produced show the distribution of the known Roman villas in the Canton of Bern, with particular attention to their surrounding environment. The Roman villas are analyzed with regard to their location in space and in relation to specific environmental features. Thus, relying on this exploratory data analysis, new assumptions were formulated and projected towards those areas of the region that, sharing similar environmental attributes, could potentially hide still unknown Roman villas.

This work has moreover shed some light on the issues that have to be dealt with when working with institutional archaeological databases and modern environmental data for the development of APM.

As introduced at the beginning of Section I and deepened in Chap. 4, cultural heritage management in Switzerland is a very complex matter because of the political decentralization of the Country, which may result in a multiplicity of methods and approaches applied in archaeological practices, in storing and recording archaeological information. As mentioned, the archaeological data modeled within this Thesis are essentially based on the information and the records stored within different regional archives through decades and until 2015. Thus, the extant archaeological data varied considerably in structure and quantity from one region to another. In order to comply with the modeling procedure requirements, a first step consisted in compiling a new georeferenced database for each of the regions analyzed and to finally merge them into a single one to provide with a more comprehensive perspective of the site distribution at a supra-regional scale.

Establishing the structure of the database and its data-fields was performed at the outset of this research (Chap. 5). Creating a unified and standardized geospatial database was performed by reorganizing the data both diachronically (i.e., per period) and hierarchically (i.e., per size category). By disentangling the original variegate typological classifications, interpretative biases and variations in definition, the sites were compared and the new database structure was defined on the basis of common formal attributes, such as the end use of a place, of the structure and of the period. Firstly, the site period was chosen (Roman) to further target the modeling procedure, secondly the typology was defined by assigning universal definitions across the different regions. A standard classification of 'settlement' was assigned to those evidences originally defined as: building, house, dwelling, vicus, village, city, settlement, production/industry place, villa, agglomeration, establishment, etc. In the same way, the standard classification of 'single finds' was assigned to those evidences defined as: artifacts, coin, coins, single finds, scatter finds, other single finds, treasure, material, depot, etc.

Moreover this database architecture was designed in order to minimize the potential errors during new data entry or subsequent modification, to maximize the database flexibility and especially its potential for further research based on the same data.

A further step concerning the database consisted in the exploration of suitable solutions for managing and incorporating data uncertainty within the APM procedure (Chap. 5). As seen in Chaps. 3 and 4, the data holding a certain degree of uncertainty was abundant in the databases of the Cantons of Aargau and Geneva. In both cases the uncertainty was expressed through linguistic variables. In the case of Geneva, a settlement could have been defined as potential, suspected or sure with regard to its typology. In the case of Aargau, the uncertainty analyzed and computed referred to single finds. They were defined as unsure or unknown, with respect both to the chronology and the typology itself. A single find could be either difficult to recognize as such or difficult to assign to a precise time period.

Thus, instead of simply deleting the uncertain information from the database and drastically reduce the amount of data to model, it was decided to explore alternative solutions to quantify the uncertainty and to incorporate it into the Machine Learning modeling procedure. The literature provides with valuable examples as to how to address this specific research needs [7–9, 10]).

However the solution proposed in this Thesis does not supply an automated mechanism for dealing with unreliable data, it offers a possibility to include and visualize the uncertainty in predictive modeling maps. The approach developed in this Study, based on the fuzzy logic structure, allowed to assign quantitative uncertainty values to qualitative linguistic labels (such as "unsure" or "unknown" etc.) in a systematic manner. Once obtained, the quantitative values were algorithmically processed as part of the predictive modeling procedure.

In the database of Geneva, the sites classified as *Presumed* or *Suspected* (when the chronology was provided and ascertained) were assigned a value of 1, while the sites defined as sure, were assigned a value of 2 (Table 5.4). In the database of Aargau, the single finds classified as typologically sure were sometimes chronologically unknown or *viceversa,* and were thus assigned a value of 3, 2, or 1 (Table 5.3).

At the same time, as described in Chap. 5, fourteen environmental features were selected and computed as influencing factors of archaeological settlement location using Geographic Information Systems (GIS). A Digital elevation model—DEM was used to derive additional features such as the terrain orientation (Easteness and Northness) and the steepness (Slope), computed at the same spatial resolution as the DEM (100 m cell size). The proximity of each settlement to water ways, such as the main rivers and main lakes, was calculated using the Euclidean distance function. Rather than considering all the modern rivers and lakes of the Country, a preselection was made to exclude modern dam lakes and isolate only those rivers carrying a certain water capacity, which could have served as water supply and transportation routes during Roman times. The agricultural suitability was considered to further assess the soils that had higher probability to be exploited for agricultural activities. The productive potential (e.g. good, very good, sufficient) of each soil unit is established on the basis of other important qualities of the soil, namely fertility (depending on the availability of nutrients and minerals, and the drainage status of the soil) and workability (depending on skeleton and stoniness qualities) [11–13]. From the agricultural suitability map, more variables were derived, such as Water saturation, Water storage, Permeability, Depth of Vegetal soil, Nutrient storage, Skeleton of soil.

The geological map was used to provide information about the distribution of the uppermost rock or sediment strata. During this GIS preprocessing phase, a deep screening of the variables enabled to correct and eliminate construction errors (e.g. no data, resampling the size to match a common spatial resolution).

At this point, once the new database was generated, embedding 1488 Roman settlements, and the geo-environmental aspects quantified and correctly computed into 14 variables, a first descriptive statistical analysis was performed, allowing to observe several trends and patterns at regional and supra-regional scales.

Although the regions analyzed are geomorphologically very different, the results of this locational preference analysis show that natural factors (i.e. altitude, slope, aspect, proximity to water) affect settlement distribution quite uniformly across the various regions. Most of the Roman settlements (85% of 1488 total) are distributed between 350–650 m a.s.l., with exception of the Cantons of Vaud and Grisons, where the sites are distributed over a wider range of heights, reaching up to 1600 m a.sl. In the Canton of Vaud, these few sites discovered at high altitude (above 1000 m.a.s.l.) represent outliers. In the Canton of Grisons, strongly characterized by high mountainous peaks, the locational analysis was performed on a small dataset of Roman settlements embedding only 14 settlements. The analysis did not return any visible trend in the site distribution over the altitude because of the small number of input data.

Concerning the distribution of sites over the slope, there is a visible common trend in all the regions analyzed. The settlements preferred terrains with a gentle slope, generally concentrating around slopes of less than 5°. The only exception is represented by the small dataset of the Canton of Grisons, where the settlements are distributed over slopes reaching up to 30°, concentrating around 13°.

In most of the regions, the sites are preferentially oriented north, respectively northwest or northeast (28% and 25,4%), with only one exception to this general rule. In the Canton of Vaud, the sites seem to prefer a southeast orientation.

When examining the proximity of the sites to the water bodies, intended as main rivers and lakes, the image that emerges is quite diversified from one region to another. With regard to the main rivers, in five out of six Cantons, the sites are mainly located within a distance of 5 km. Only in the Canton of Vaud, some sites have been discovered at distances over 15 km away from main rivers. With regard to the main lakes, the sites are mostly concentrated at distances of 10 km or less from the nearest main lake, with exception of the Canton of Grisons, where the almost total absence of big natural lakes must have guided the different choice of settlement locations.

A majority of sites across all the regions (53%) seem to have been attracted by the most suitable land for agricultural activity, in particular on soils classified as very good. However, it is interesting to observe that along with the 'very good' class, many sites in all regions fall in the class of the map identifying the urbanized areas. The spatial correlation between the modern urban agglomerations and the archaeological discoveries is a topic leading to further in-depth analysis and the possibility to further explore this topic can be listed among the outlooks of this research.

With regard to the ensemble of the other geo-environmental sub-variables (Water saturation, Water storage, Permeability, Depth of Vegetal soil, Nutrient storage, Skeleton of soil), the sites are located on slightly stony soils with good nutrient storage and drainages capacities in almost all the regions. They are identified as generally good and easily workable lands.

It seems that the choice of preferentially settling these regions in the Roman period was dictated exactly by the availability of cultivable lands more then by other constraints.

In the final section of this Thesis, several Machine learning-based models were computed in order to produce multiple predictive maps at regional scale, plus a wider scale model that extended the prediction on the other regions not directly considered in this research.

This Thesis provided a clear and narrative protocol for applying Random Forest (RF) classification and regression. Different models were realized for each of the study area. The regional datasets, combining archaeological data and the environmental variables, were used as input data in the modeling procedure.

A first action in this regard involved the creation of pseudo absence datasets, selected randomly over the study areas, where no Roman sites were discovered. To ensure a good generalization of the models and to avoid the overestimation of lower classes, a balanced number of absences i.e. in a number equal to the observed presences, was specified. Thus, the pseudo absence and the presence (the archaeological data) datasets were combined with the environmental variables to produce a spatial matrix or grid of 100 m equidistant cells for each of the study areas, where to each cell, the information about the presence or absence of a Roman site and the information concerning the environmental features (quality of soil, degree of slope, altitude values, distance to water, etc.) were associated.

To avoid spatial autocorrelation and model overfitting, i.e. introducing a bias due to similar characteristics held by observations (or cells of the raster) close to each other [14], a *k*-fold cross validation was performed. A Spatial Block function was computed in order to create several separated spatial folds based on a pre-specified distance in different number and size, containing subsets of the original dataset. This operation was reiterated hundred times to identify the random fold assignation that most effectively reduces autocorrelation, using manually defined size and number of *k* folds based on a tool specifically designed to calculate the spatial correlation range and thus to help in calibrating these parameters [15].

The *k*-folds were used to train the Random Forest algorithm. A 60% of the data was used as training dataset while 40% was used to test the model. Between 700 and 1000 decision trees were grown, as each time, a subset of archaeological data (presence-absence) is combined with a subset of geo-environmental variables (generally 4 variables randomly selected). The final prediction of new data is computed by taking the majority or the soft voting (by counting how many times each cell is classified as "presence" or "absence"), and converting it into a final probability map.

An internal testing dataset was used to assess the accuracy of these models. After computing several different models, the performance accuracies for the final models retained ranged between 68% (Aargau) and 94% (Grisons). The models run on the Canton of Zurich and Fribourg reached an accuracy of 72%, the model run for the Canton of Vaud 81% and the model run for the Canton of Geneva 87%. Several assumptions can be formulated to explain the lowest model performance accuracies: (i) the spatial autocorrelation existing between the geo-environmental variables. Indeed, sub-variables like water saturation, water storage, permeability etc. were all derived from the agricultural suitability map and did not contribute much to improve the final prediction values. (ii) The small number of sites available as input for the models. It is well known that Machine Learning techniques perform better on larger dataset.

A final model was run on the unified regional datasets in order to project the prediction of Roman settlements also in those areas that were not directly included in this research. The performance accuracy of that model is 89.1%.

The model outputs are several, such as visual graphs and maps. The final predictive maps highlight in red the areas that hold the highest probability to host Roman settlements.

As discussed in Chap. 6, no historical map was integrated in the models for availability reasons and because they are not assumed to represent a 2000 year old landscape better than modern maps. However, the model nevertheless automatically recognized the ancient marshy areas and floodplains like e.g. the Grosses Moos in the Canton of Fribourg as very low probability areas, indicating that the Random Forest model extracts information from the data in a plausible manner to compute the predictive output.

The ranking of variable importance shows the importance or the influence of each variable in computing the final prediction. According to this ranking, a general trend sees the altitude, the slope and the proximity to water resources playing a fundamental role in the settlement location as strong attractors for settling in the best areas in terms of fertility and workability. The altitude hereby plays a particularly important part.

With regard to the research questions formulated at the beginning of this Study, the following conclusions can be drawn:

This study has demonstrated that, if approached in a proper way, data from institutional collections can disclose important information about settlement patterns and about the principal processes that determined these patterns. These archaeological data are of evident value and the databases can also be merged for modeling purposes, but they require necessary preprocessing work to meet the minimum standard of interoperability and transferability. To this effect, this study has presented a method which not only offers a way to formally tackle research questions using institutional data collections, but also provides widely-applicable means to deal with uncertainty and biases affecting these data. Moreover, it has been shown how uncertain and biased data can be incorporated into machine learning applications if an uncertainty quantification procedure is considered and undertaken at the beginning of the procedure.

Locational analysis for the entire areas has shown that Roman settlements tend to cluster on natural resources similar to those of today. A large majority of sites indeed fall within those classes of the agricultural suitability map and its sub-categories that define modern soils as suitable for agriculture. This highly evident site distribution pattern over modern soil variables strongly supports the conclusion that areas suitable for agriculture today were already good arable lands during the Roman period, evidencing a certain spatio-temporal continuity in the anthropogenic land use. It thus appears that modern data can indeed be used to model the location of ancient sites, at least to a certain extent.

Furthermore, these analyses have emphasized the determining role of environmental natural factors in the Roman settlement location strategies in Switzerland. When looking more carefully at the site distribution, at regional and supra-regional scales, it is possible to recognize strong similar trends: a remarkable preference for relatively flat lands and gentle slopes, for more fertile soils and in close proximity to the main rivers and lakes.

The results provided by the Machine Learning models offer a wider range of interesting results and a more accurate comprehensive framework of the environmental variable importance.

The locational analysis and the Machine Learning procedures developed and tested both contributed substantially to enhance the current knowledge in archaeological research. At a wider scale, they provide new information about what kind of resources were considered most valuable in the settlement choices of past populations and offer a better understanding of local variations in settlement preferences. The Archaeological Site Modeling procedure, as understood here as a quantitative, objectified approach, has facilitated an unprecedented view on the archaeological phenomena investigated in the case studies and moving from the local to the supra-regional level.

7.2 Research Perspectives

There are many avenues for future research arising from the completion of this Thesis. The results obtained within this Study are very promising in terms of technological innovation and relevance of the acquired new knowledge, as well for archaeological research as for cultural heritage management (CHM).

The methodological procedure can easily find application in the cultural heritage management departments that provided with the basic data for the research. Archaeological models can be used to comply with inventory requirements, forecast effects, and make proactive streamlined management decisions regarding where to place CHM efforts. The locational preference analysis and the predictive maps developed could represent useful support tool in the decision making process. In particular, the possibility to rely on maps that highlight the areas holding the highest probability to discover archaeological sites could certainly help to improve the assignment of human and financial resources in CHM.

Although the criticisms towards the efficiency of these maps stands on a long tradition, in the last decades, numerous investigations have explored methods for improving the strength and utility of models by applying a broad array of inductive and deductive techniques to model site location in diverse contexts. Most recent applications have seen the use of more sophisticated technologies borrowed from sister disciplines like Machine Learning, which is constantly improving the way of doing archaeological research. Such kind of procedures could be continuously updated and tested in the field. The simplest method to update the model(s) is with the data collected from other surveys in the future. However, to make a better use of it, the data should ideally fulfill a minimum standard of precision and be comparable with other new data collected from across the regions.

Furthermore, the first innovative application of spatial statistics and machine learning techniques to perform site locational analysis and to develop archaeological predictive maps, at regional and supra-regional level in Switzerland, offers interesting new perspectives and strengthens future collaborations between academia and governmental institutions as well as between archaeological research and sister disciplines.

Further investigations that move from this research results could include for example: a reassessment of the archaeological site types which could be categorized into more specific typologies, allowing different analyses focusing on e.g. routes, graves, production centers, religious places, etc. and including cultural factors, as the distance to the main urban centers, to other sites and other particular attractors. The reproducible methodology deployed in this Study permits to model any kind of archaeological evidence and to obtain various thematic and predictive maps at different scales. An assessment of the site size, e.g. defining site catchment areas prior to the modeling procedure would also be an interesting outlook, useful to provide with more accurate information about the extension of the high probability areas expressed in the final predictive maps

Likewise, the use of additional environmental variables such as paleo-climate data, carpological and faunistic data, if available, would help to model possible distinct geographical areas, interested by different past subsistence economies. This in-depth analyses would furthermore help to reconstruct land use exploitation systems over time and space.

Ultimately, the use of high resolution satellite images, combined with the predictive map results, would offer the possibility to remotely detect the presence of any archaeological remain still undisclosed in those zones defined as 'high probability'. This outlook, by combining predictive modeling and remote sensing, would furthermore offer a new and interesting perspective to assess once more the prediction accuracy of the modeling procedure and to further strengthen interdisciplinary collaborations.

Archaeological Site Modeling represents a powerful tool that offers an unlimited range of possible applications with which to test new lines of research and stimulate archaeologists to formulate new narratives.

References

1. Burrough PA (1986) Principles of geographical information systems for land resources assessment. Clarendon Press. https://doi.org/10.1080/10106048609354060
2. Altschul J (1988) Models and the modeling process. In: Judge W, Sebastian L (eds) Quantifying the past and predicting the past: theory, method, and application of archaeological predictive modeling. US Bureau of Land Management, Denver (CO), pp 61–96
3. Ebert JI, Kohler TA (1988) The theoretical basis of predictive modeling and a consideration of appropriate data-collection methods. In: Judge WJ, Sebastian L (eds) Quantifying the present and predicting the past: theory, method, and application of archaeological predictive modeling. US Department of the Interior, Bureau of Land Management Service Center, Denver, pp 97–171
4. Lock G (ed) (20140 Using computers in archaeology, using computers in archaeology. https://doi.org/10.4324/9780203451076
5. Herzog I (2018) Least-cost networks. Archaeology in the digital era. (Pampus 1998), pp 237–248. https://doi.org/10.1515/9789048519590-026
6. Fonte J, Parcero-Oubiña C, Costa-García JM (2017) A GIS-based analysis of the rationale behind Roman roads. The case of the so-called via XVII (NW Iberian Peninsula). Mediterr Archaeol Archaeom 17(3):163–189
7. Martin-Rodilla P, Pereira-Farĩna M, Gonzalez-Perez (C 2019) Qualifying and quantifying uncertainty in digital humanities: a fuzzy-logic approach. In: ACM international conference proceeding series. pp 788–794. https://doi.org/10.1145/3362789.3362833
8. Martin-Rodilla P, Gonzalez-Perez C (2019) Conceptualization and non-relational implementation of ontological and epistemic vagueness of information in digital humanities. Informatics 6(2). https://doi.org/10.3390/informatics6020020.
9. Gonzalez-Perez C (2018) Information modelling for archaeology and anthropology. Inf Model Archaeol Anthropol . https://doi.org/10.1007/978-3-319-72652-6
10. Niccolucci F (2006) Managing uncertainty in archaeological GIS applications, Reading historical spatial information from around the world studies of culture and civilization based on geographic information systems data.
11. White KD (1970) Roman farming. Thames and Hudson, London
12. Frayn JM (1979) Subsistence farming in Roman Italy. Centaur Press Ltd., Fontwell
13. Spurr MS (1986) Arable cultivation in Roman Italy c. 200 B.C.—c. A.D. 100. J Roman Stud Monogr. Society for the Promotion of Roman Studies, London
14. Getis A (2008) A history of the concept of spatial autocorrelation: a geographer's perspective. Geogr Anal 40(3):297–309
15. Valavi R, Elith J, Guillera-Arroita G (2019) BlockCV: an r package for generating spatially or environmentally separated folds for k-fold cross-validation of species distribution models. Methods Ecol Evol 10(2):225–232. https://doi.org/10.1111/2041-210X.13107

Appendix

A.1 Canton of Aargau

A.1.1 Database Reclassification

See Table A.1.

Table A.1 Roman site typology classification for Aargau

Roman site typology database Aargau		
Class	English classification	Original classification
Bridges	River crossing	Flussübergang
Fortifications	(Military) camp	(Militär) Lager
	Castella	Burg, Schloss, Kastell
	Other fortifications	Andere Befestigung
	Refuge, earthwork	Refugium, Erdwerk
	Tower, watchtower	Wachturm, Hochwacht
Graves	Cemetery (group or field of graves)	Gräbergruppe, -feld
	Flat grave	Flachgrab
	Flat grave (1–2)	Flachgrab (1–2)
	Flat grave group	Flachgräbergruppe
	Grave	Grab
	Grave (1–2)	Grab (1–2)
	Grave hill	Grabhügel

(continued)

M. E. Castiello, *Computational and Machine Learning Tools for Archaeological Site Modeling*, Springer Theses, https://doi.org/10.1007/978-3-030-88567-0

Table A.1 (continued)

Roman site typology database Aargau		
Class	English classification	Original classification
Others	Ditch, wall	Graben, wall
	Form: others	Form: andere
	Hint from local name	Hinweis aus Flurname
	Others	Andere
	Pit, shaft, holes	Grube, Schacht, Mulde
	Various	Diverses
	Wall, border ditch	Weidmauern, Grenzgraben
Quarry	Quarry	Abbau, Steinbruch
Religious sites	Religious building—Temple, Church, Chapel	Kultbau—Tempel, Kirche, Kapelle
	Religious site—offertory	Kultplatz, Opferstelle
Roads	Road	Strasse
	Road network (supposedly)	Strassennetz (angeblich)
	Traffic	Verkehr
Settlements	(Single) Building, house, dwelling tower, basement	(Einzel-)Gebäude, Haus, Wohnturm, Keller
	Camp, village with walls	Lagerplatz, Dorf, mit Umwehrung
	Camp, village, Vicus	Lagerplatz, Dorf, Vicus
	City (unfortified)	Stadt (unbefestigt)
	Construction stones, ceramic, wood	Bausteine, -keramik, -hölzer
	Construction structures (wood and stone)	Baustrukturen (Holz und Stein)
	Court, farm, installation (villa)	Hof, Gutshof, Anlage
	Crafts (oven)	Verarbeitung, Raffinerie
	Occupation layer, stone concentration	Kulturschicht, Steinkonzentration
	Other installations	Sonstige Anlagen
	Production, industry, crafts	Produktionsplatz, Industrie, Gewerbe
	Settlement	Ansiedlung
	Settlement	Siedlungsstelle
	Structures and observations (unidentified)	Festgestellte Strukturen und Beobachtungen
Single finds	Artifact	Artefakt
	Coin	Münze
	Coin deposit	Münz-Depot
	Coins	Münzen

(continued)

Table A.1 (continued)

Roman site typology database Aargau		
Class	English classification	Original classification
	Concentration of single finds (ceramic)	Konzentration von Einzelfunden
	Material, deposit	Material, Depot
	Other single finds	Anderer Einzelfund
	Single finds, scattered finds	Einzelfund, Lesefund, Streufund
	Treasure	Schatz
Unknown	Unknown	Unbekannt
Water infrastructure	Canalization	Kanalisation
	Water distribution	Wasserversorgung
	Well, water canalization	Sodbrunnen, Wasserleitungen

A.1.2 Environmental Variables

See Figs. A.1, A.2 and A.3.

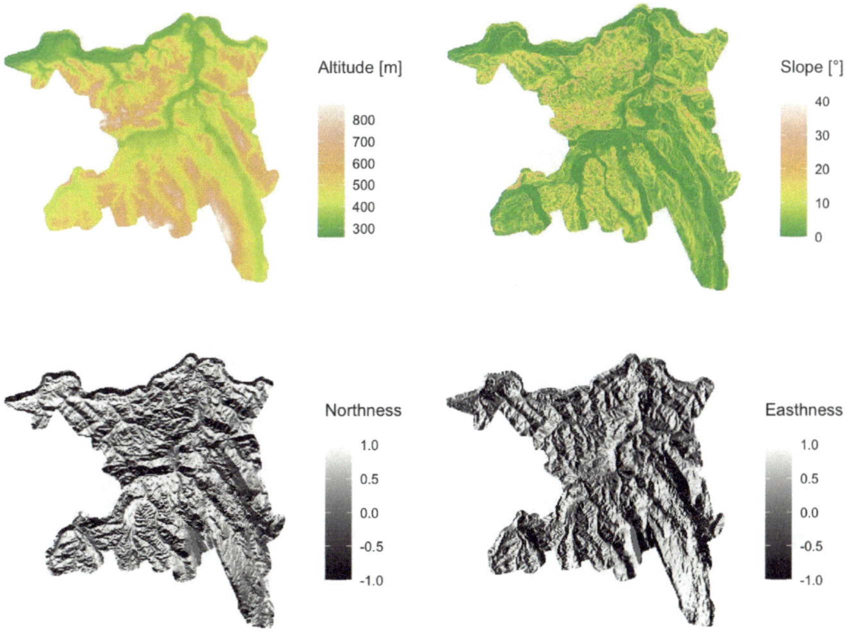

Fig. A.1 Environmental variables Aargau (part 1)

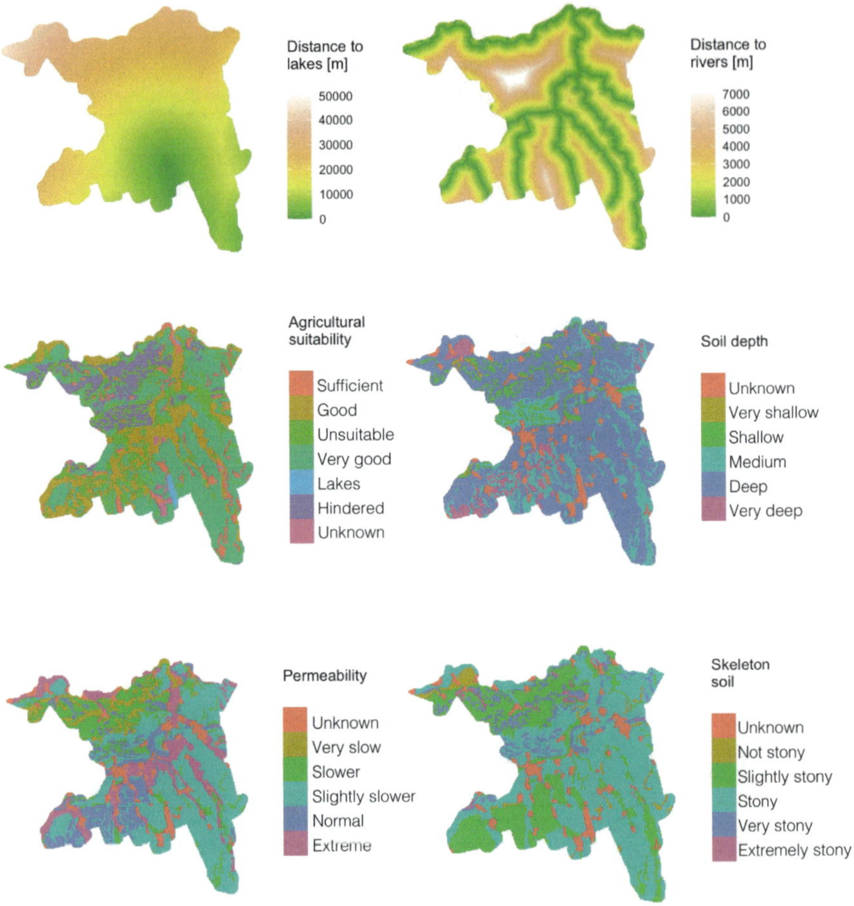

Fig. A.2 Environmental variable Aargau (part2)

Fig. A.3 Environmental variables Aargau. For geology classes see Table 5.6 (part 3)

A.1.3 Locational Preference Analysis

See Figs. A.4, A.5, A.6 and A.7.

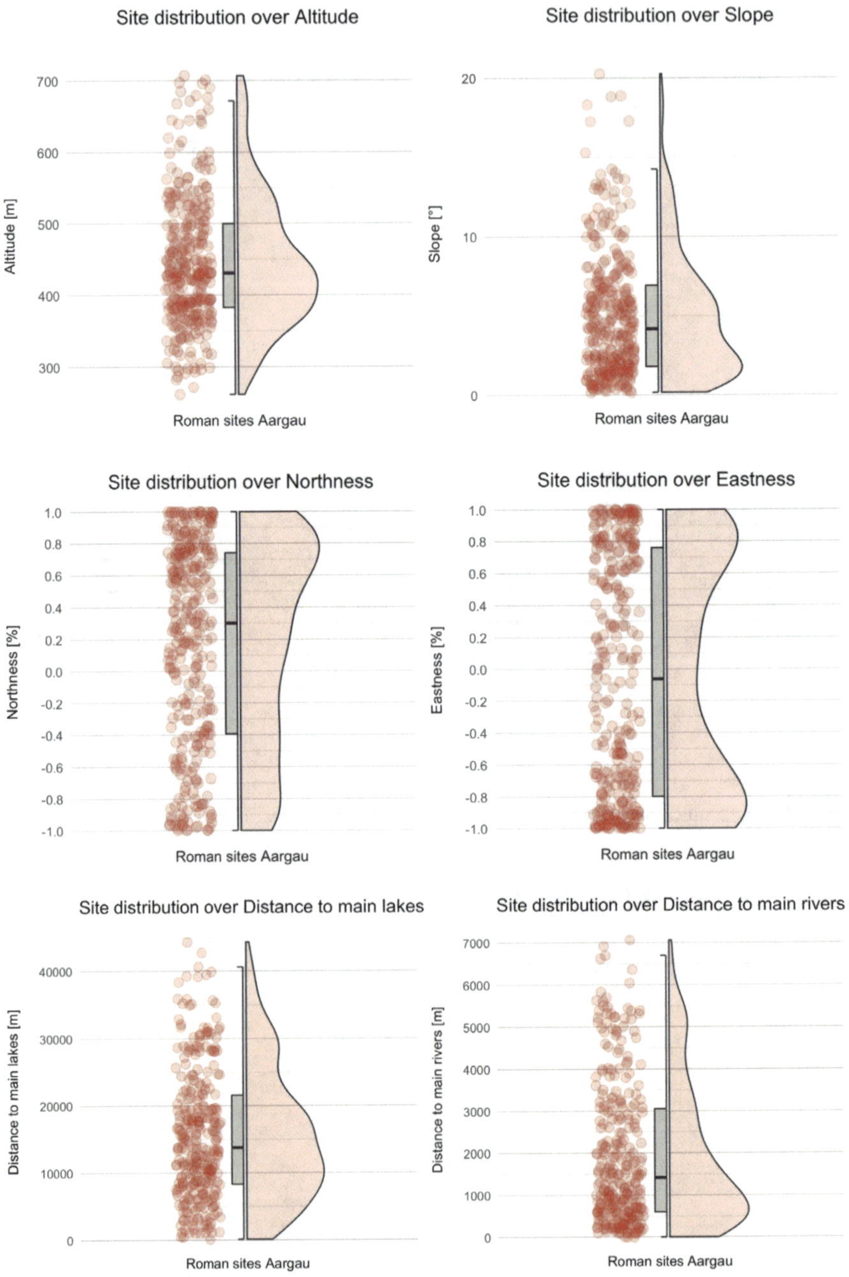

Fig. A.4 Locational preference analysis Aargau. The point jitters show all occurrences. The boxplot indicate the median value and the interquartile range IQR (25th (Q1) to 75th (Q3) percentile (half of all occurrences lie within the gray box), the whiskers show the "minimum" (Q1 − 1.5 * IQR) and "maximum" (Q3 + 1.5 * IQR) values. The colored area indicates the concentration of occurrences (Kernel probability density)

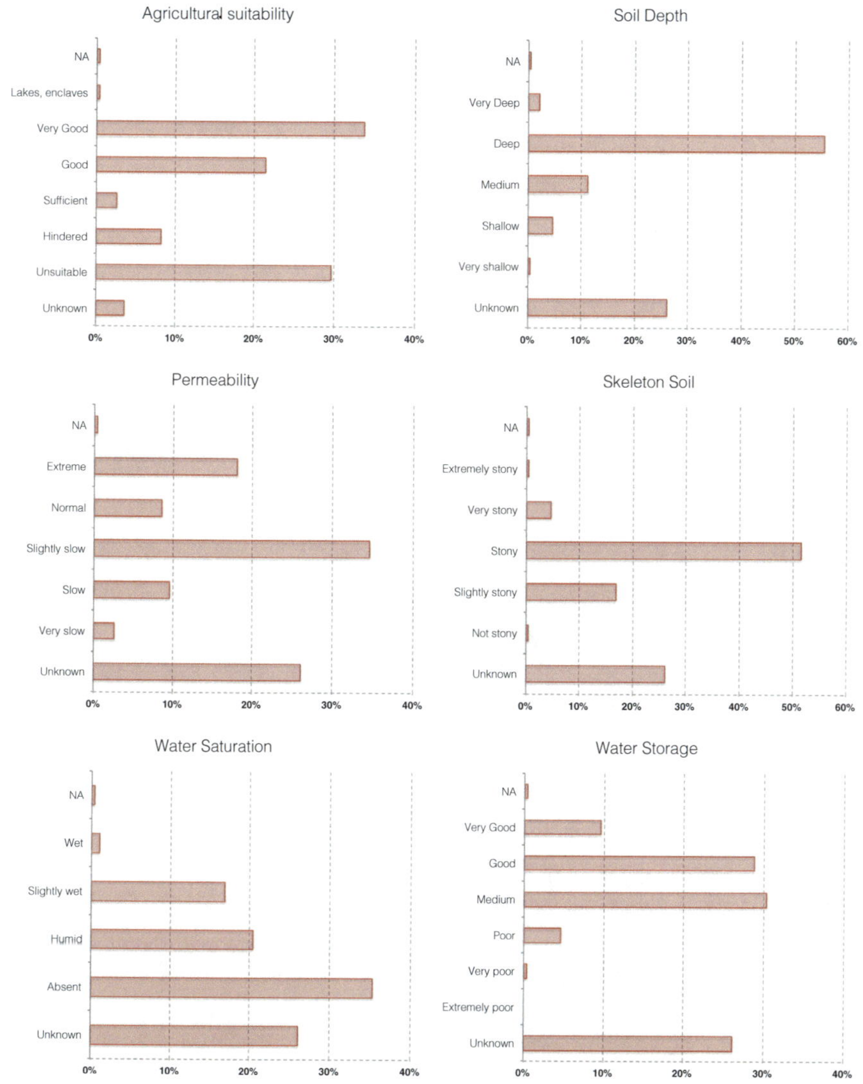

Fig. A.5 Locational preference analysis Aargau. Distribution of the Roman settlements over the agricultural suitability and the soil property classes (part 1)

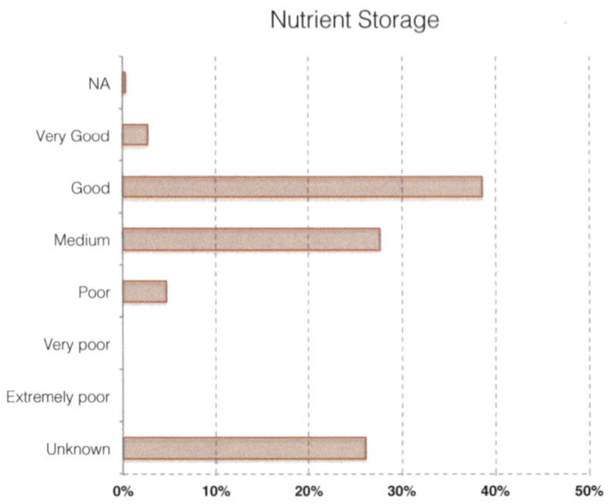

Fig. A.6 Locational preference analysis Aargau. Distribution of the Roman settlements over the agricultural suitability and the soil property classes (part 2)

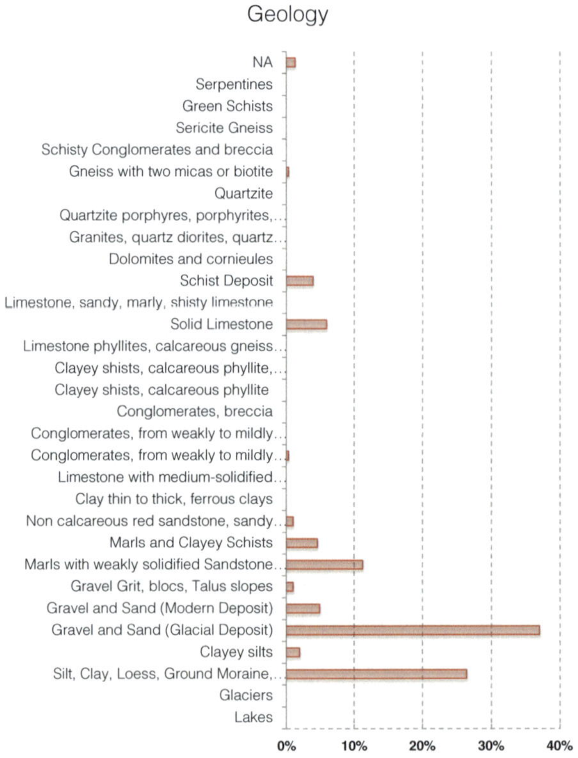

Fig. A.7 Locational preference analysis Aargau. Distribution of the Roman settlements over the geology classes

A.1.4 RF Classification Model Results

See Figs. A.8, A.9, A.10, A.11 and A.12.

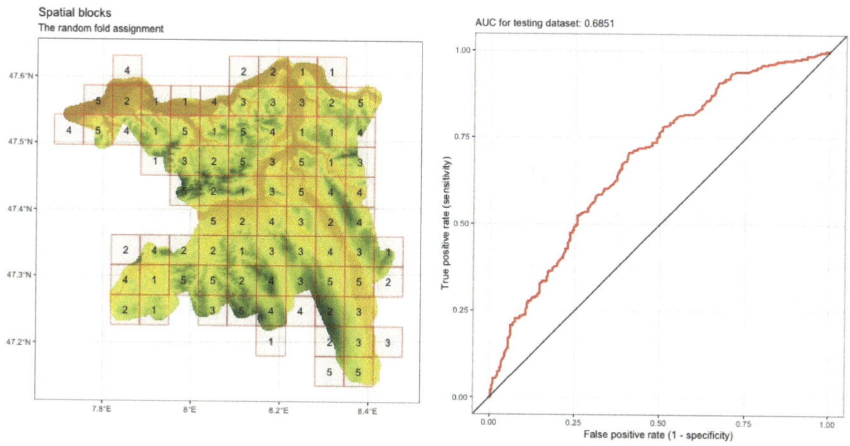

Fig. A.8 Left: Spatial blocks for *k-folds* cross validation Aargau. Number of folds: 5, size of blocks: 5000 m. Right: ROC curve and AUC (0.68) for Aargau RF classification model

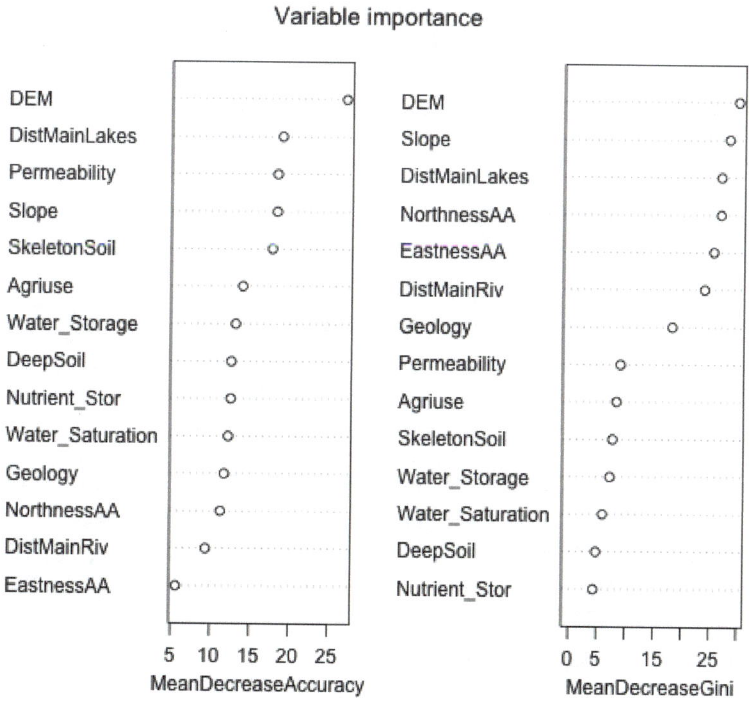

Fig. A.9 Variable importance for RF classification model Aargau

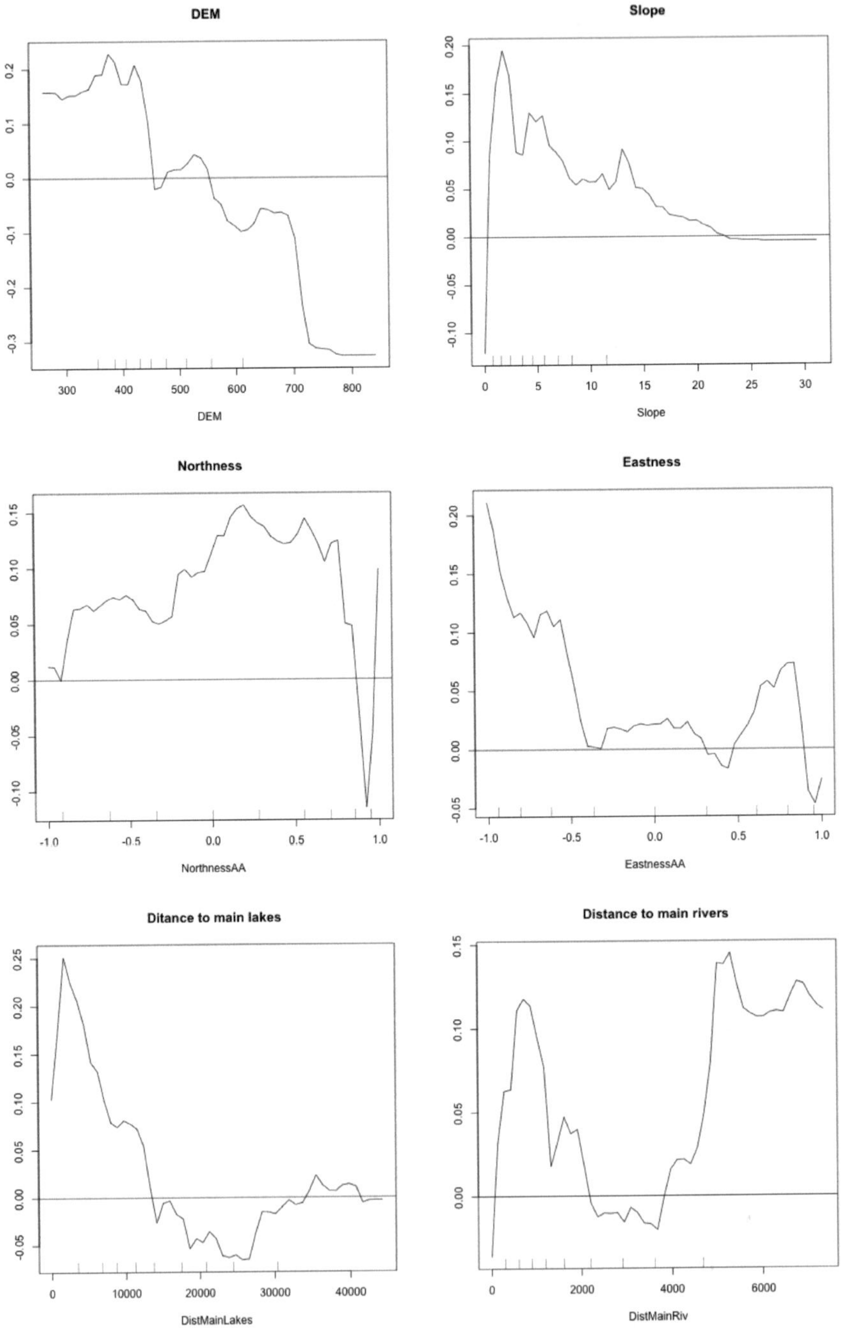

Fig. A.10 Partial dependence plots for RF classification model Aargau (part 1)

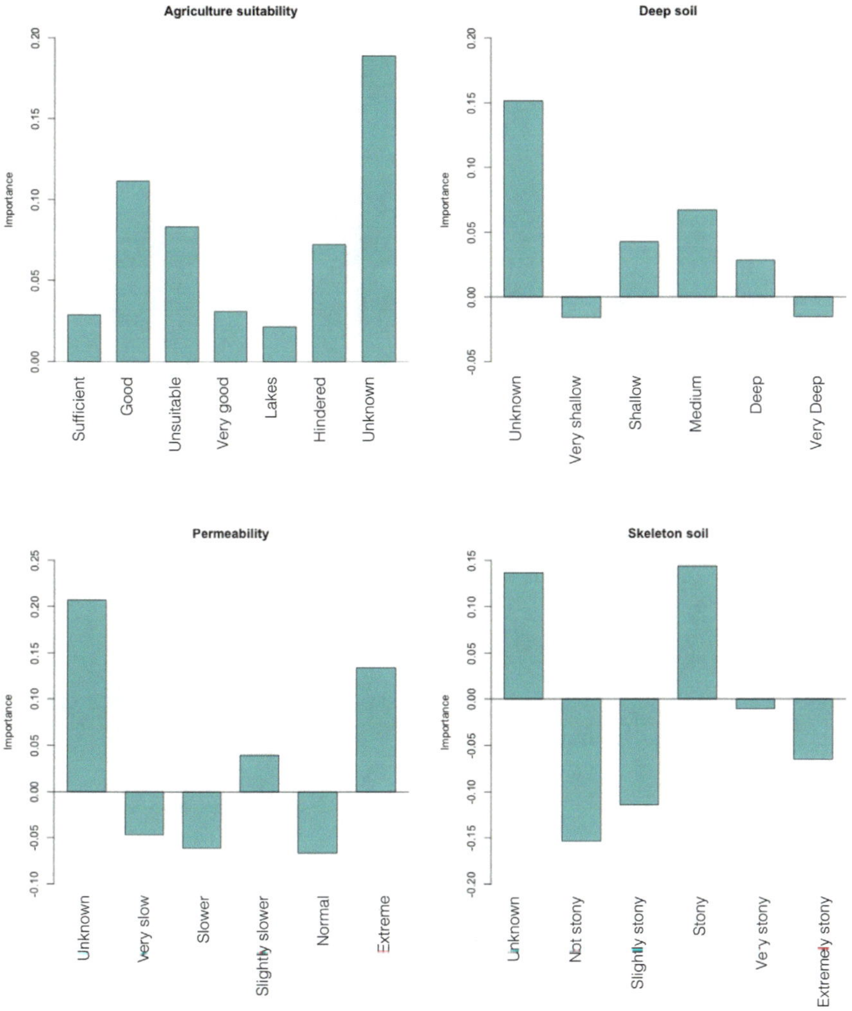

Fig. A.11 Partial dependence plots for RF classification model Aargau (part 2)

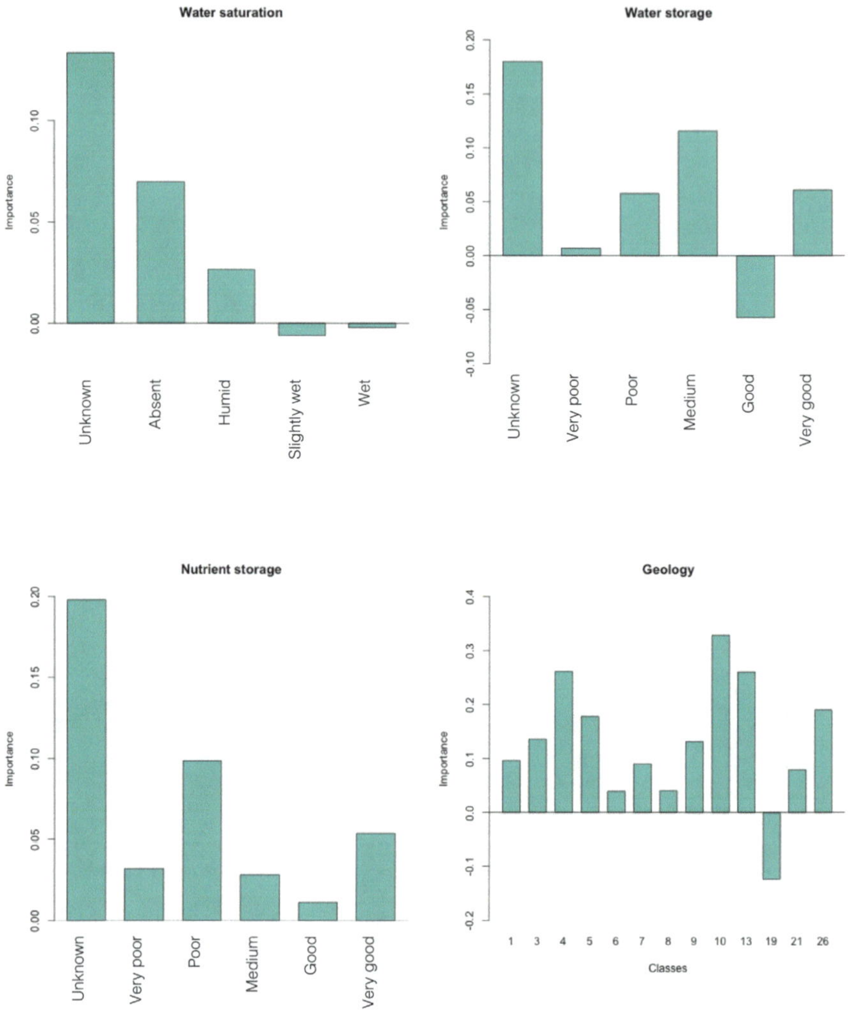

Fig. A.12 Partial dependence plots for RF classification model Aargau (part 3). For geology classes, see Table 5.6

A.1.5 Comparative Analysis of Predicted High Probability Areas—RF Classification

See Figs. A.13, A.14, A.15 and A.16.

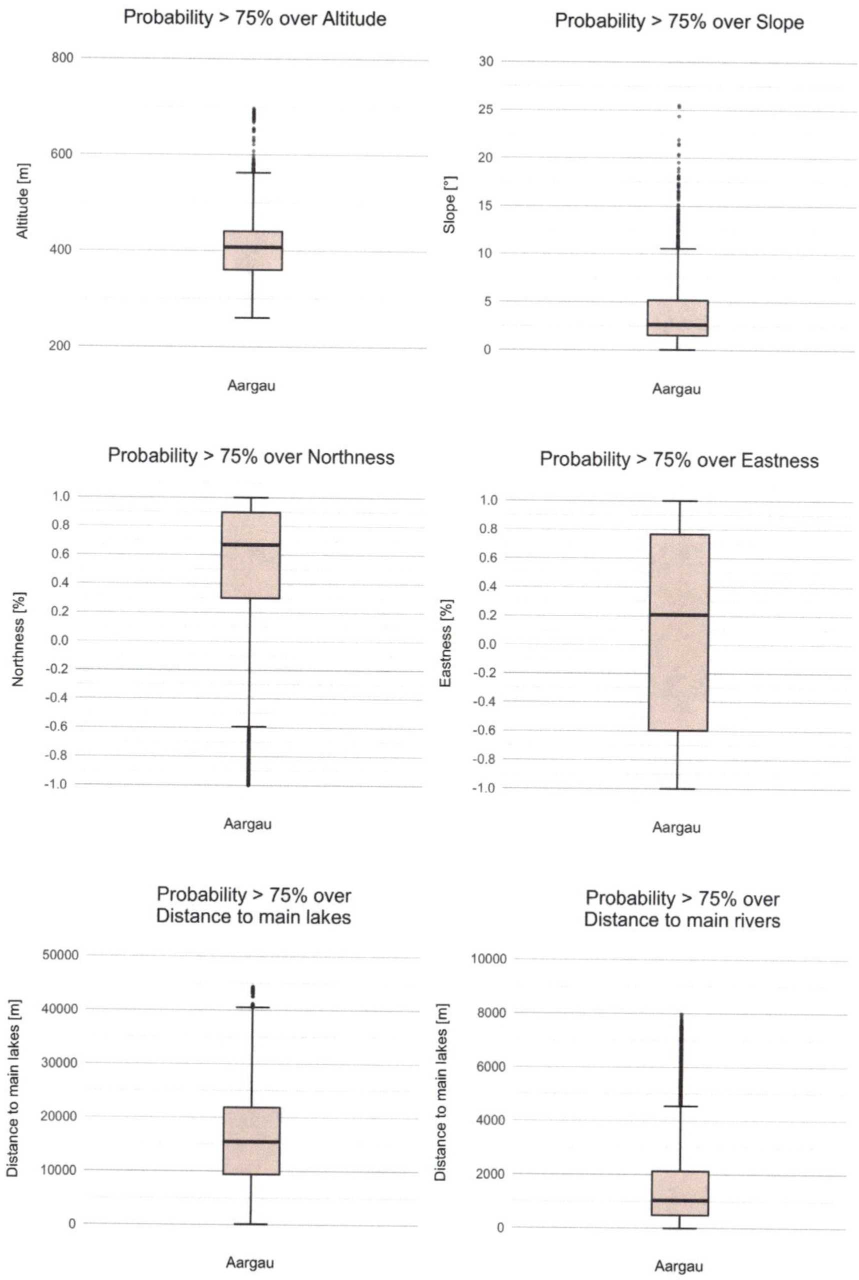

Fig. A.13 Comparative analysis of predicted high probability (>75%) areas using RF classification in Aargau (part 1). The boxplot indicate the median value and the interquartile range IQR (25th (Q1) to 75th (Q3) percentile (half of all occurrences lie within the box), the whiskers show the "minimum" (Q1 − 1.5 * IQR) and "maximum" (Q3 + 1.5 * IQR) values. Points show outliers

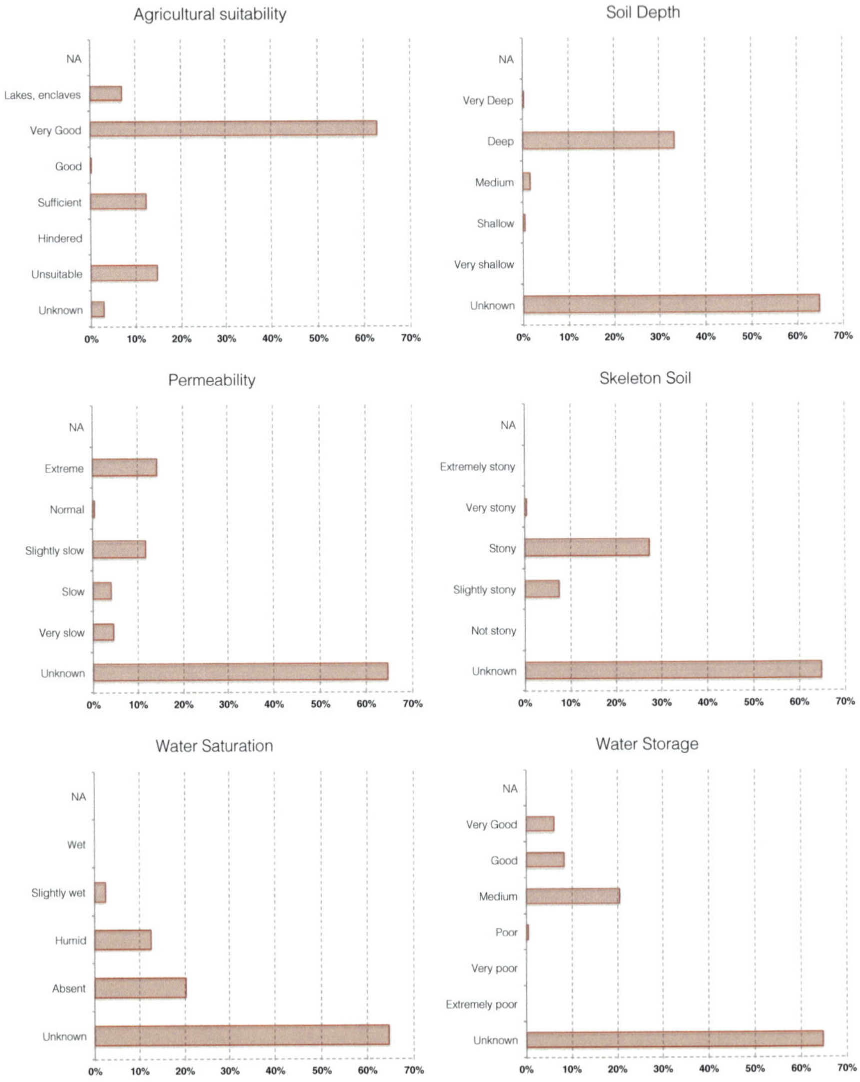

Fig. A.14 Comparative analysis of predicted high probability (>75%) areas using RF classification in Aargau (part 2)

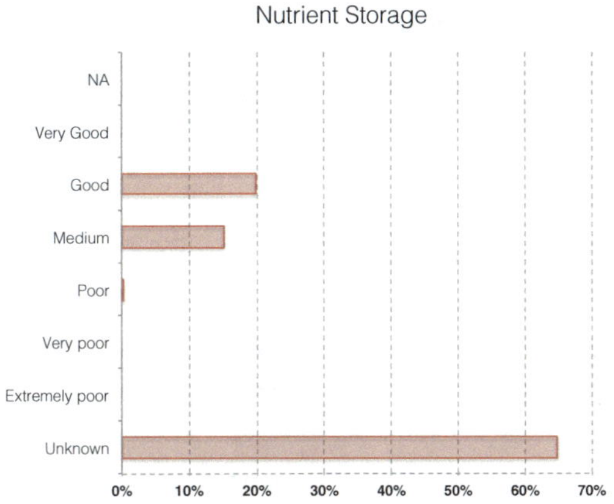

Fig. A.15 Comparative analysis of predicted high probability (>75%) areas using RF classification in Aargau (part 3)

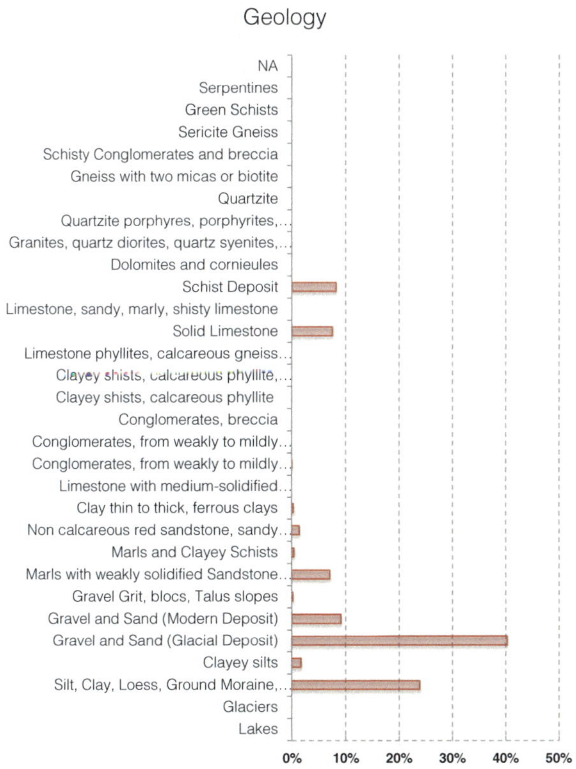

Fig. A.16 Comparative analysis of predicted high probability (>75%) areas on geology classes using RF classification in Aargau

A.1.6 RF Regression Model Results (Single Finds)

See Fig. A.17.

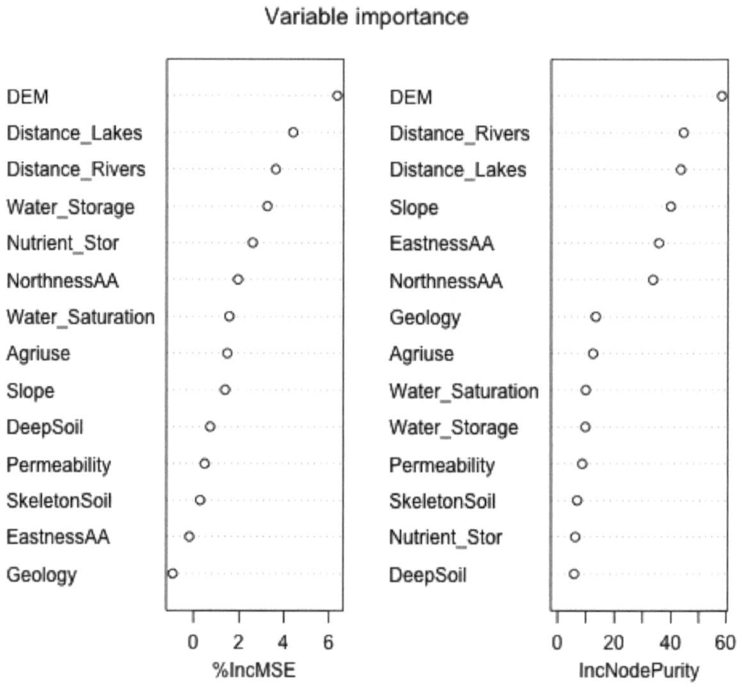

Fig. A.17 Variable importance for RF regression model Aargau

A.2 Canton of Fribourg

A.2.1 Database Reclassification

See Table A.2.

Table A.2 Roman site typology classification for Fribourg

Roman site typology database Fribourg		
Class	English classification	Original classification
Bridges	Bridge	Pont
	Bridge, deposit	Pont, dépôt
	Bridge, dwelling	Pont, habitat
Fanum	Fanum	Fanum
	Mausoleum	Mausolé
Graves	Funerary stele	Stèle funéraire
	Necropolis	Nécropole
	Necropolis, mausoleum	Nécropole, mausolé
	Necropolis, occupation	Nécropole, occupation
	Necropolis?	Nécropole?
Roads	Road	voie
	Road?	voie?
Settlements	Agglomeration	Agglomération
	Altitude site?	Site de hauteur?
	Baths	Bains
	Baths?	Bains?
	Crafts	Artisanat
	Crafts, dwelling, road	Artisanat, habitat, voie
	Crafts, road	Artisanat, voie
	Crafts?	Artisanat?
	Establishment	Etablissement
	Landing stage	Débarcadère
	Occupation	Occupation
	Occupation, fanum?	Occupation, fanum?
	Occupation, road	Occupation, voie
	Villa	Villa
	Villa, basement	Villa, cave
	Wall, fence	Mur d'enclos
Single finds	Coin deposit	Dépôt monétaire
	Cultural deposit	Dépôt culturel
	Deposit	Dépôt
	Single finds	Trouvaille isolée
	Single finds, road	Trouvaille isolée, voie
Unknown	Undetermined	Indéterminé
Water infrastructures	Aqueduct	Aqueduc
	Canalization	Canalisation

(continued)

Table A.2 (continued)

Roman site typology database Fribourg		
Class	English classification	Original classification
	Spring catchment	Captage de source?
	Springs	Sources
	Wells	Puits

A.2.2 Environmental Variables

See Figs. A.18, A.19 and A.20.

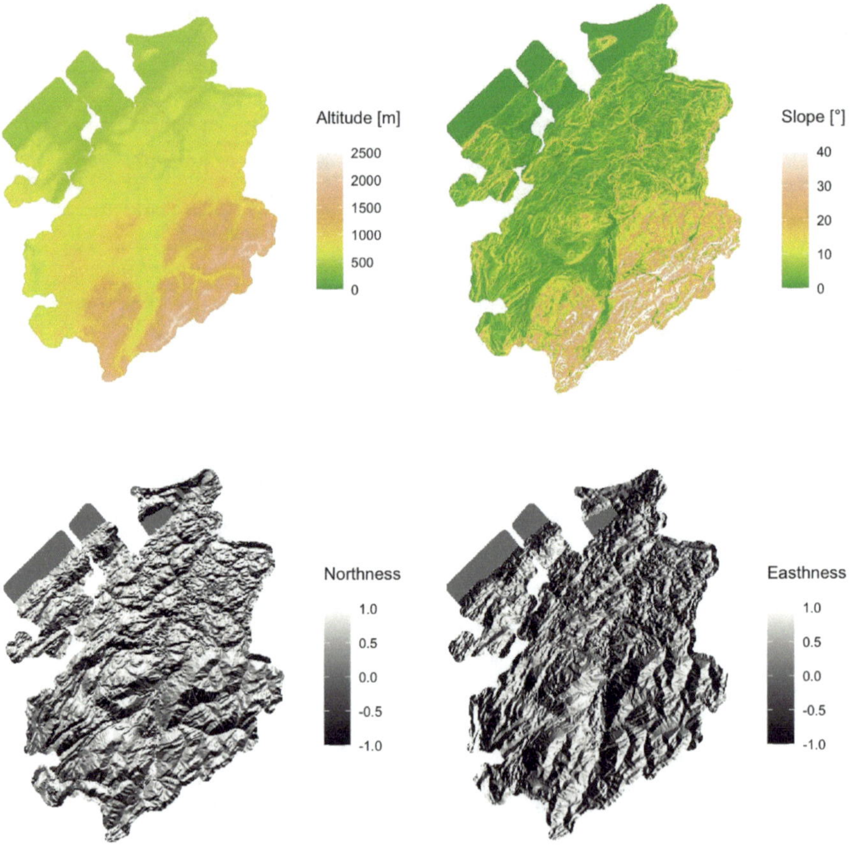

Fig. A.18 Environmental variables Fribourg (part 1)

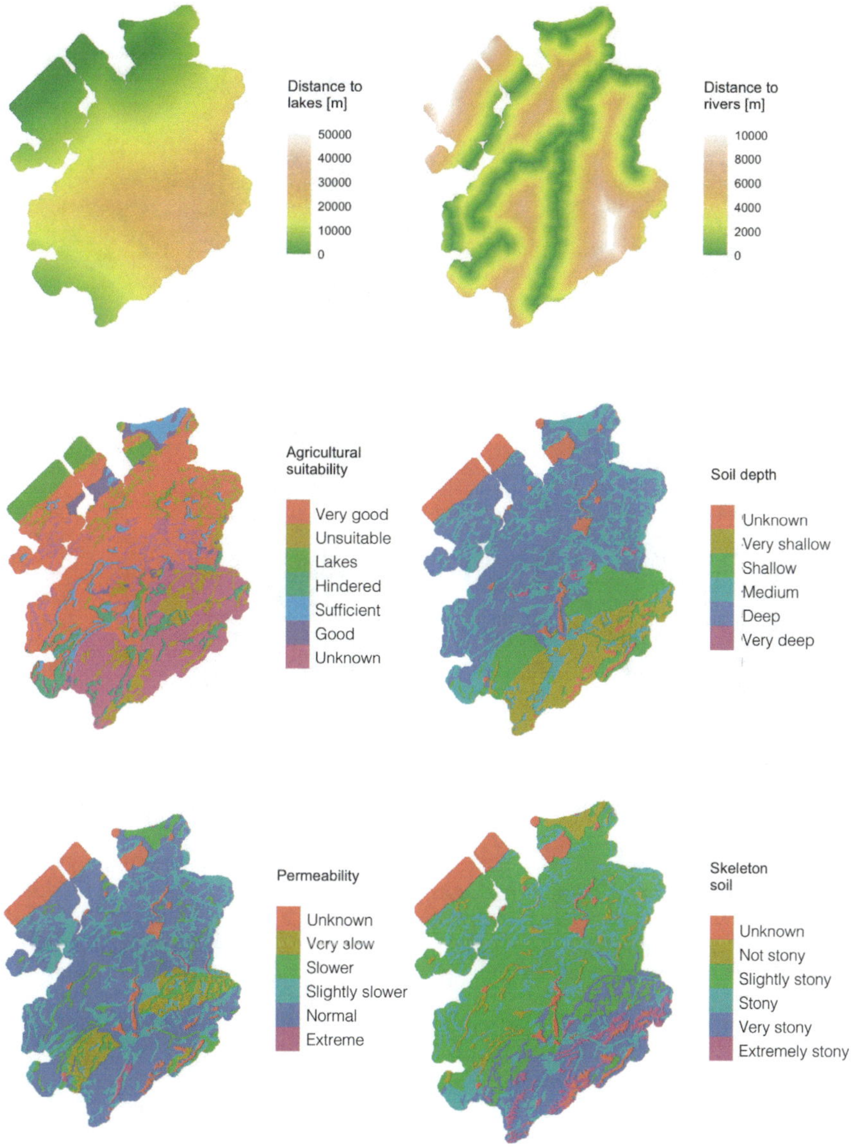

Fig. A.19 Environmental variables Fribourg (part 2)

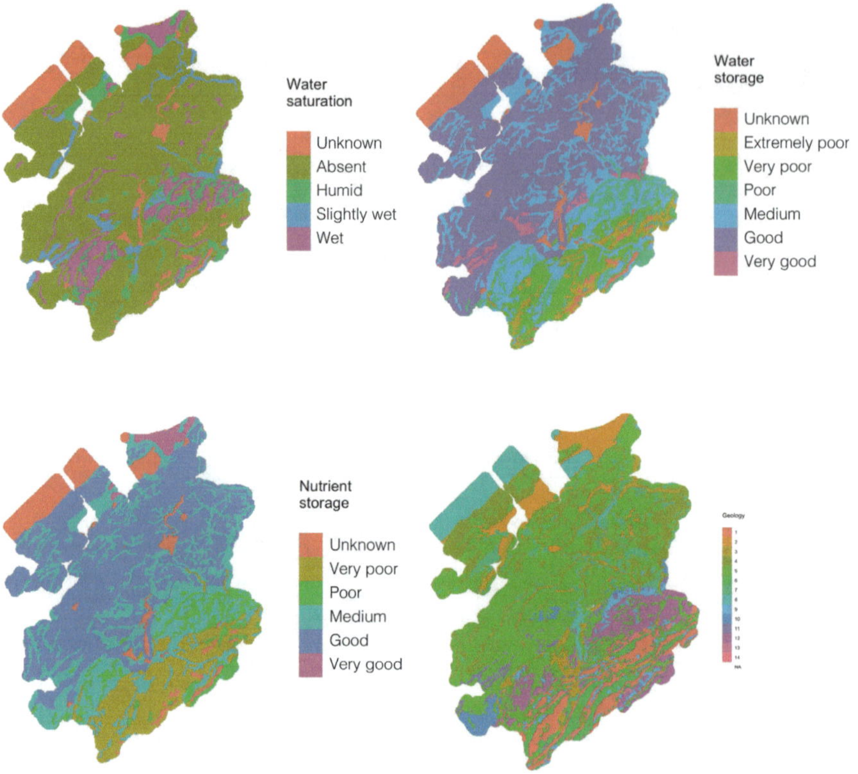

Fig. A.20 Environmental variables Fribourg. For geology classes see Table 5.6 (part3)

A.2.3 Locational Preference Analysis

See Figs. A.21, A.22, A.23 and A.24.

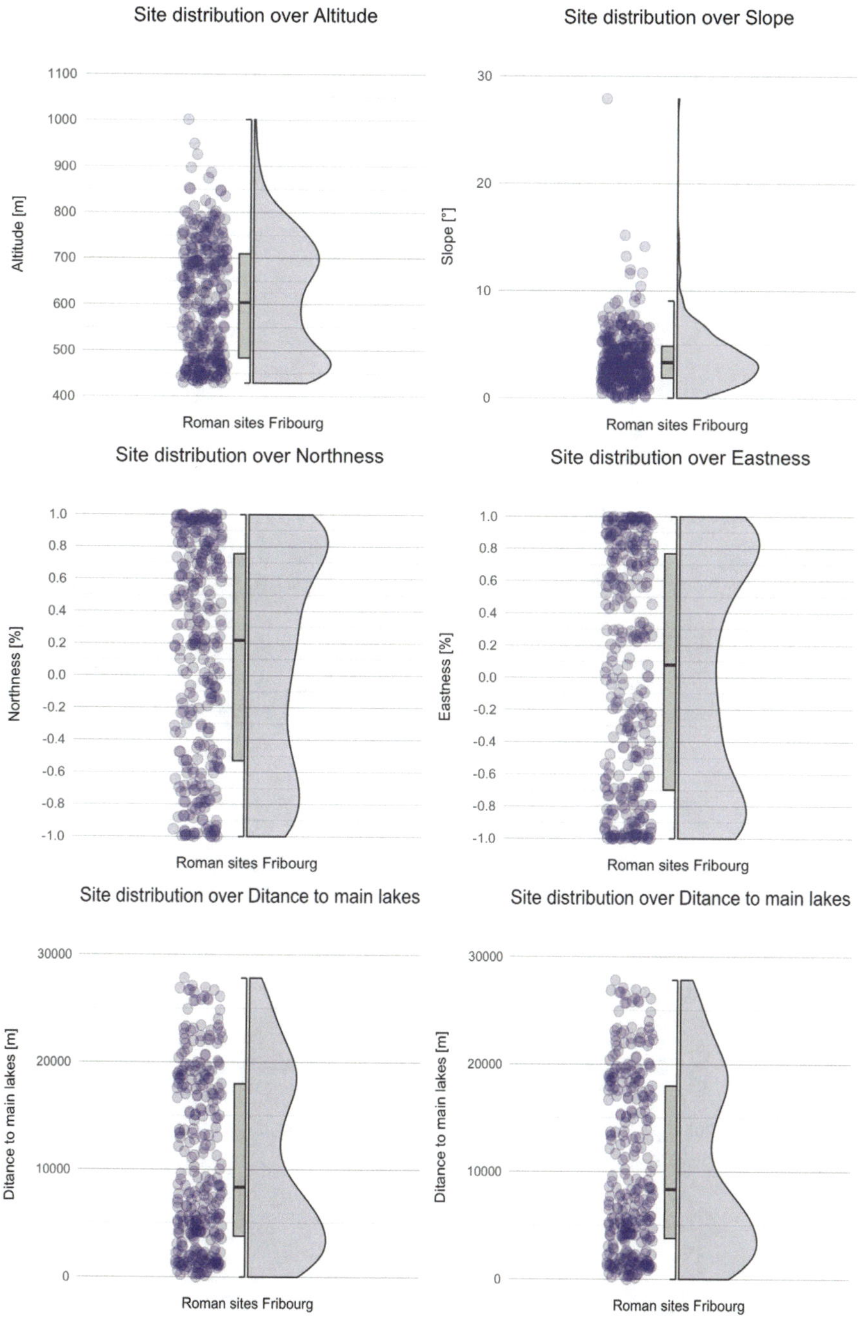

Fig. A.21 Locational preference analysis Fribourg. The point jitters show all occurrences. The boxplot indicate the median value and the interquartile range IQR (25th (Q1) to 75th (Q3) percentile (half of all occurrences lie within the gray box), the whiskers show the "minimum" (Q1 – 1.5 * IQR) and "maximum" (Q3 + 1.5 * IQR) values. The colored area indicates the concentration of occurrences (Kernel probability density)

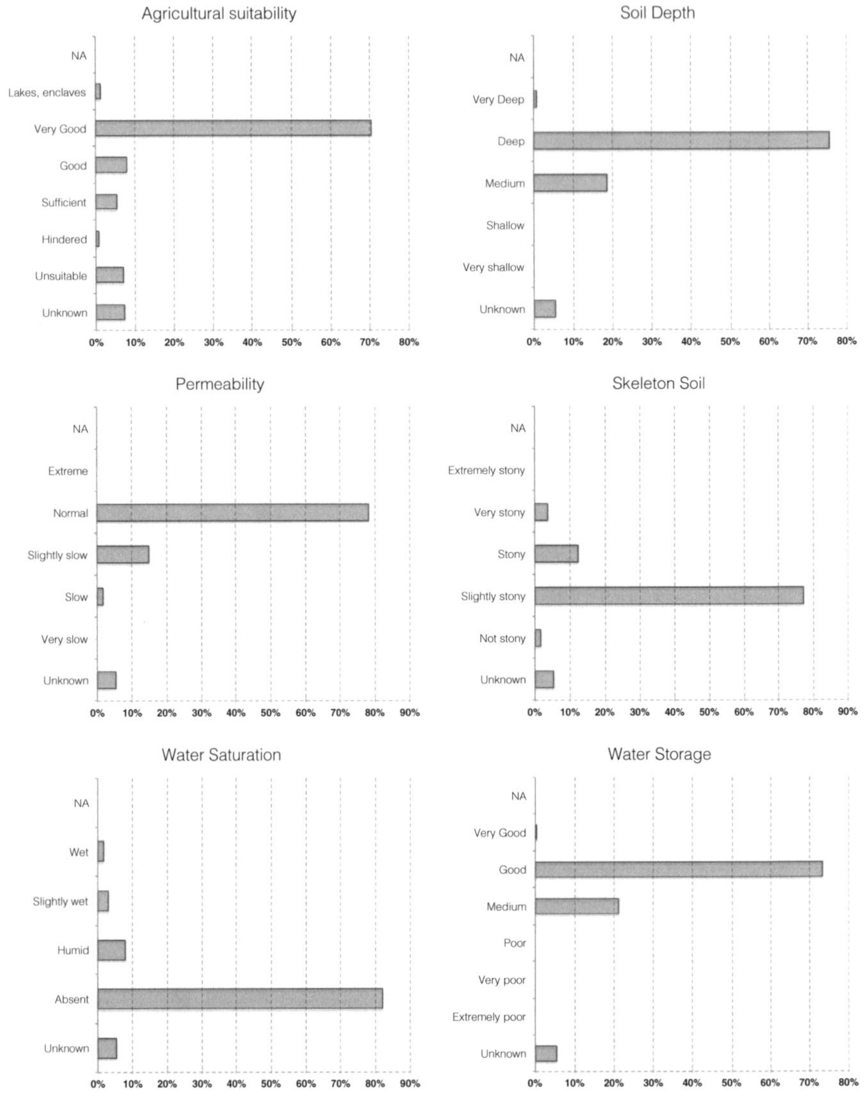

Fig. A.22 Locational preference analysis Fribourg. Distribution of the Roman settlements over the agricultural suitability and the soil property classes (part 1)

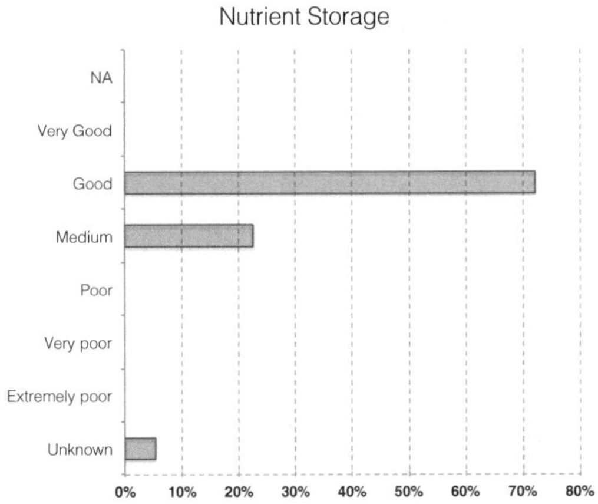

Fig. A.23 Locational preference analysis Fribourg. Distribution of the Roman settlements over the agricultural suitability and the soil property classes (part 2)

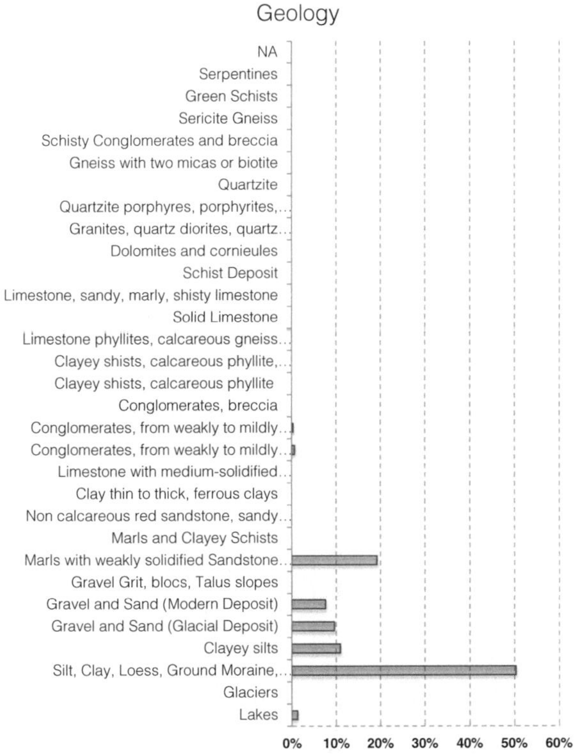

Fig. A.24 Locational preference analysis Fribourg. Distribution of the Roman settlements over the geology classes

A.2.4 RF Classification Model Results

See Figs. A.25, A.26, A.27, A.28 and A.29.

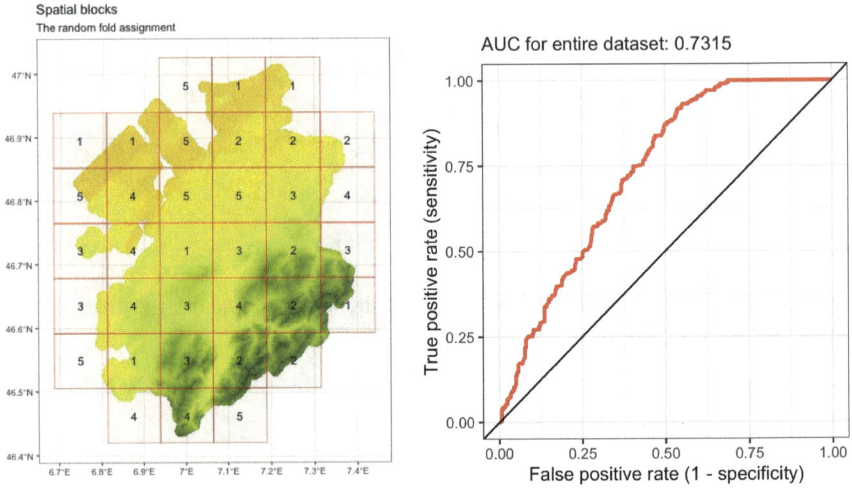

Fig. A.25 Left: Spatial blocks for k-folds cross validation Fribourg. Number of folds: 5, size of blocks: 9639 m. Right: ROC curve and AUC (0.73) for Fribourg RF classification model

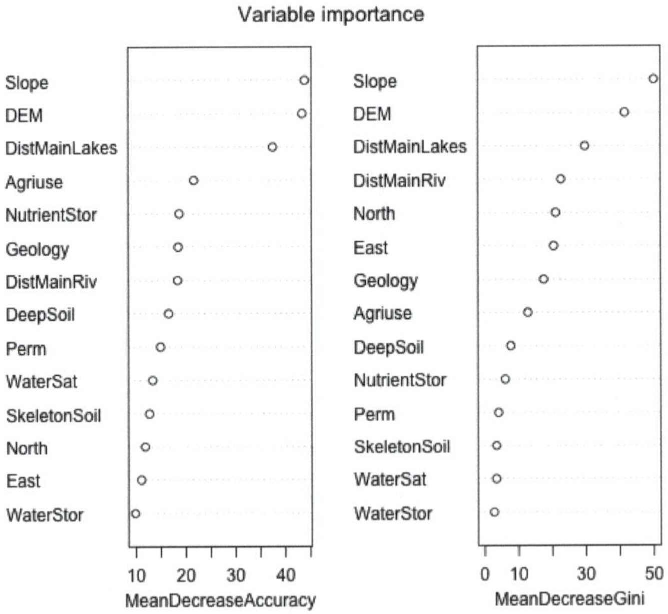

Fig. A.26 Variable importance for RF classification model Fribourg

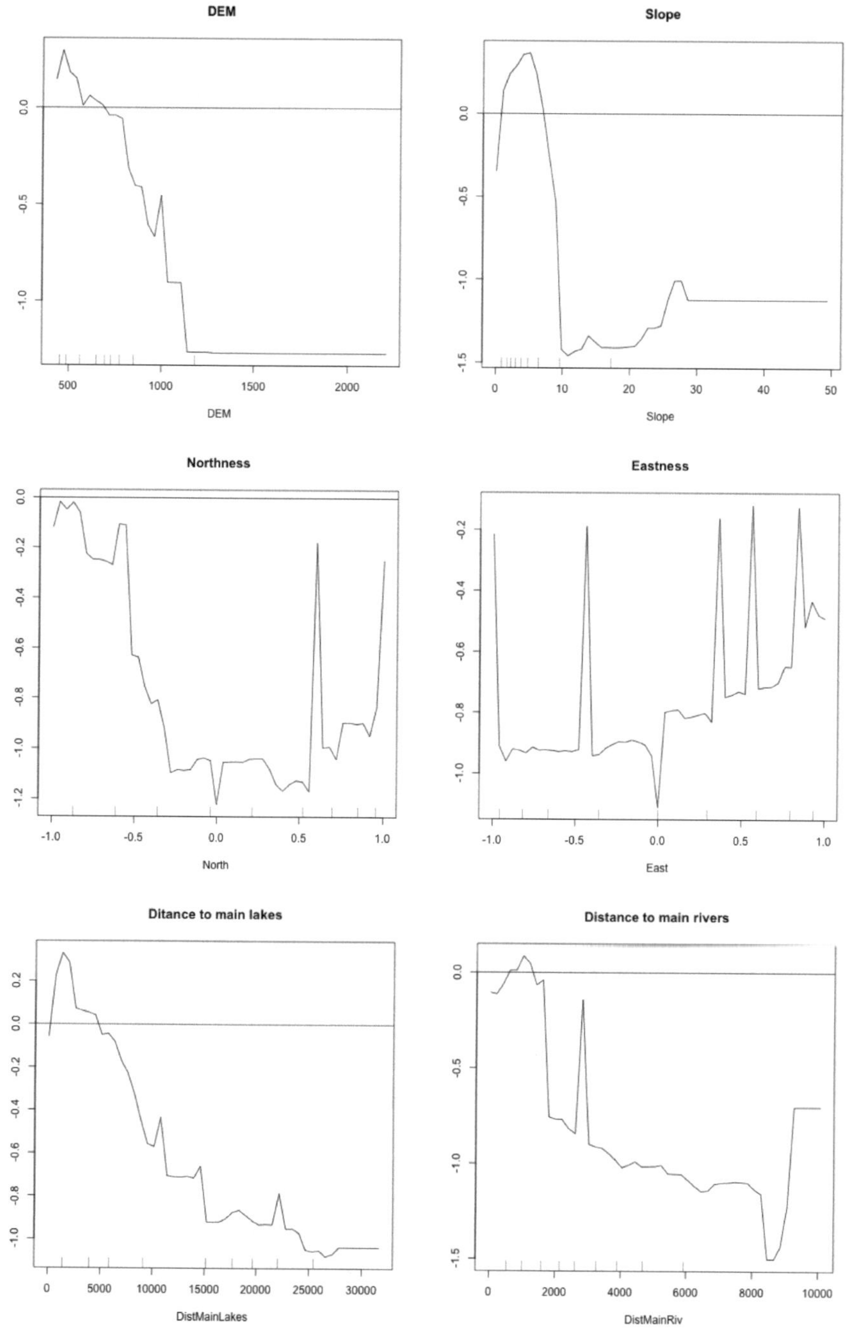

Fig. A.27 Partial dependence plots for RF classification model Fribourg (part 1)

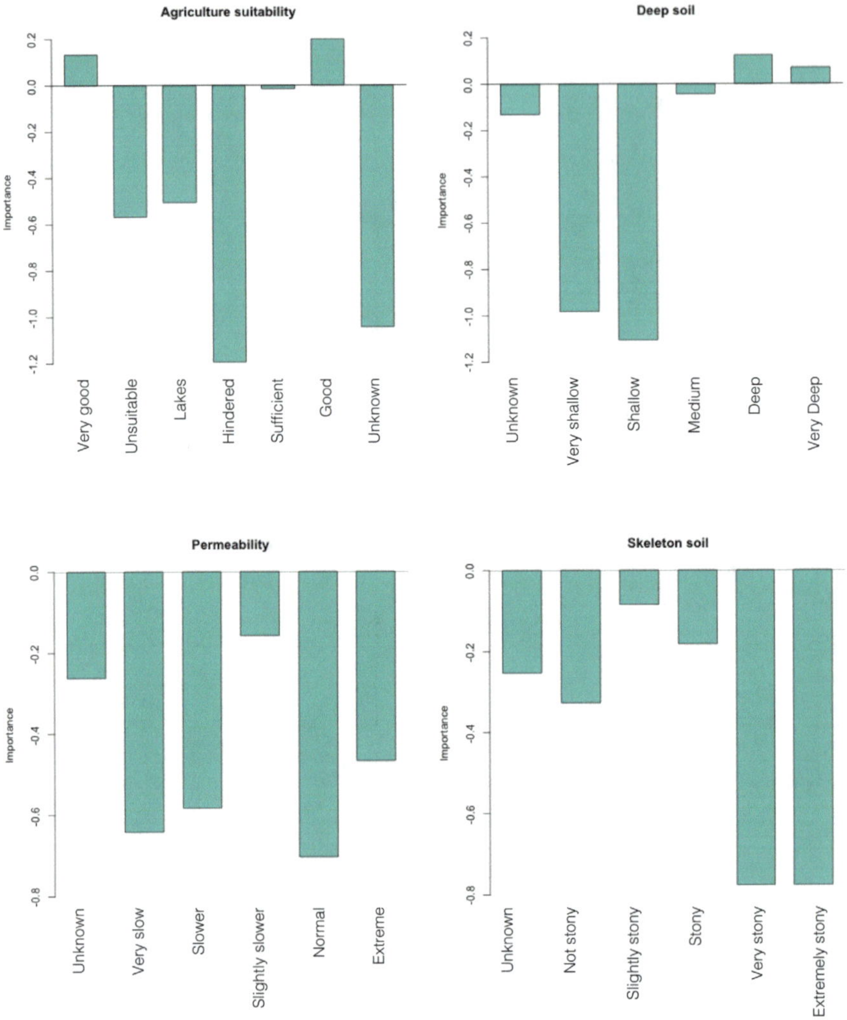

Fig. A.28 Partial dependence plots for RF classification model Fribourg (part 2)

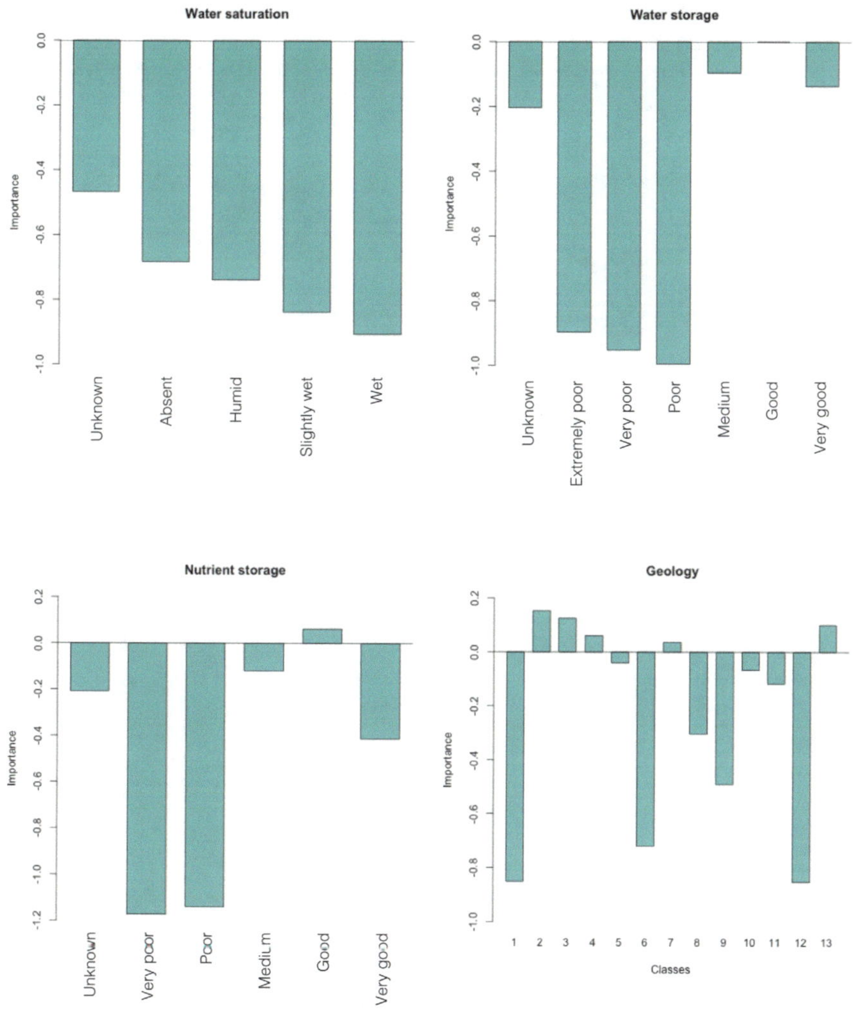

Fig. A.29 Partial dependence plots for RF classification model Fribourg (part 3). For geology classes, see Table 5.6

A.2.5 Comparative Analysis of Predicted High Probability Areas—RF Classification

See Figs. A.30, A.31, A.32 and A.33.

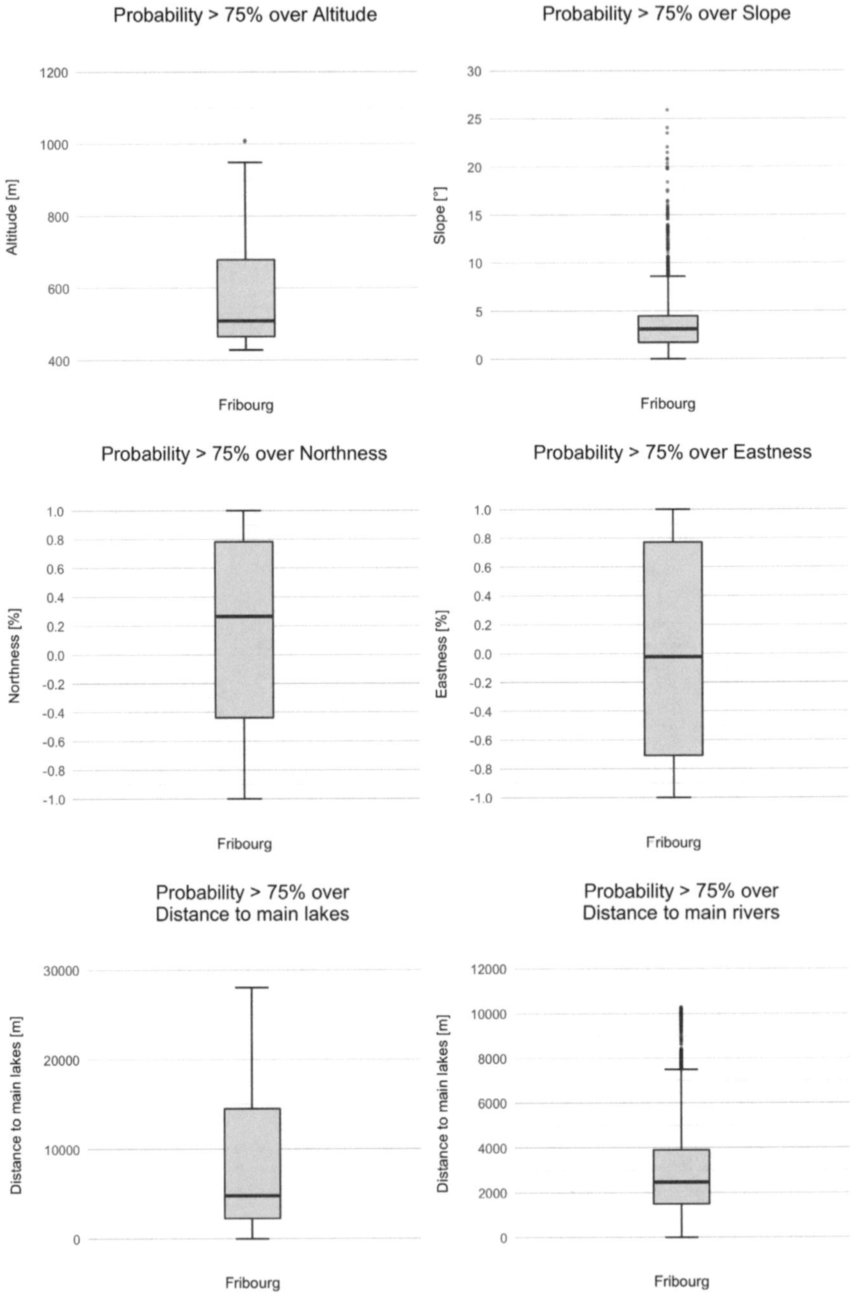

Fig. A.30 Comparative analysis of predicted high probability (>75%) areas using RF Classification in Fribourg (part 1). The boxplot indicate the median value and the interquartile range IQR (25th (Q1) to 75th (Q3) percentile (half of all occurrences lie within the box), the whiskers show the "minimum" (Q1 − 1.5 * IQR) and "maximum" (Q3 + 1.5 * IQR) values. Points show outliers

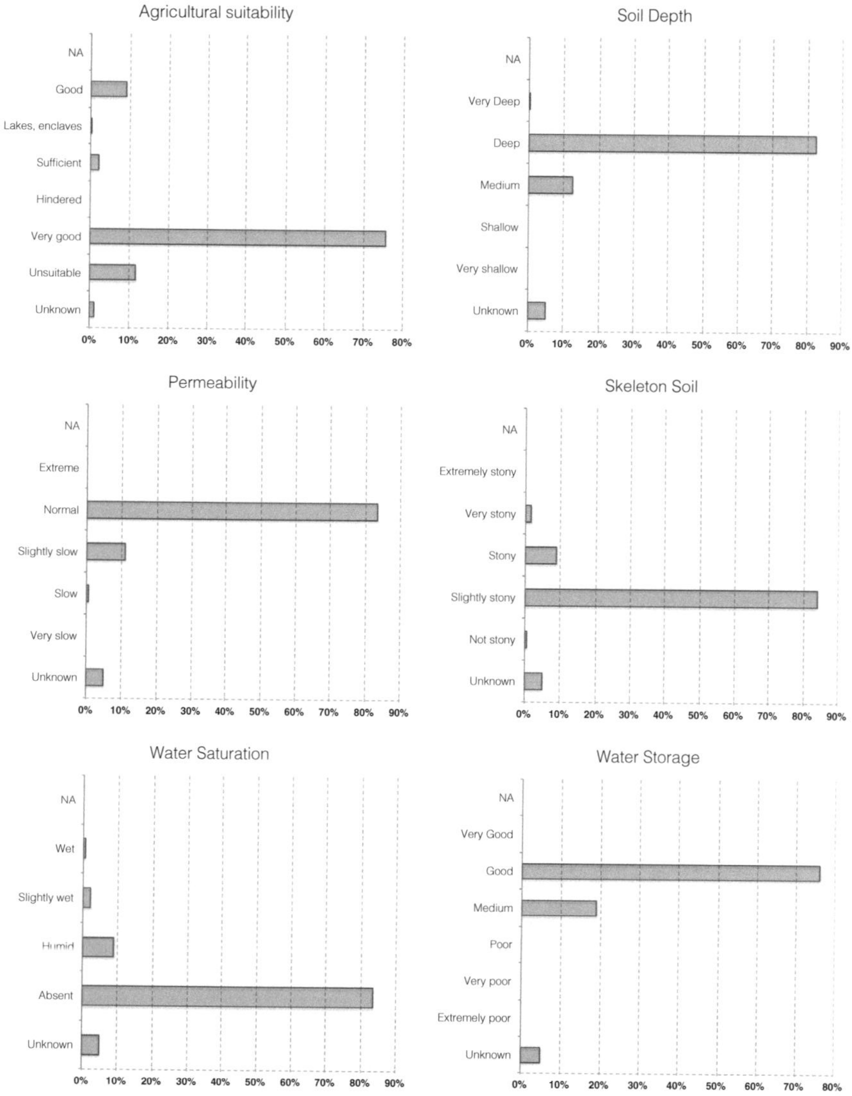

Fig. A.31 Comparative analysis of predicted high probability (>75%) areas using RF classification in Fribourg (part 2)

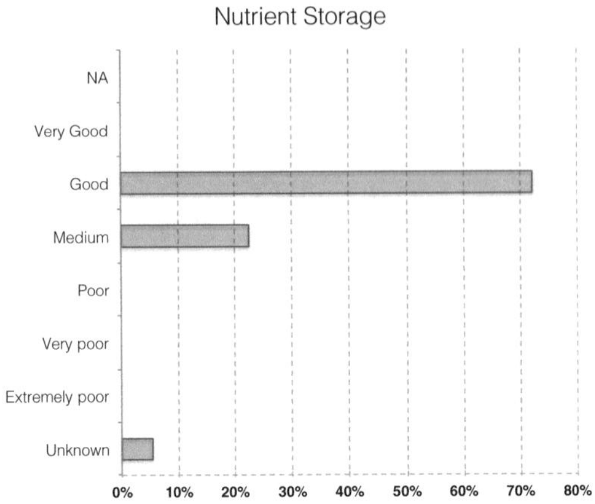

Fig. A.32 Comparative analysis of predicted high probability (>75%) areas using RF classification in Fribourg (part 3)

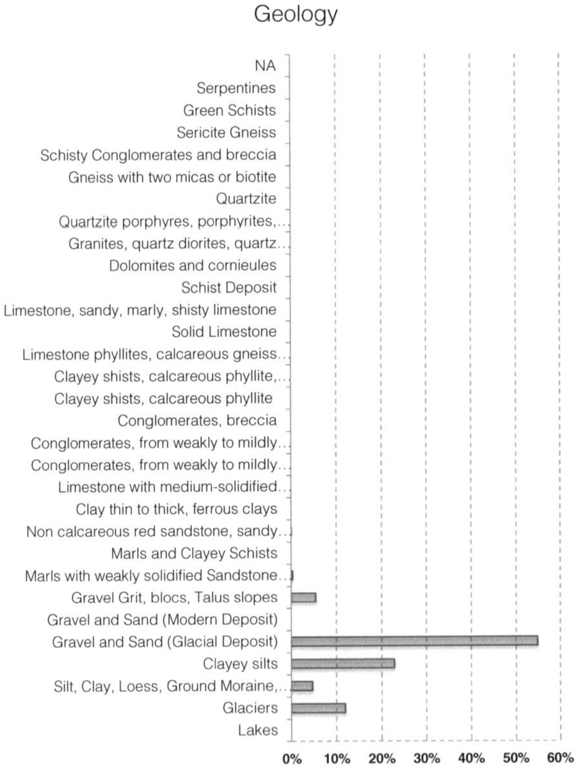

Fig. A.33 Comparative analysis of predicted high probability (>75%) areas on Geology classes using RF classification in Fribourg

A.3 Canton of Geneva

A.3.1 Database Reclassification

See Table A.3.

Table A.3 Roman site typology classification for Geneva

Roman site typology database Geneva		
Class	English classification	Original classification
Fortifications	Castle, fortified town, etc	Château/Bourg fortifié/etc
Graves	Necropolis	Nécropole/Cimetière
Others	Ancient land limits	Parcellaire ancien
	Ditch	Fossé
Religious sites	Religious site	Lieu de culte
Settlements	Crafts, industry	Structure artisanale ou industrielle
	Dwelling, villa	Habitat/Villa
Unknown	Blanks (tiles)	Blanks

A.3.2 Environmental Variables

See Figs. A.34, A.35 and A.36.

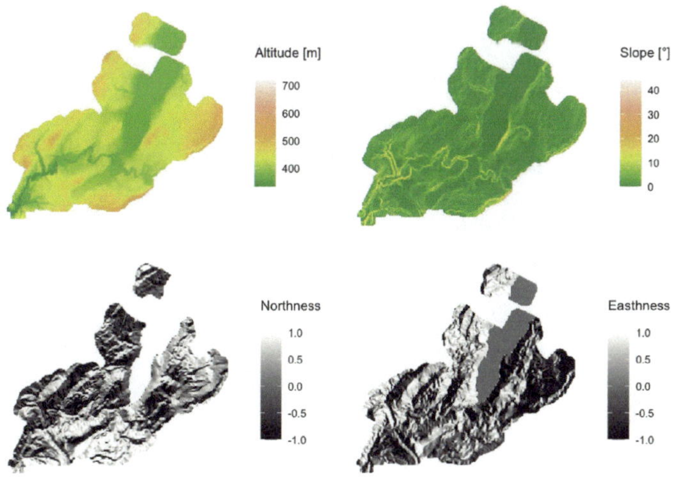

Fig. A.34 Environmental variables Geneva (part 1)

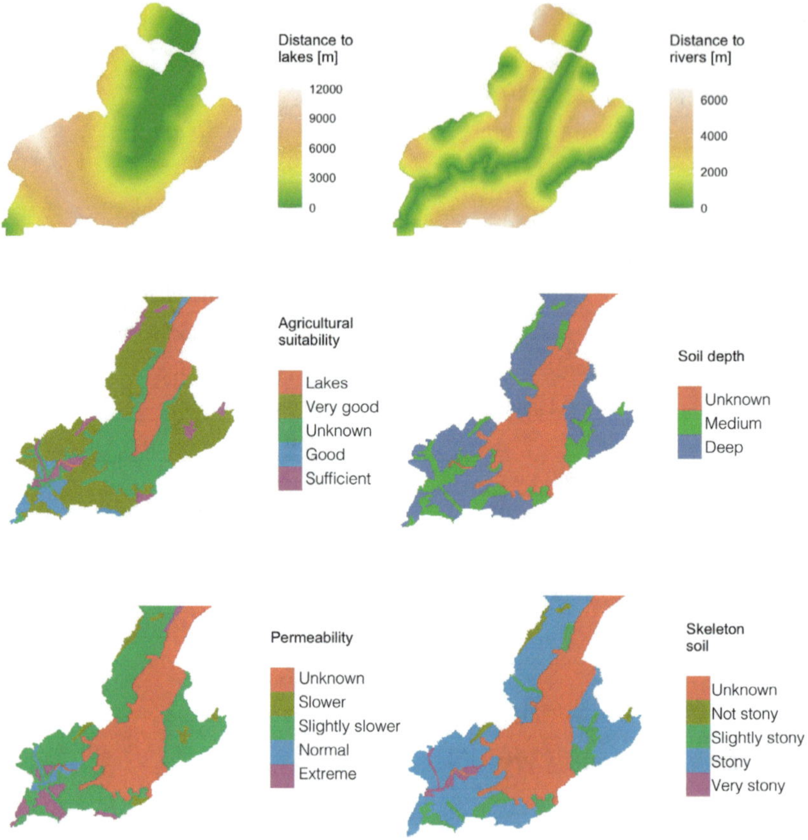

Fig. A.35 Environmental variable Geneva (part2)

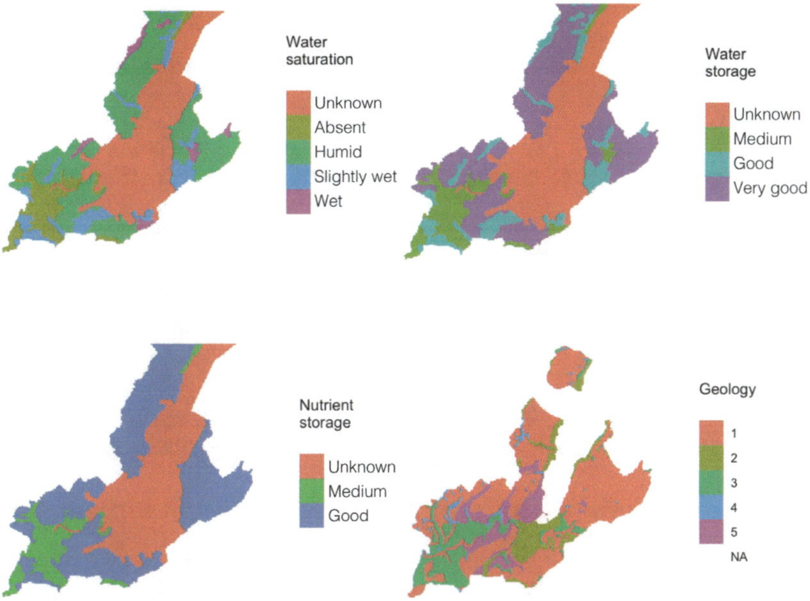

Fig. A.36 Environmental variables Geneva. For Geology classes see Table 5.6 (part3)

A.3.3 Locational Preference Analysis

See Figs. A.37, A.38, A.39 and A.40.

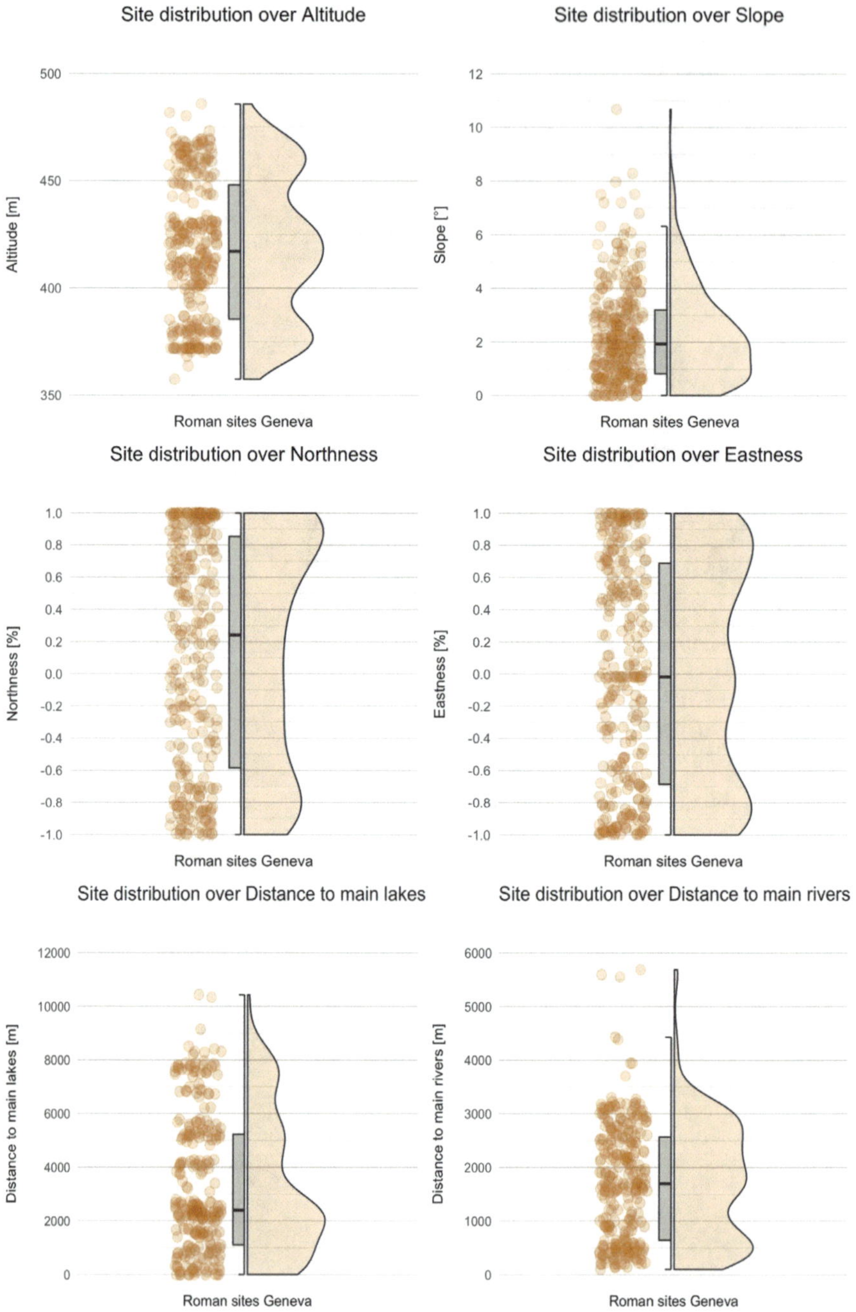

Fig. A.37 Locational preference analysis Geneva. The point jitters show all occurrences. The boxplot indicate the median value and the interquartile range IQR (25th (Q1) to 75th (Q3) percentile (half of all occurrences lie within the gray box), the whiskers show the "minimum" (Q1 − 1.5 * IQR) and "maximum" (Q3 + 1.5 * IQR) values. The colored area indicates the concentration of occurrences (Kernel probability density)

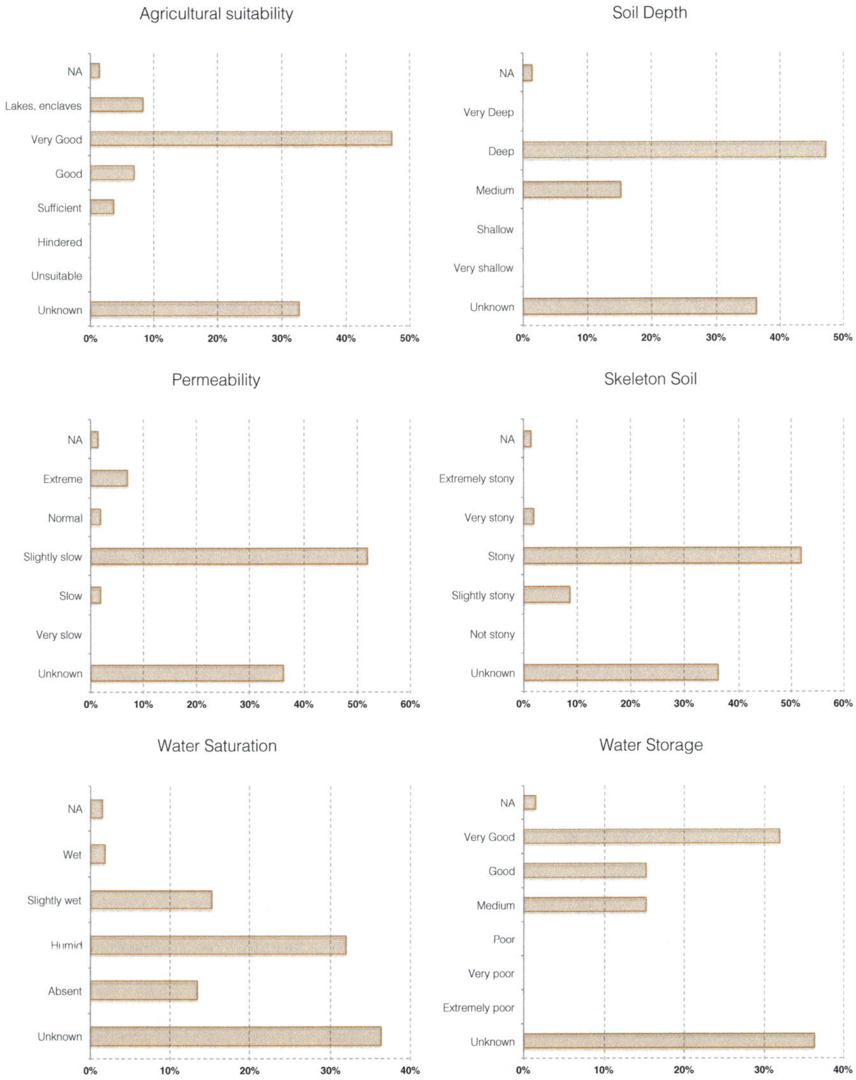

Fig. A.38 Locational preference analysis Geneva. Distribution of the Roman settlements over the agricultural suitability and the soil property classes (part 1)

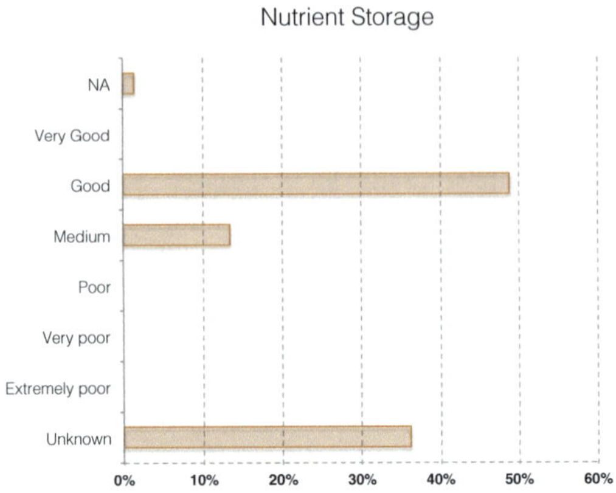

Fig. A.39 Locational preference analysis Geneva. Distribution of the Roman settlements over the agricultural suitability and the soil property classes (part 2)

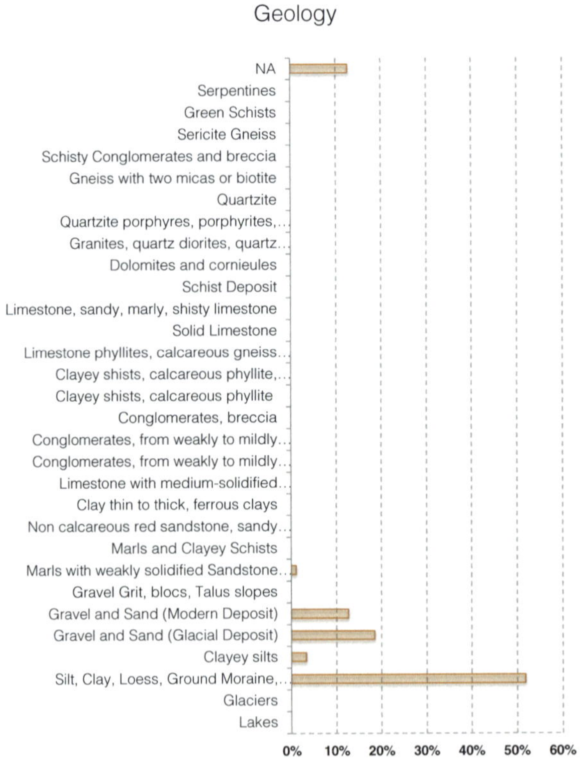

Fig. A.40 Locational preference analysis Geneva. Distribution of the Roman settlements over the geology classes

A.3.4 RF Classification Model Results

See Figs. A.41, A.42, A.43, A.44 and A.45.

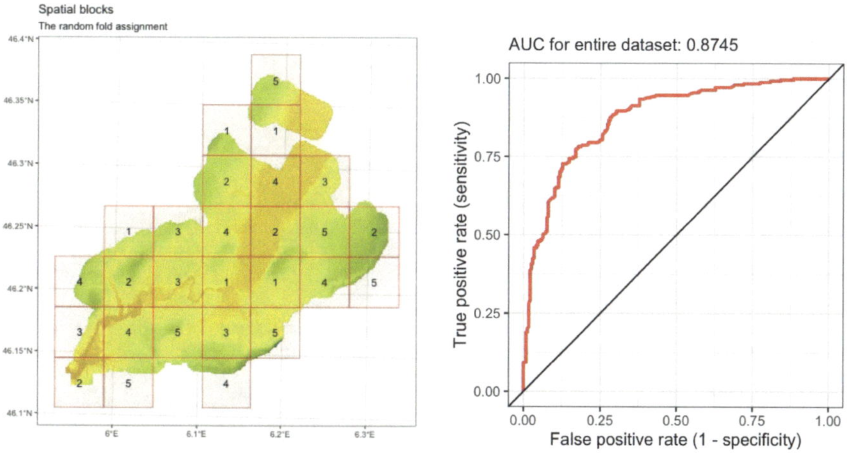

Fig. A.41 Left: Spatial blocks for *k-folds* cross validation Geneva. Number of folds: 5, size of blocks: 4500 m. Right: ROC curve and AUC (0.87) for Geneva RF classification model

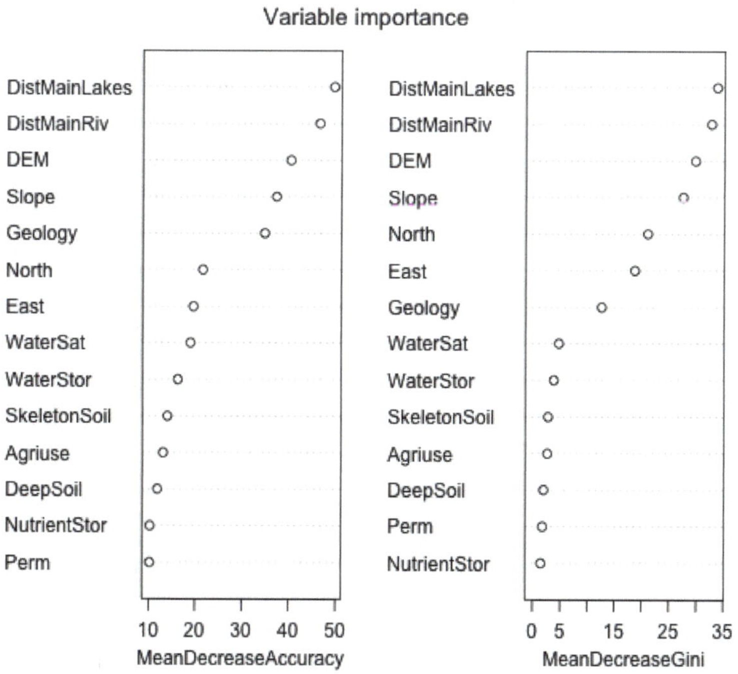

Fig. A.42 Variable importance for RF classification model Geneva

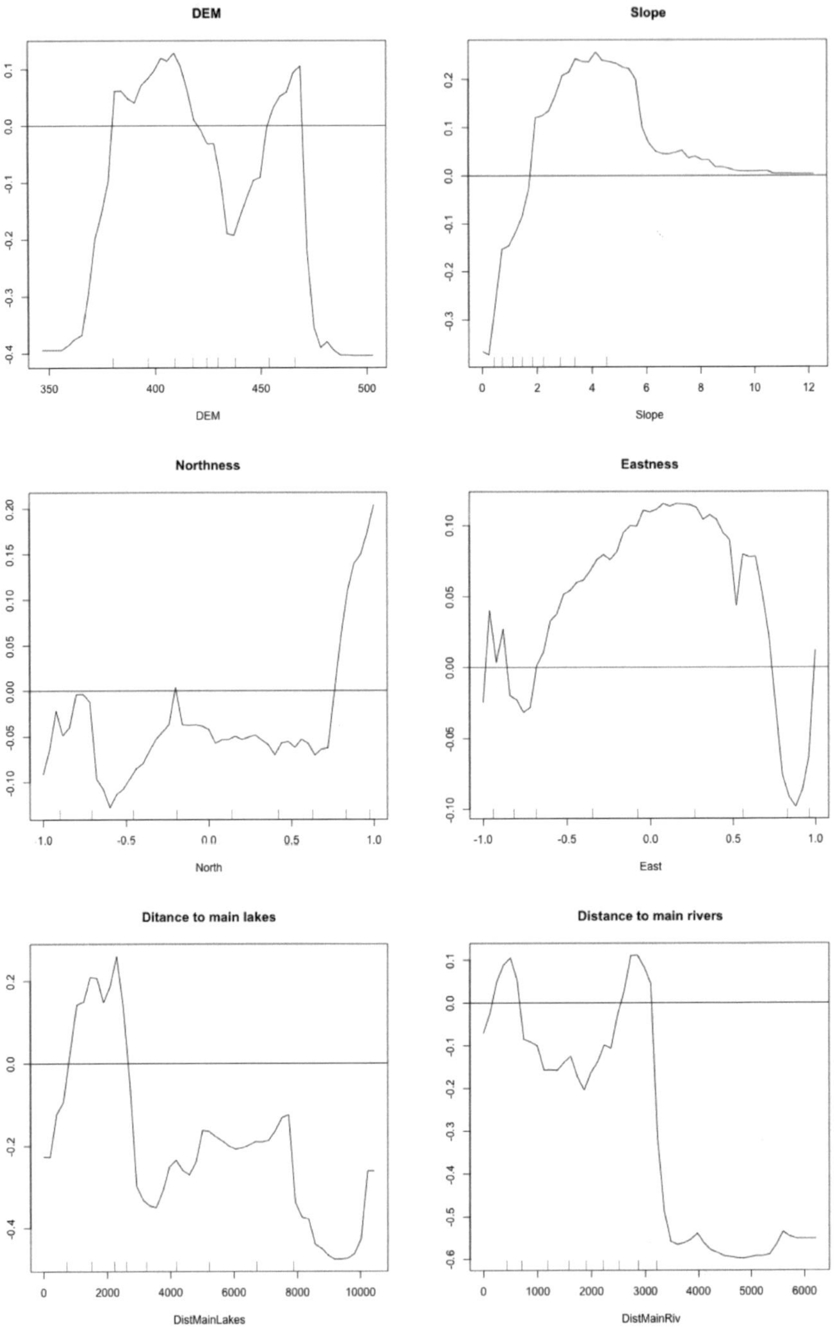

Fig. A.43 Partial dependence plots for RF classification model Geneva (part 1)

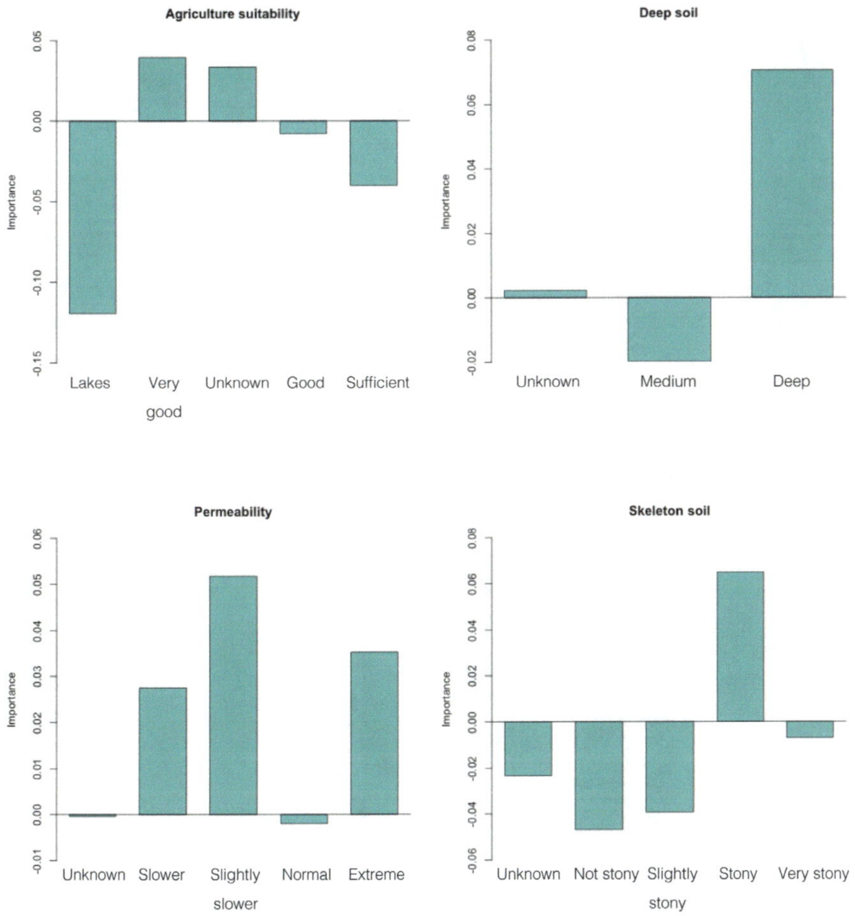

Fig. A.44 Partial dependence plots for RF classification model Geneva (part 2)

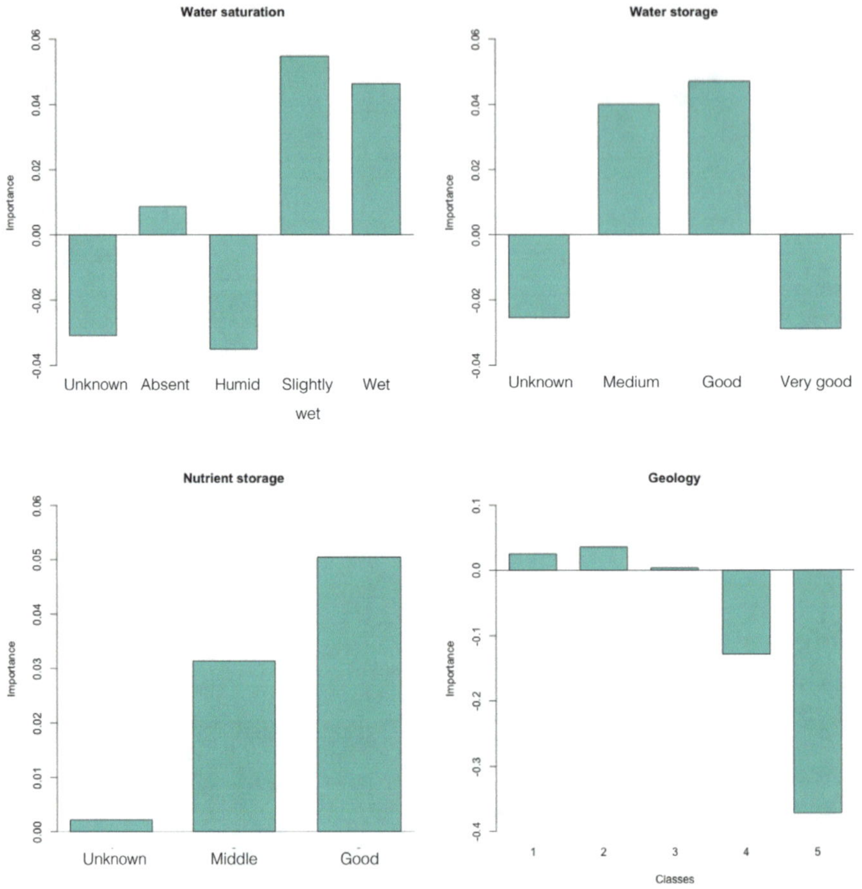

Fig. A.45 Partial dependence plots for RF classification model Geneva (part 3). For geology classes, see Table 5.6

A.3.5 Comparative Analysis of Predicted High Probability Areas—RF Classification

See Figs. A.46, A.47, A.48 and A.49.

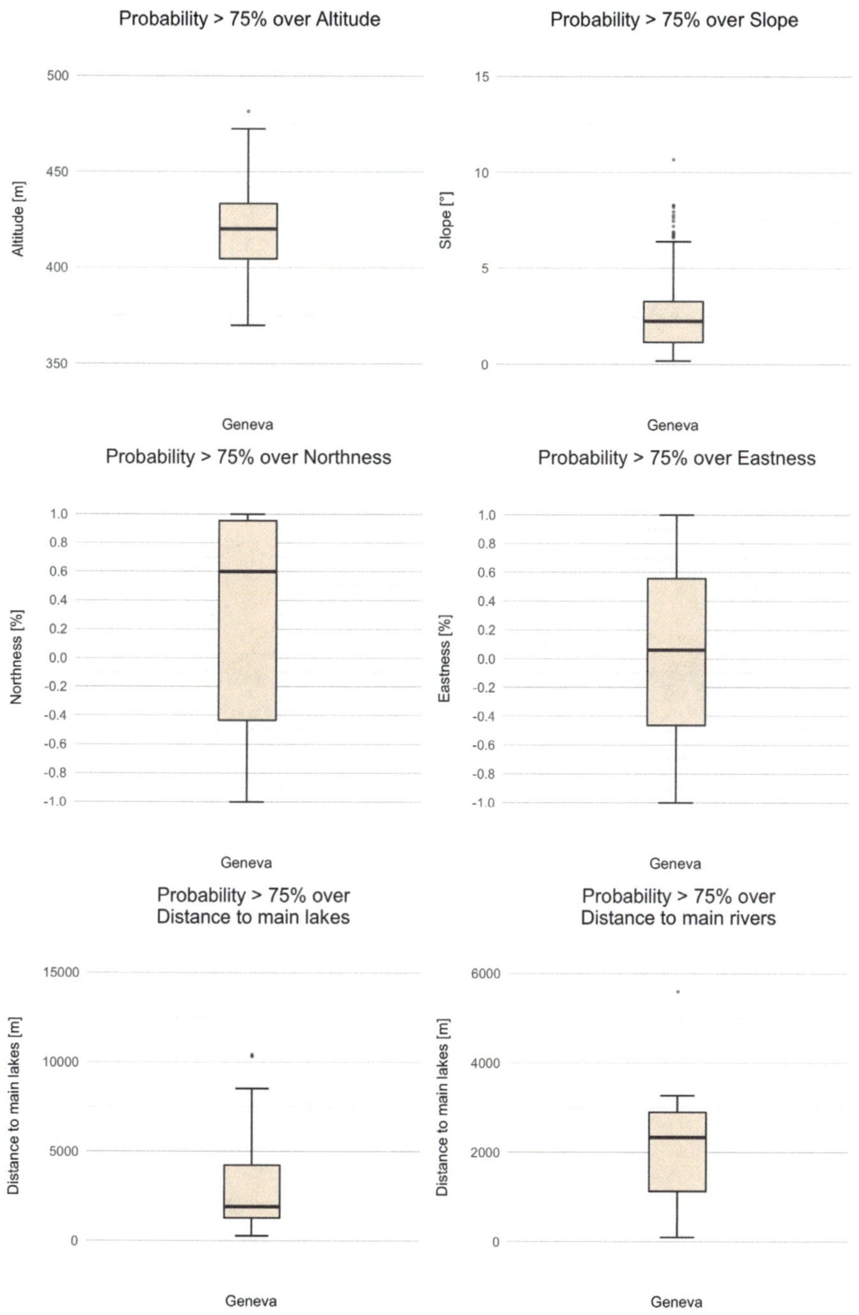

Fig. A.46 Comparative analysis of predicted high probability (>75%) areas using RF classification in Geneva (part 1). The boxplot indicate the median value and the interquartile range IQR (25th (Q1) to 75th (Q3) percentile (half of all occurrences lie within the box), the whiskers show the "minimum" (Q1 − 1.5 * IQR) and "maximum" (Q3 + 1.5 * IQR) values. Points show outliers

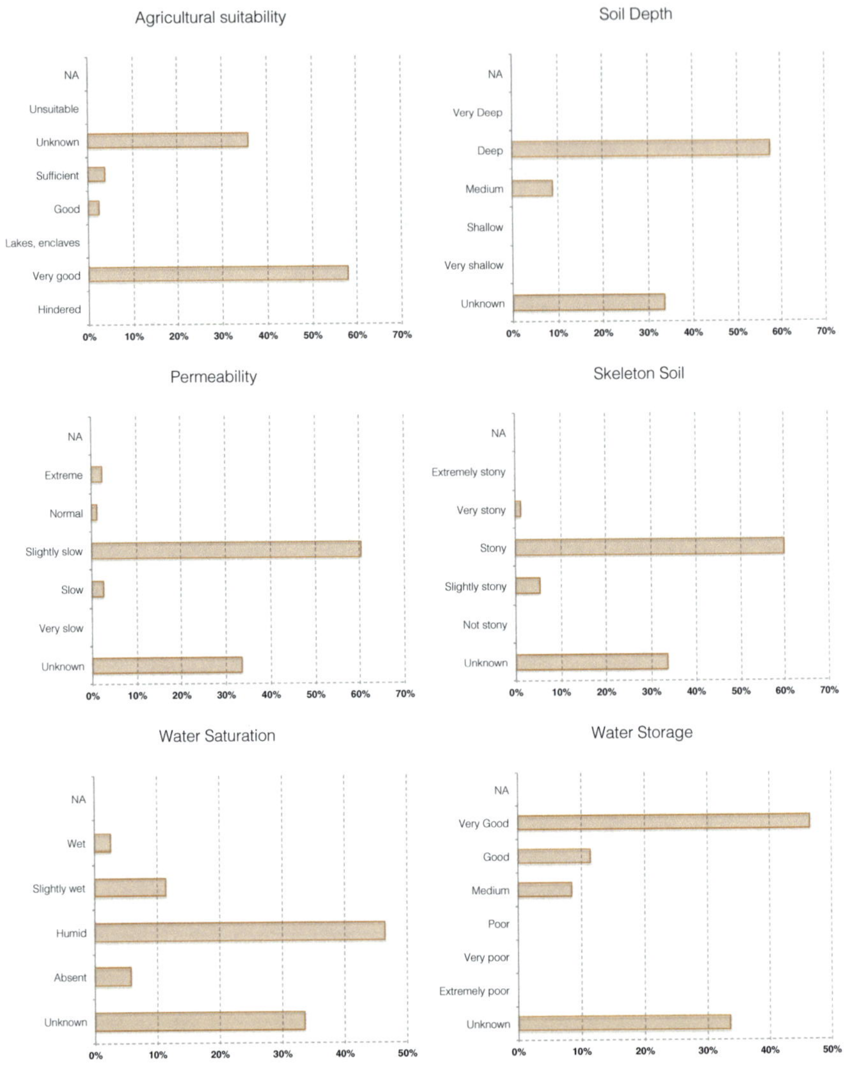

Fig. A.47 Comparative analysis of predicted high probability (>75%) areas using RF classification in Geneva (part 2)

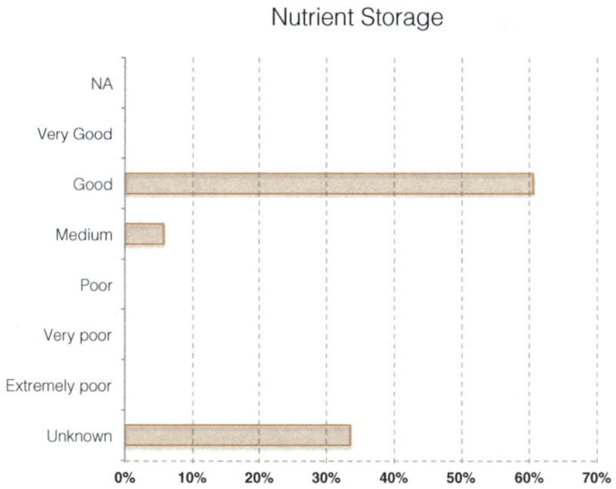

Fig. A.48 Comparative analysis of predicted high probability (>75%) areas using RF classification in Geneva (part 3)

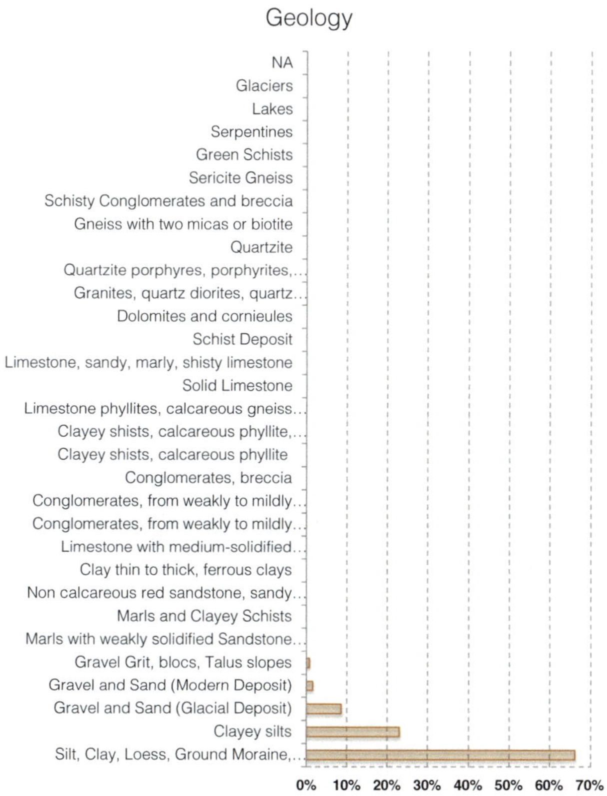

Fig. A.49 Comparative analysis of predicted high probability (>75%) areas on geology classes using RF classification in Geneva

A.3.6 RF Regression Model Results

See Fig. A.50.

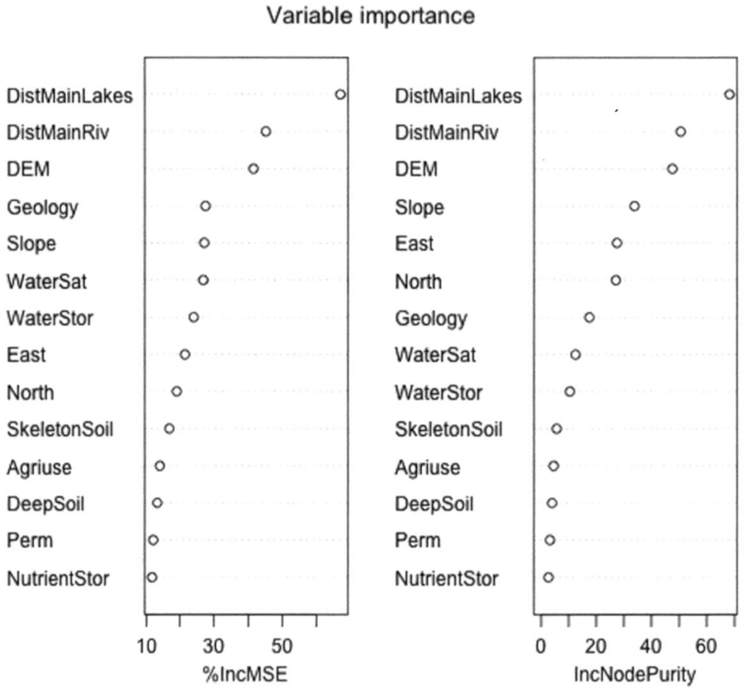

Fig. A.50 Variable importance for RF regression model Geneva

A.3.7 Comparative Analysis of Predicted High Probability Areas—RF Regression

See Figs. A.51, A.52, A.53 and A.54.

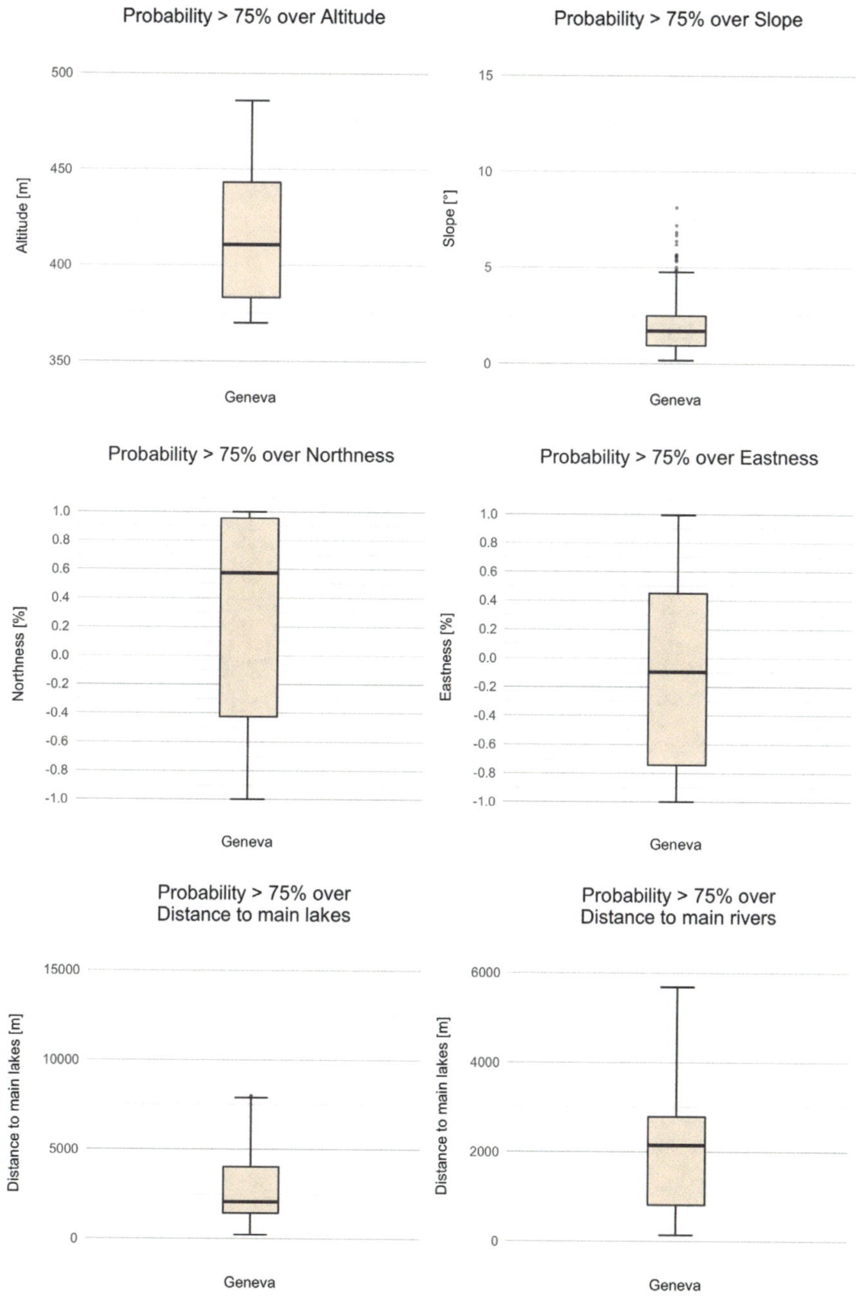

Fig. A.51 Comparative analysis of predicted high probability (>75% or >1.5) areas using RF regression in Geneva (part 1). The boxplot indicate the median value and the interquartile range IQR (25th (Q1) to 75th (Q3) percentile (half of all occurrences lie within the box), the whiskers show the "minimum" (Q1 − 1.5 * IQR) and "maximum" (Q3 + 1.5 * IQR) values. Points show outliers

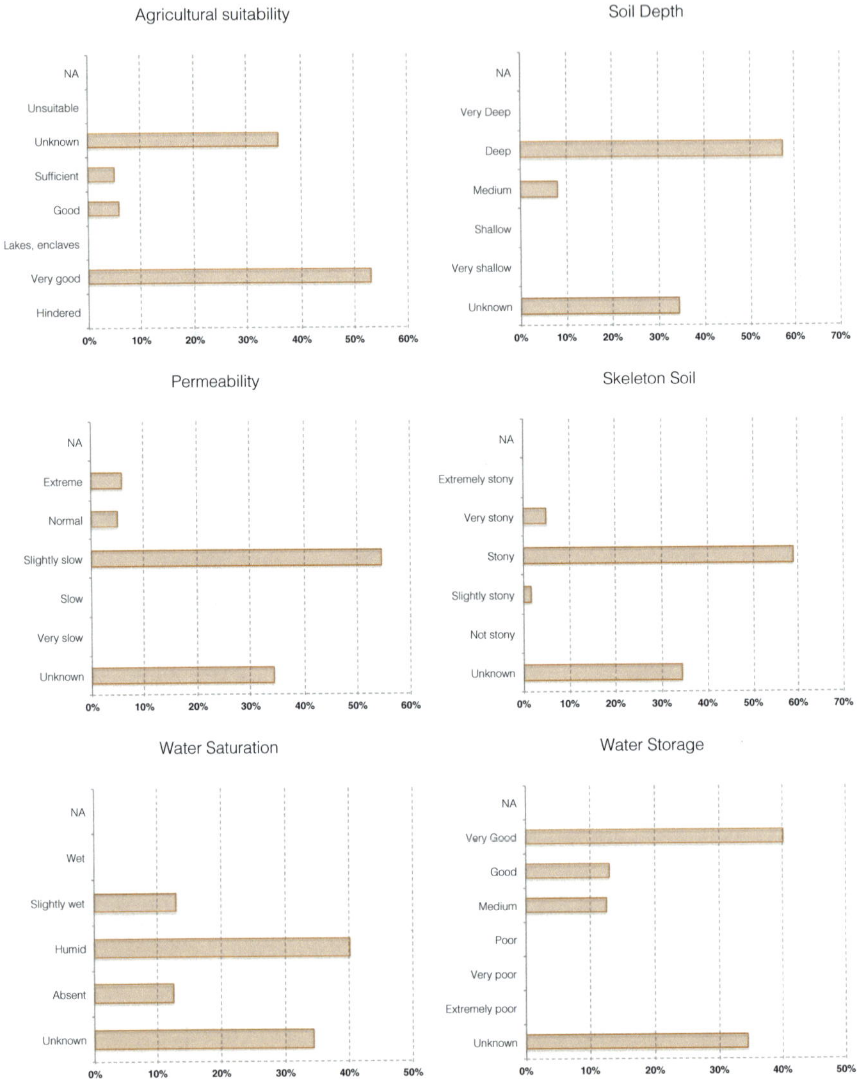

Fig. A.52 Comparative analysis of predicted high probability (>75% or >1.5) areas using RF regression in Geneva (part 2)

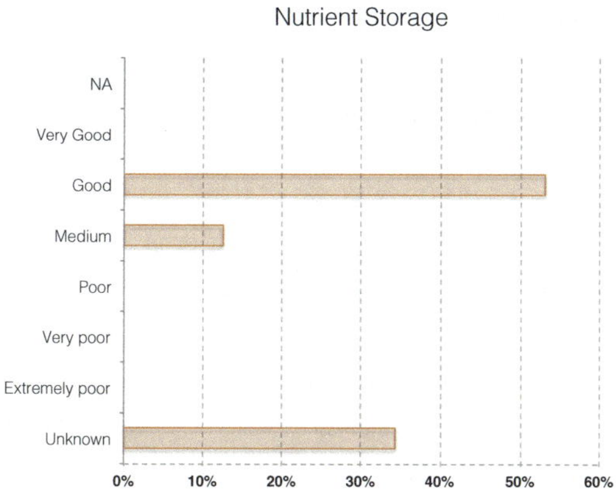

Fig. A.53 Comparative analysis of predicted high probability (>75% or >1.5) areas using RF regression in Geneva (part 3)

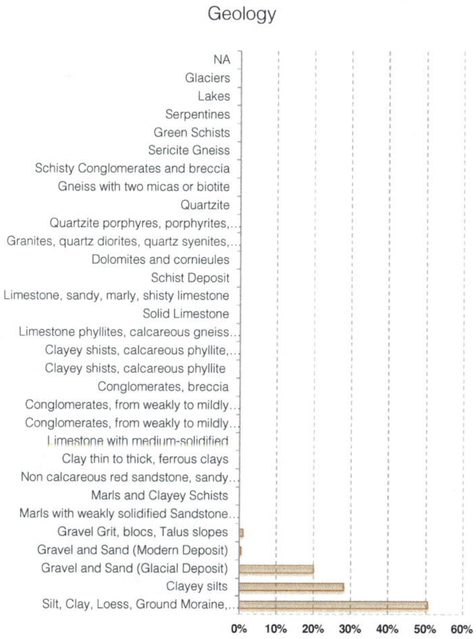

Fig. A.54 Comparative analysis of predicted high probability (>75% or >1.5) areas on geology classes using RF regression in Geneva

A.4 Canton of Grisons

A.4.1 Database Reclassification

See Table A.4.

Table A.4 Roman site typology classification for Grisons

Roman site typology database Graubünden		
Class	English classification	Original classification
Fortifications	Fortification, bunker	Wehranlage/Bunker
Graves	Grave, burial	Grab/Bestattungen
Others	Other	Anderes
Religious sites	Religious site, holy site	Kultplatz/Heiligtum
Roads	Traffic	Verkehr
Settlements	Settlement	Siedlung
Single finds	Single finds	Einzelfund
	Treasure deposit	Hort/Depot
Unknown	Blanks	Blanks
	Undetermined	Unbestimmt

A.4.2 Environmental Variables

See Figs. A.55, A.56 and A.57.

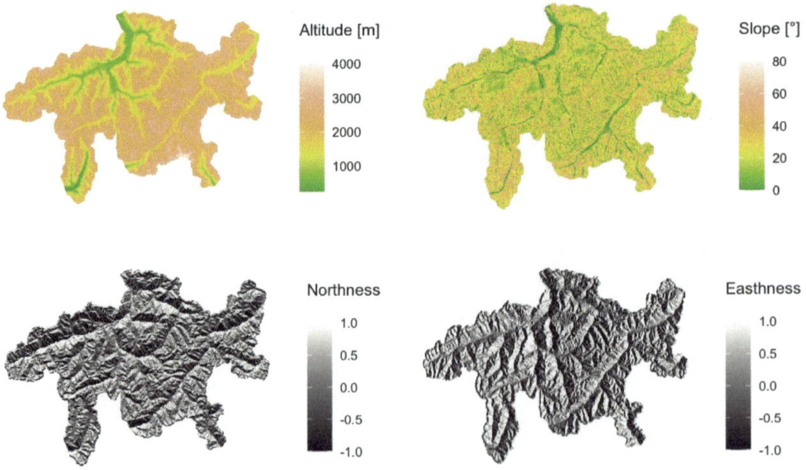

Fig. A.55 Environmental variables Grisons (part 1)

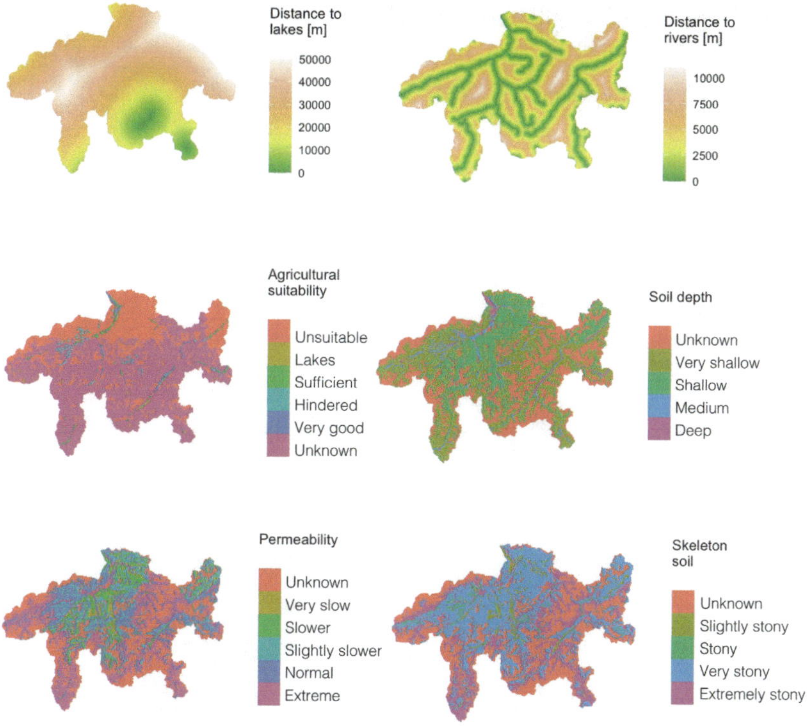

Fig. A.56 Environmental variable Grisons (part 2)

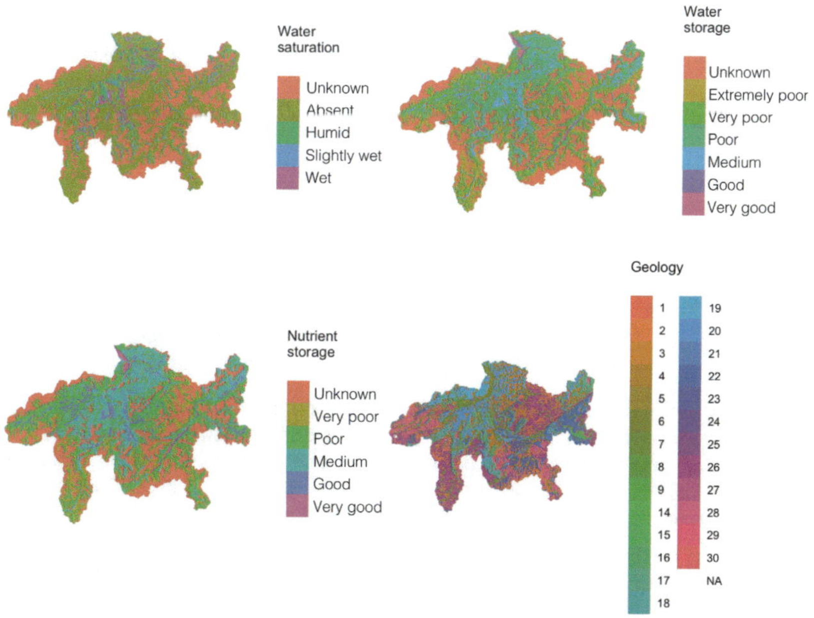

Fig. A.57 Environmental variables Grisons. For geology classes see Table 5.6 (part 3)

A.4.3 Locational Preference Analysis

See Figs. A.58, A.59, A.60 and A.61.

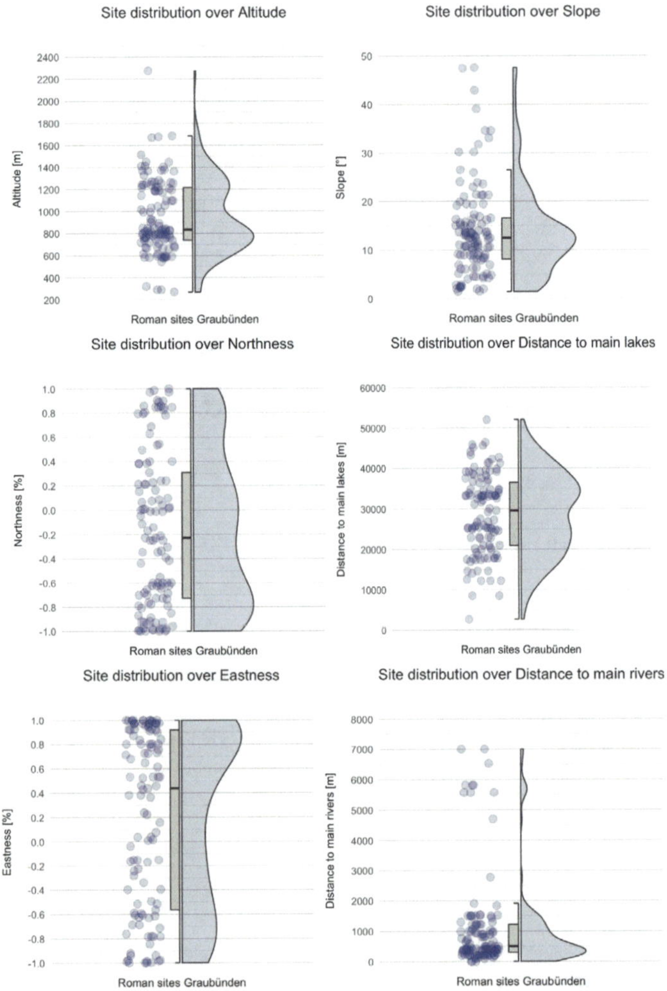

Fig. A.58 Locational preference analysis Grisons (all pre-medieval settlements). The point jitters show all occurrences. The boxplot indicate the median value and the interquartile range IQR (25th (Q1) to 75th (Q3) percentile (half of all occurrences lie within the gray box), the whiskers show the "minimum" (Q1 − 1.5 * IQR) and "maximum" (Q3 + 1.5 * IQR) values. The colored area indicates the concentration of occurrences (Kernel probability density)

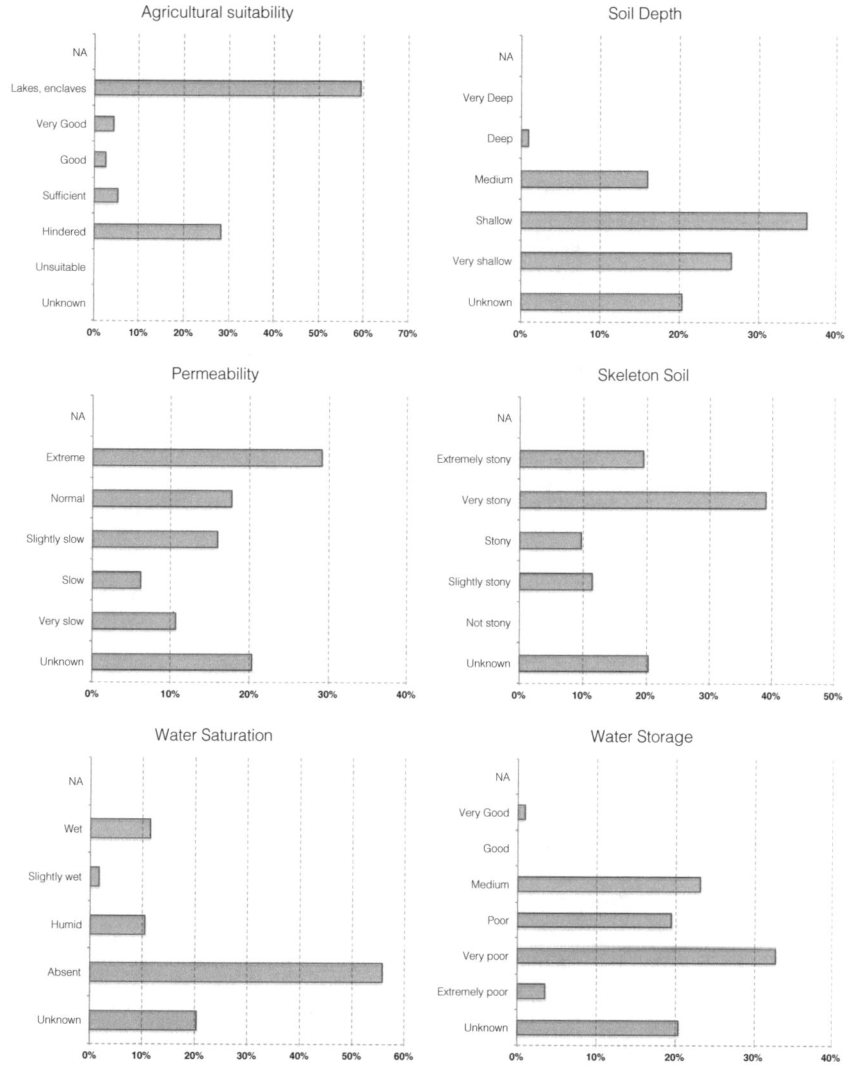

Fig. A.59 Locational preference analysis Grisons. Distribution of the Roman settlements over the agricultural suitability and the soil property classes (part 1)

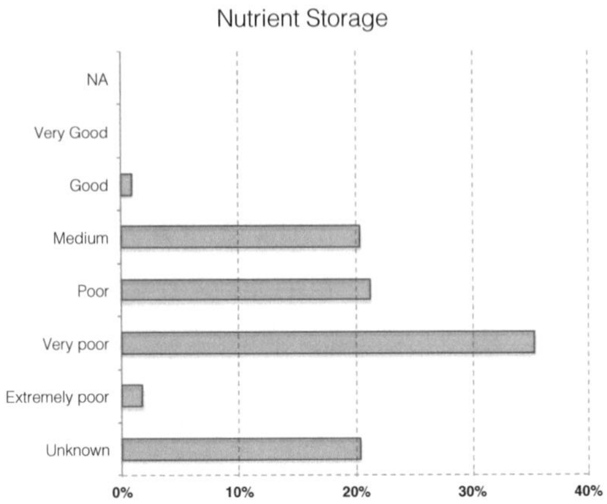

Fig. A.60 Locational preference analysis Grisons. Distribution of the Roman settlements over the agricultural suitability and the soil property classes (part 2)

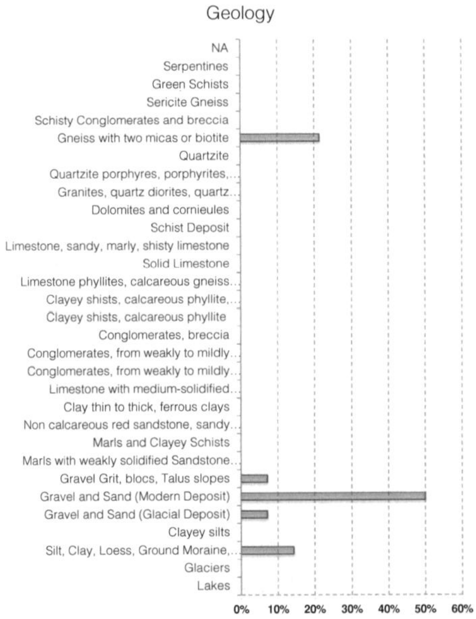

Fig. A.61 Locational preference analysis Grisons. Distribution of the Roman settlements over the geology classes

A.4.4 RF Classification Model Results

See Figs. A.62, A.63, A.64, A.65 and A.66.

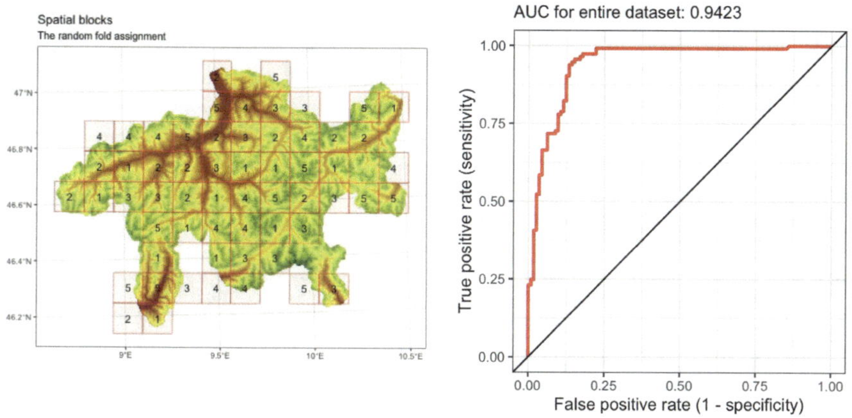

Fig. A.62 Left: Spatial blocks for k-folds cross validation Grisons. Number of folds: 5, size of blocks: 12000 m. Right: ROC curve and AUC (0.94) for Grisons RF classification model

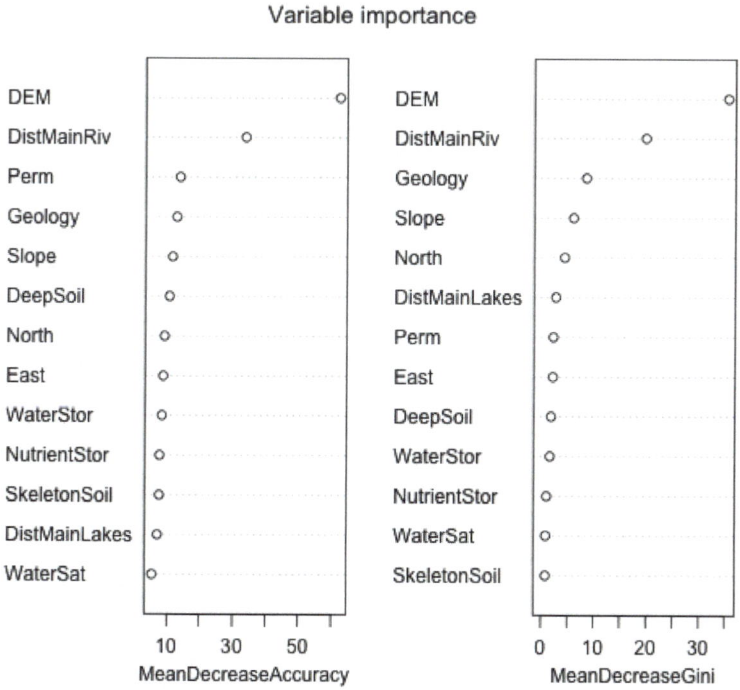

Fig. A.63 Variable importance for RF classification model Grisons

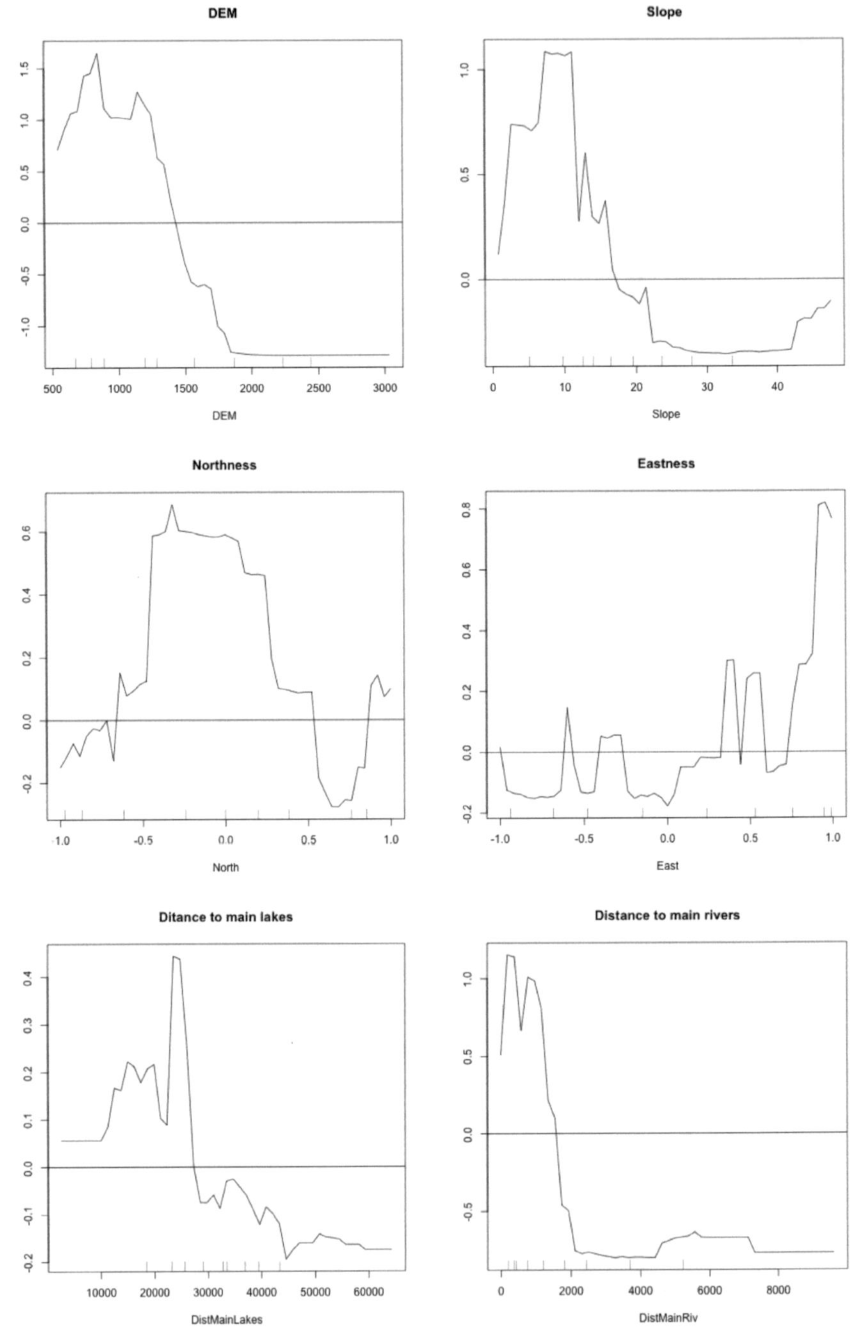

Fig. A.64 Partial dependence plots for RF classification model Grisons (part 1)

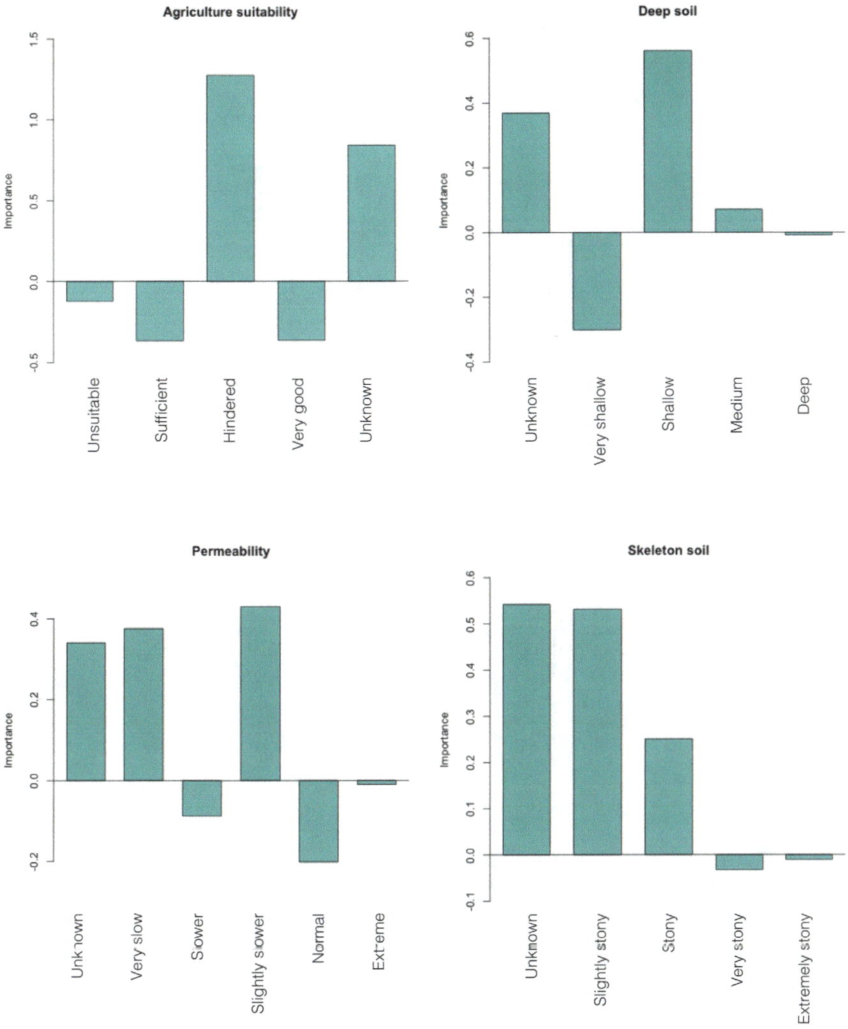

Fig. A.65 Partial dependence plots for RF classification model Grisons (part 2)

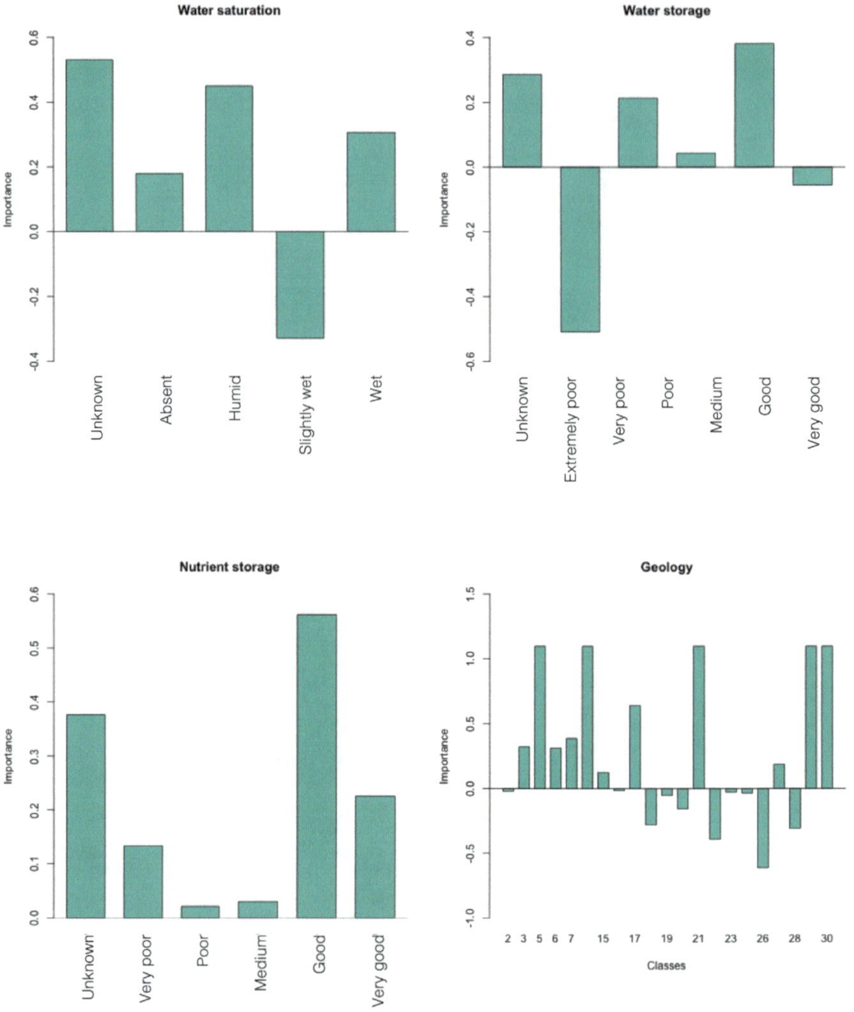

Fig. A.66 Partial dependence plots for RF classification model Grisons (part 3). For geology classes, see Table 5.6

A.4.5 Comparative Analysis of Predicted High Probability Areas—RF Classification

See Figs. A.67, A.68, A.69 and A.70.

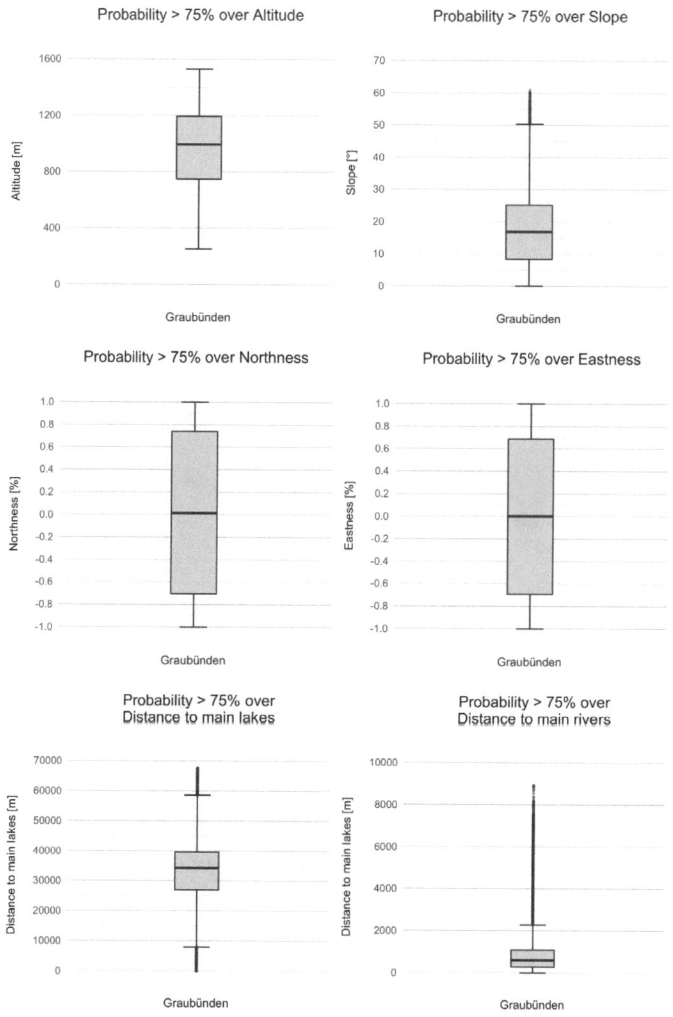

Fig. A.67 Comparative analysis of predicted high probability (>75%) areas using RF Classification in Grisons (part 1). The boxplot indicate the median value and the interquartile range IQR (25th (Q1) to 75th (Q3) percentile (half of all occurrences lie within the box), the whiskers show the "minimum" (Q1 − 1.5 * IQR) and "maximum" (Q3 + 1.5 * IQR) values. Points show outliers

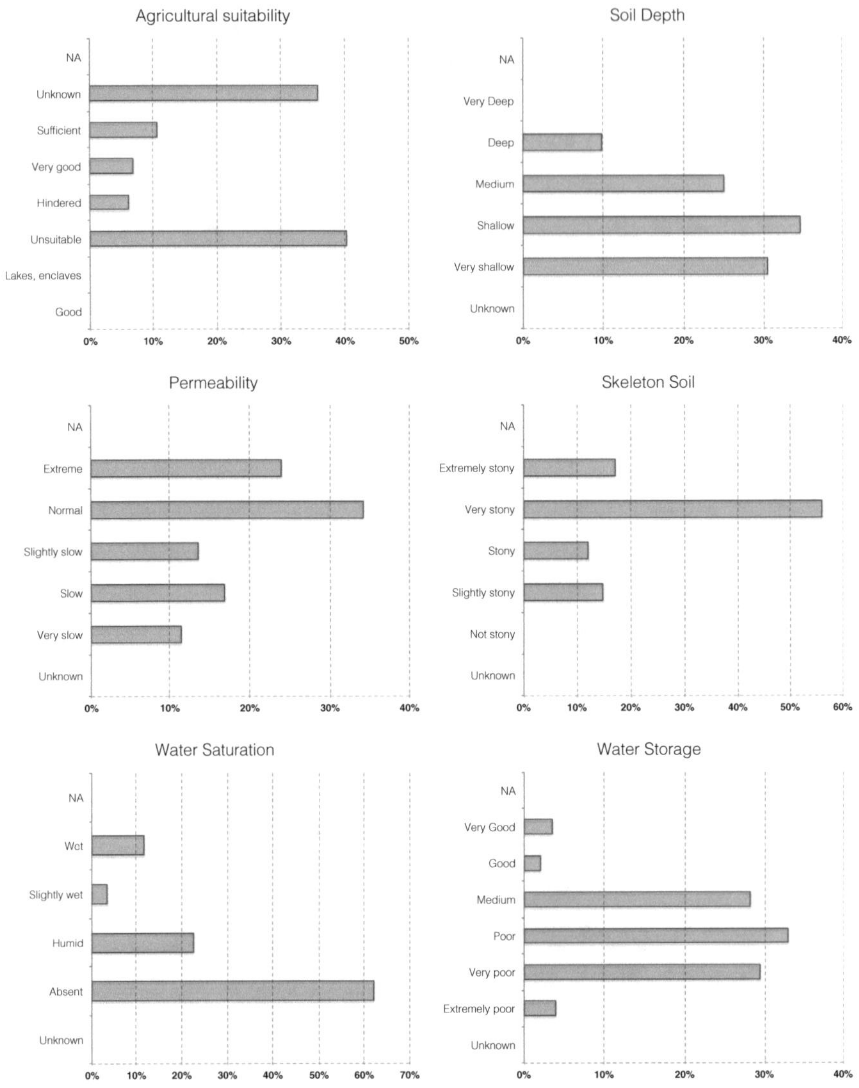

Fig. A.68 Comparative analysis of predicted high probability (>75%) areas using RF classification in Grisons (part 2)

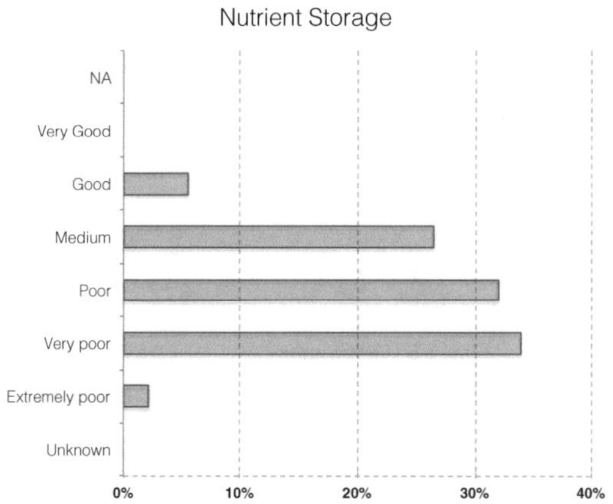

Fig. A.69 Comparative analysis of predicted high probability (>75%) areas using RF classification in Grisons (part 3)

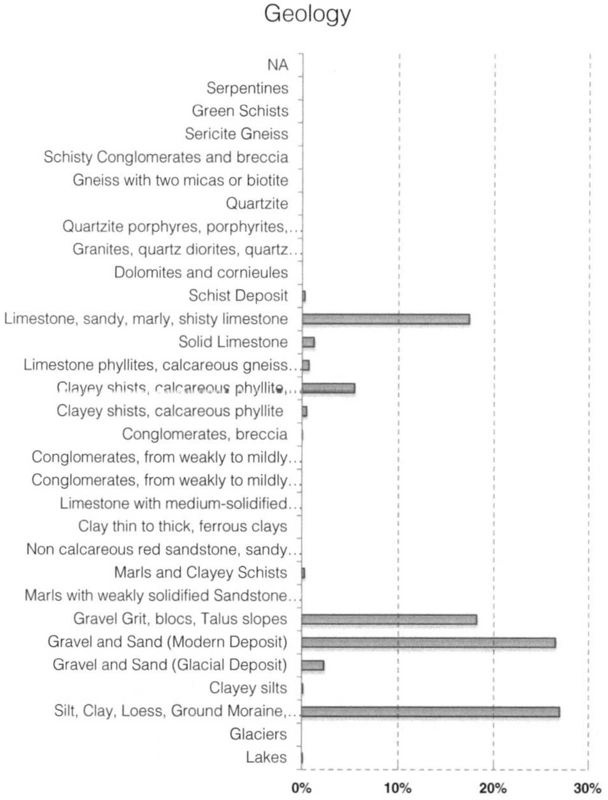

Fig. A.70 Comparative analysis of predicted high probability (>75%) areas on geology classes using RF classification in Grisons

A.5 Canton of Vaud

A.5.1 Database Reclassification

See Table A.5.

Table A.5 Roman site typology classification for Vaud

Roman site typology database Vaud		
Class	English classification	Original classification
Fortifications	Defensive	Défensive
Graves	Funerary	Funéraire
Others	Other	Autre
Religious sites	Religious	Religieuse
Roads	Communication	Communication
Settlements	Crafts	Artisanat
	Dwelling	Habitat
Single finds	Undetermined	Indéterminée
Water infrastructures	Hydraulic	Hydraulique

A.5.2 Environmental Variables

See Figs. A.71, A.72 and A.73.

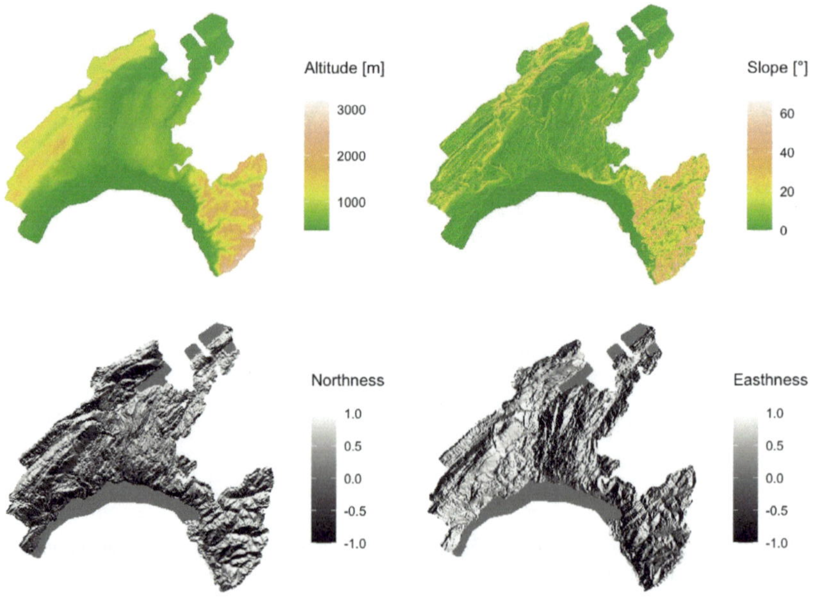

Fig. A.71 Environmental variables Vaud (part 1)

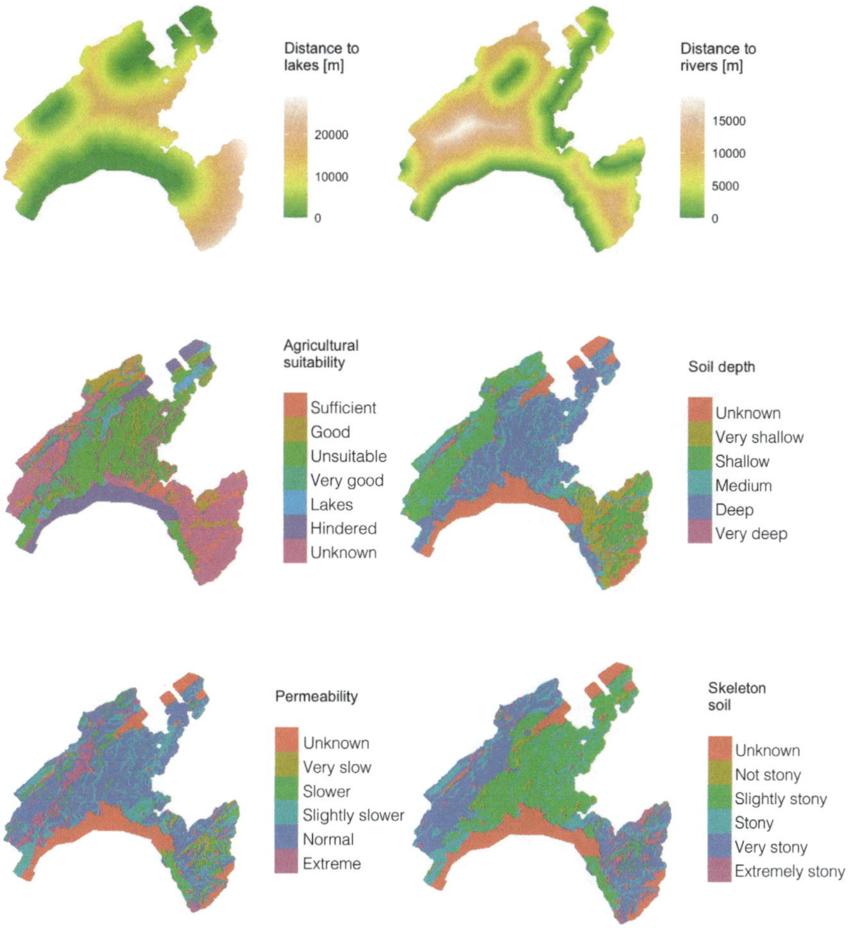

Fig. A.72 Environmental variable Vaud (part 2)

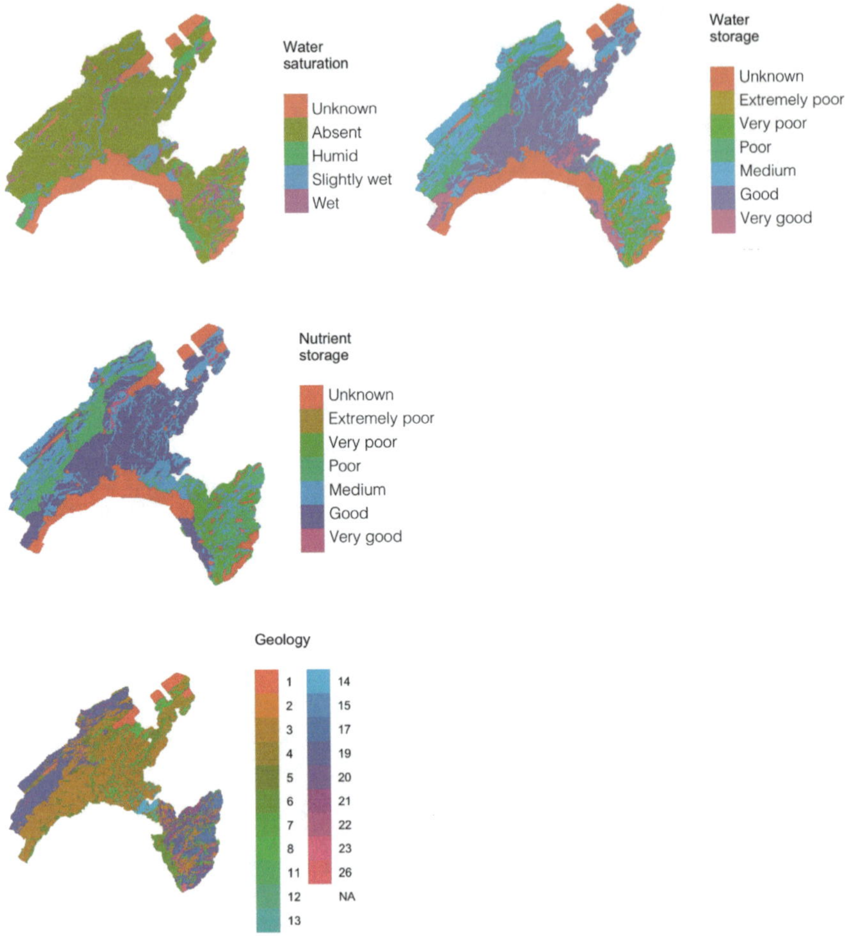

Fig. A.73 Environmental variables Vaud. For geology classes see Table 5.6 (part 3)

A.5.3 *Locational Preference Analysis*

See Figs. A.74, A.75, A.76 and A.77.

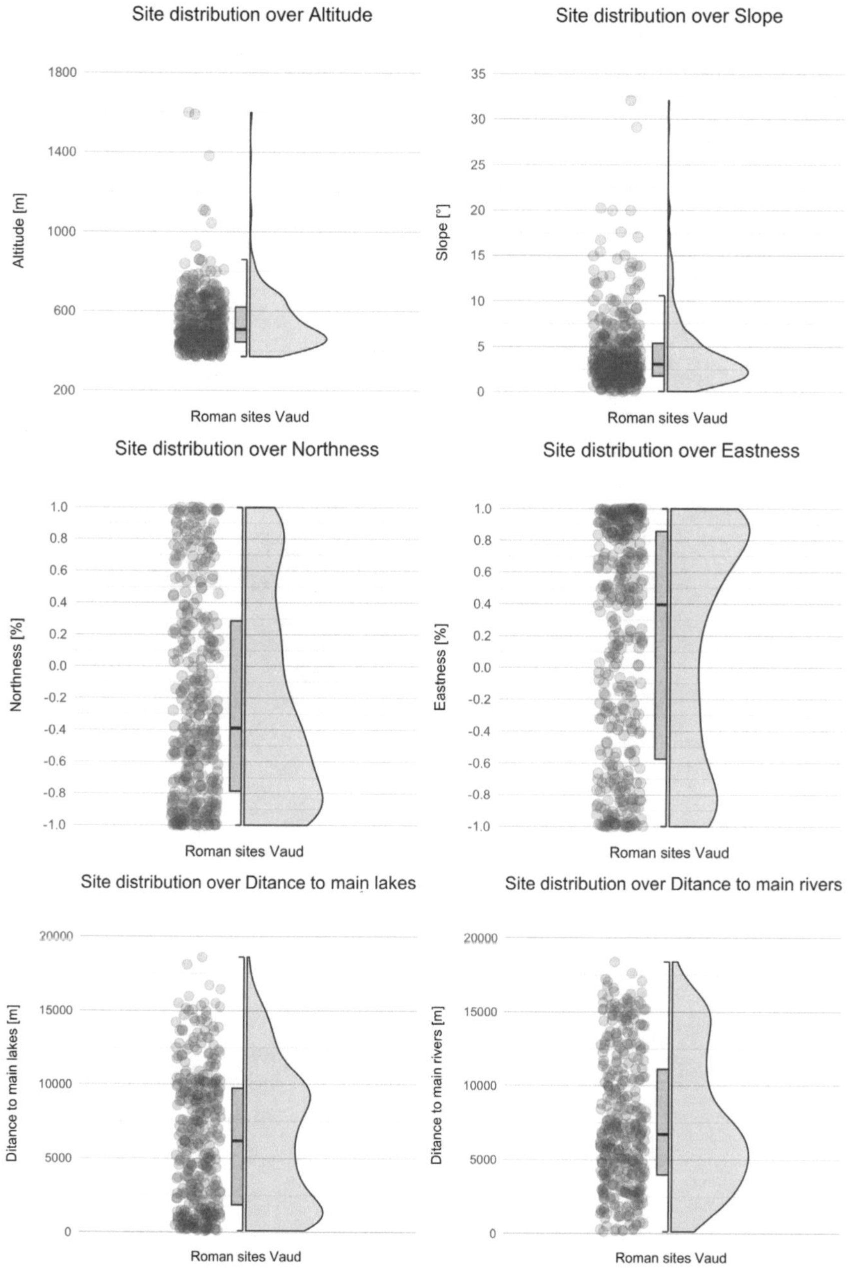

Fig. A.74 Locational preference analysis Vaud. The point jitters show all occurrences. The boxplot indicate the median value and the interquartile range IQR (25th (Q1) to 75th (Q3) percentile (half of all occurrences lie within the gray box), the whiskers show the "minimum" (Q1 − 1.5 * IQR) and "maximum" (Q3 + 1.5 * IQR) values. The colored area indicates the concentration of occurrences (Kernel probability density)

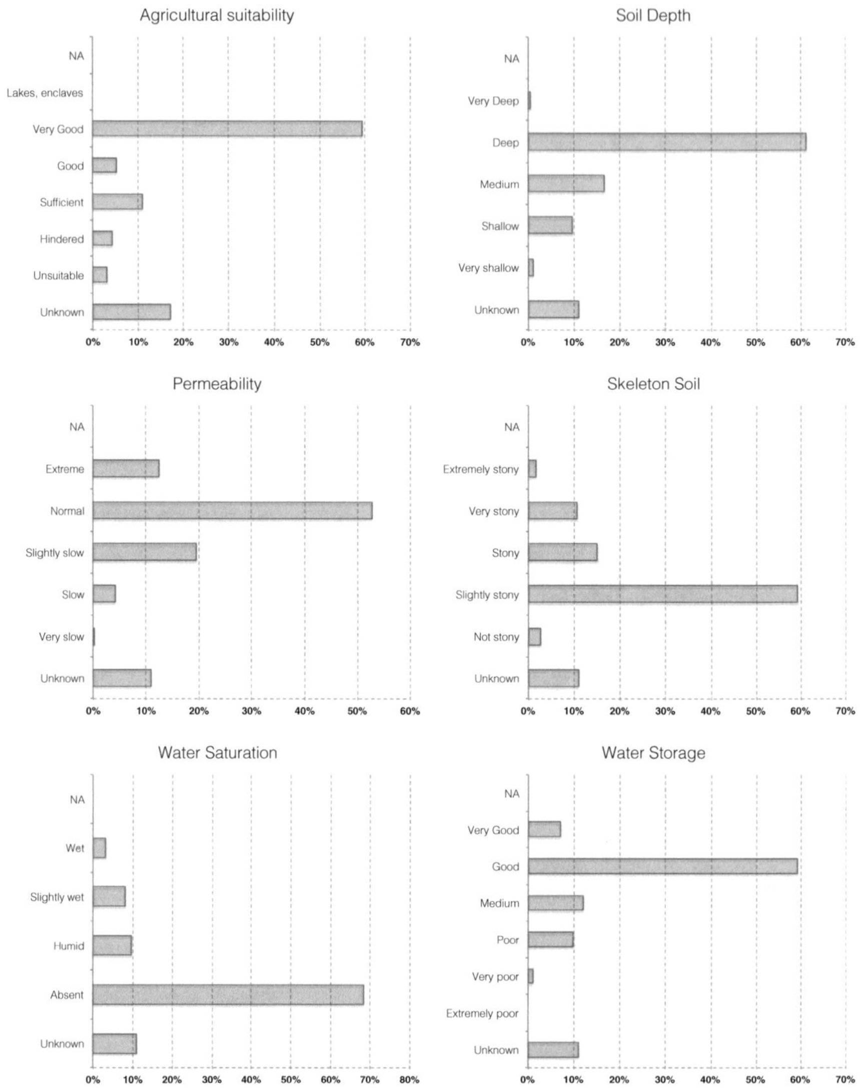

Fig. A.75 Locational preference analysis Vaud. Distribution of the Roman settlements over the agricultural suitability and the soil property classes (part 1)

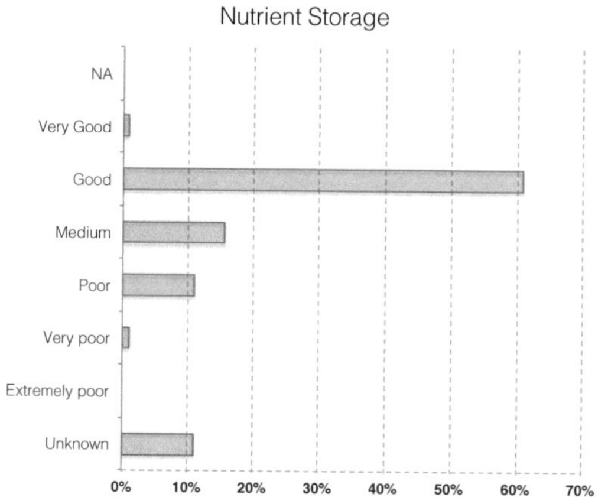

Fig. A.76 Locational preference analysis Vaud. Distribution of the Roman settlements over the agricultural suitability and the soil property classes (part 2)

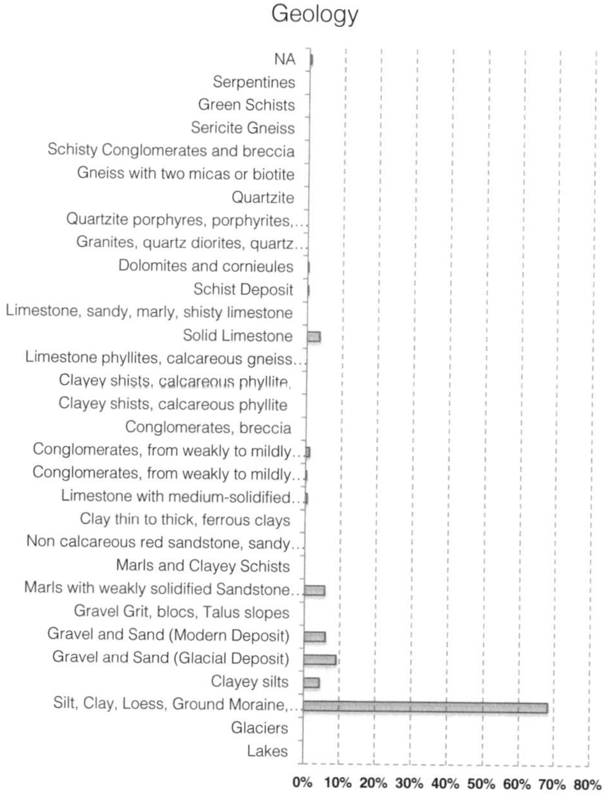

Fig. A.77 Locational preference analysis Vaud. Distribution of the Roman settlements over the geology classes

A.5.4 RF Classification Model Results

See Figs. A.78, A.79, A.80, A.81 and A.82.

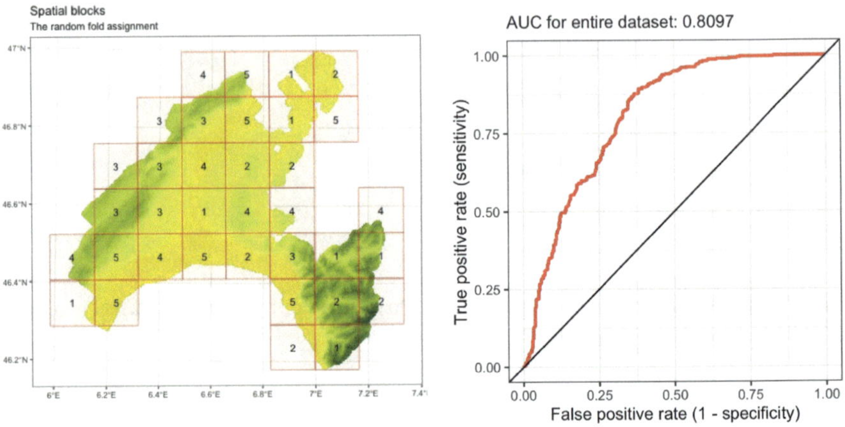

Fig. A.78 Left: Spatial blocks for *k-folds* cross validation Vaud. Number of folds: 5, size of blocks: 13000 m. Right: ROC curve and AUC (0.80) for Vaud RF classification model

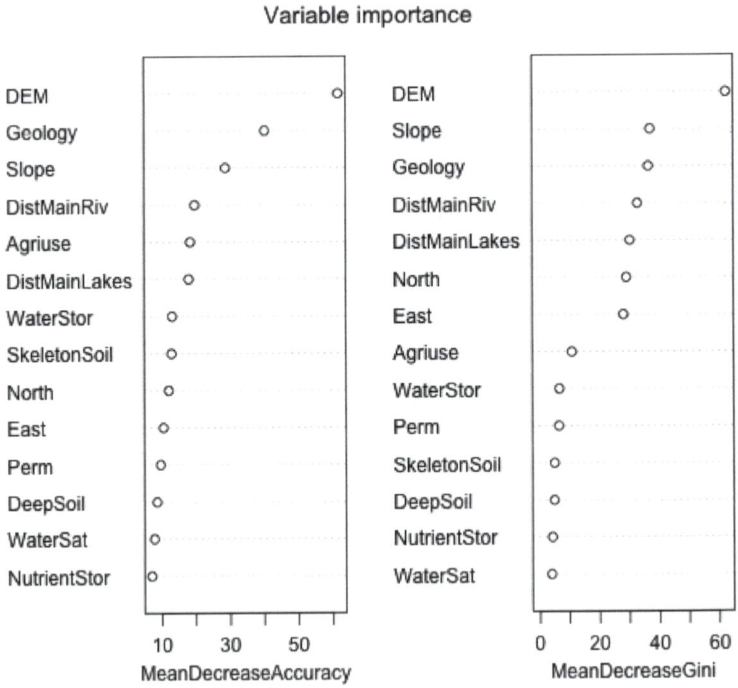

Fig. A.79 Variable importance for RF classification model Vaud

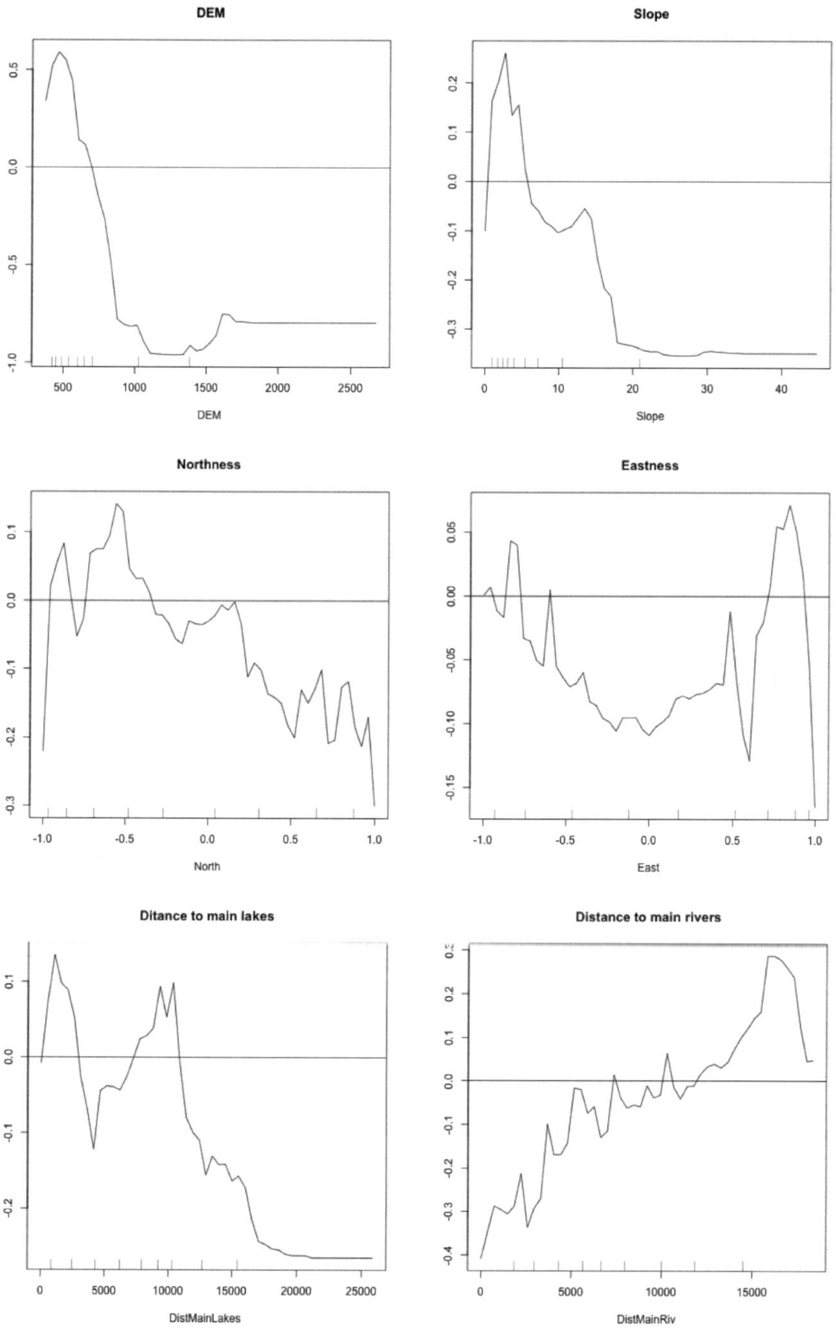

Fig. A.80 Partial dependence plots for RF classification model Vaud (part 1)

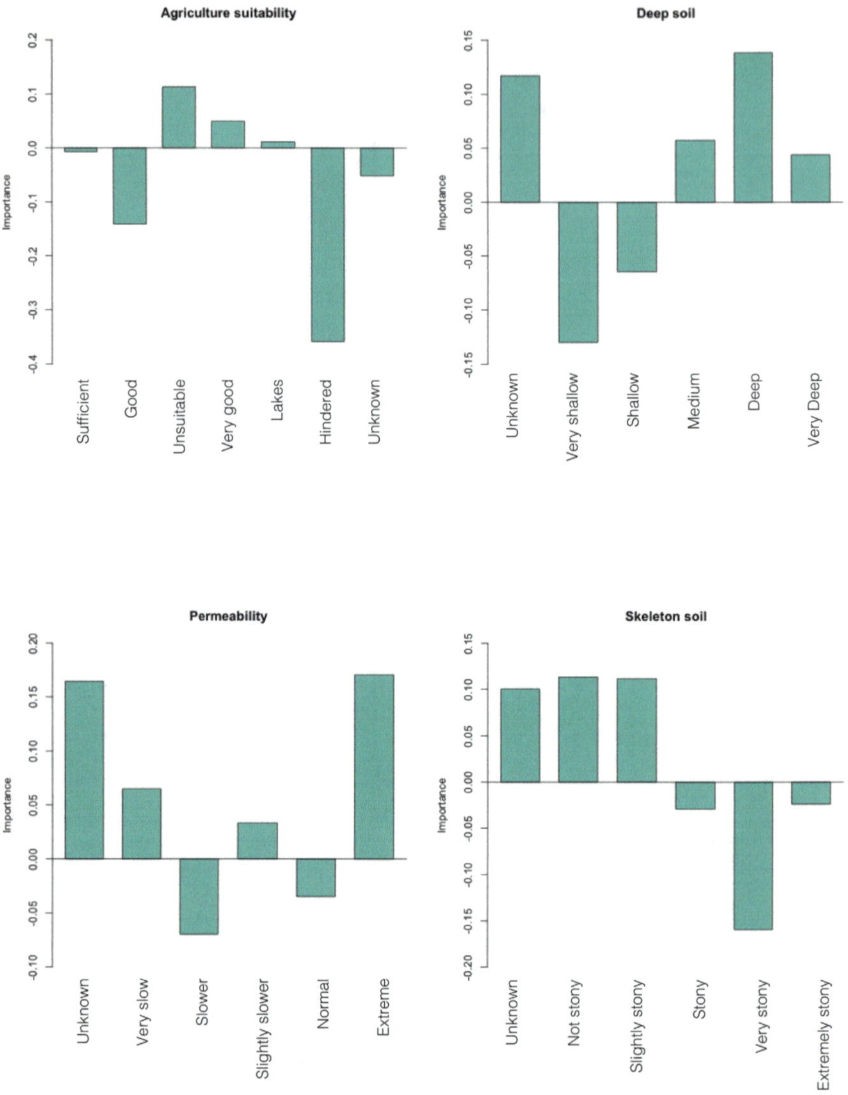

Fig. A.81 Partial dependence plots for RF classification model Vaud (part 2)

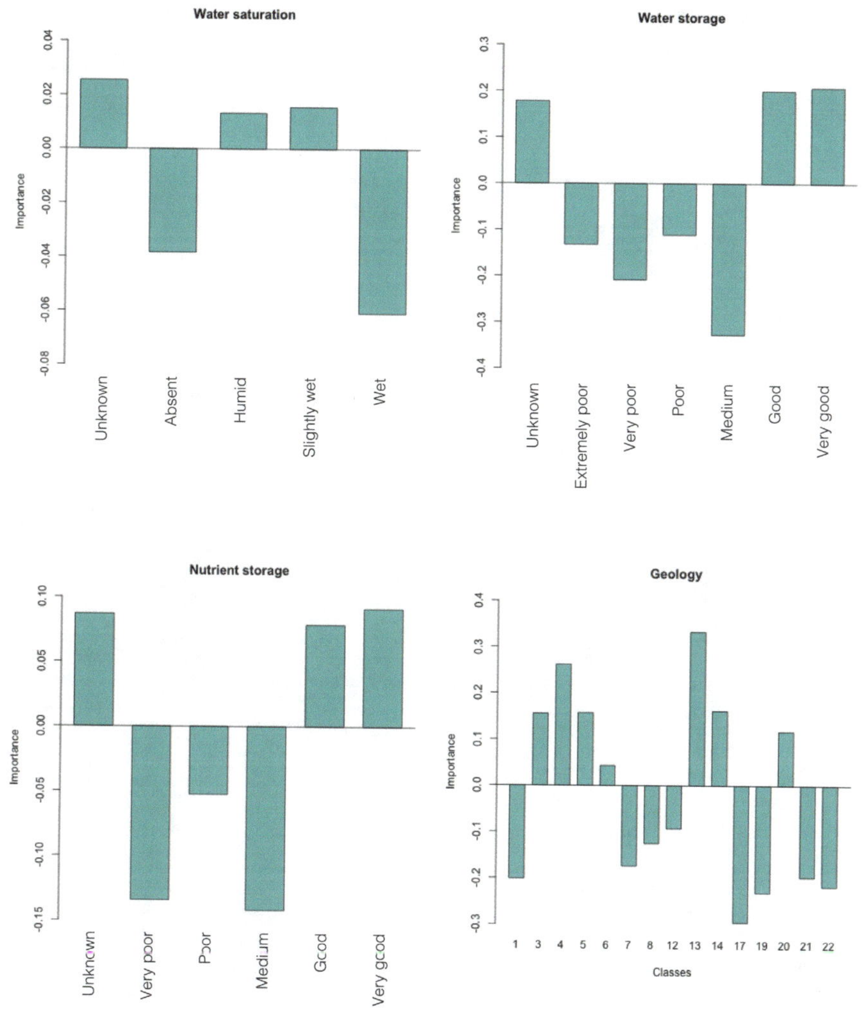

Fig. A.82 Partial dependence plots for RF classification model Vaud (part 3). For geology classes, see Table 5.6

A.5.5 Comparative Analysis of Predicted High Probability Areas—RF Classification

See Figs. A.83, A.84, A.85 and A.86.

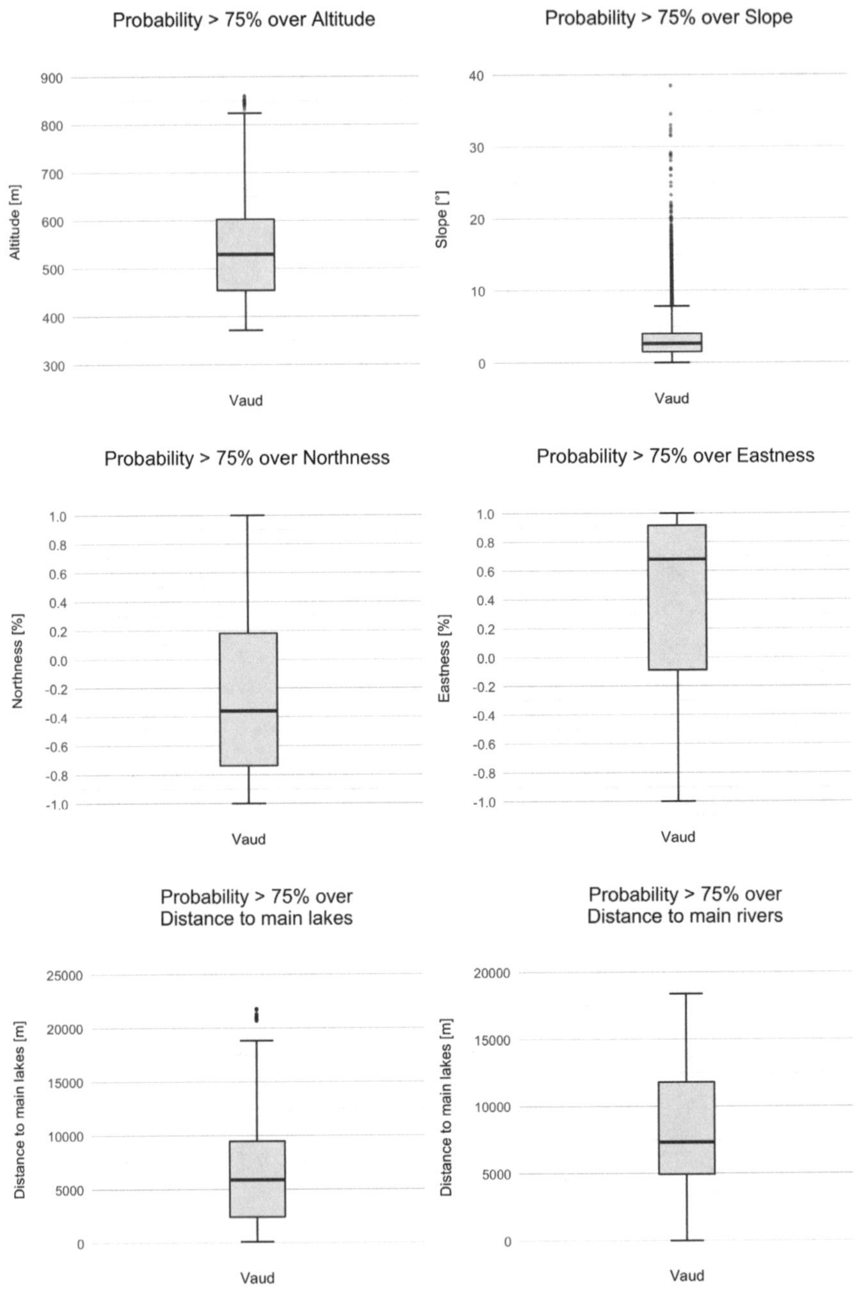

Fig. A.83 Comparative analysis of predicted high probability (>75%) areas using RF Classification in Vaud (part 1). The boxplot indicate the median value and the interquartile range IQR (25th (Q1) to 75th (Q3) percentile (half of all occurrences lie within the box), the whiskers show the "minimum" (Q1 − 1.5 * IQR) and "maximum" (Q3 + 1.5 * IQR) values. Points show outliers

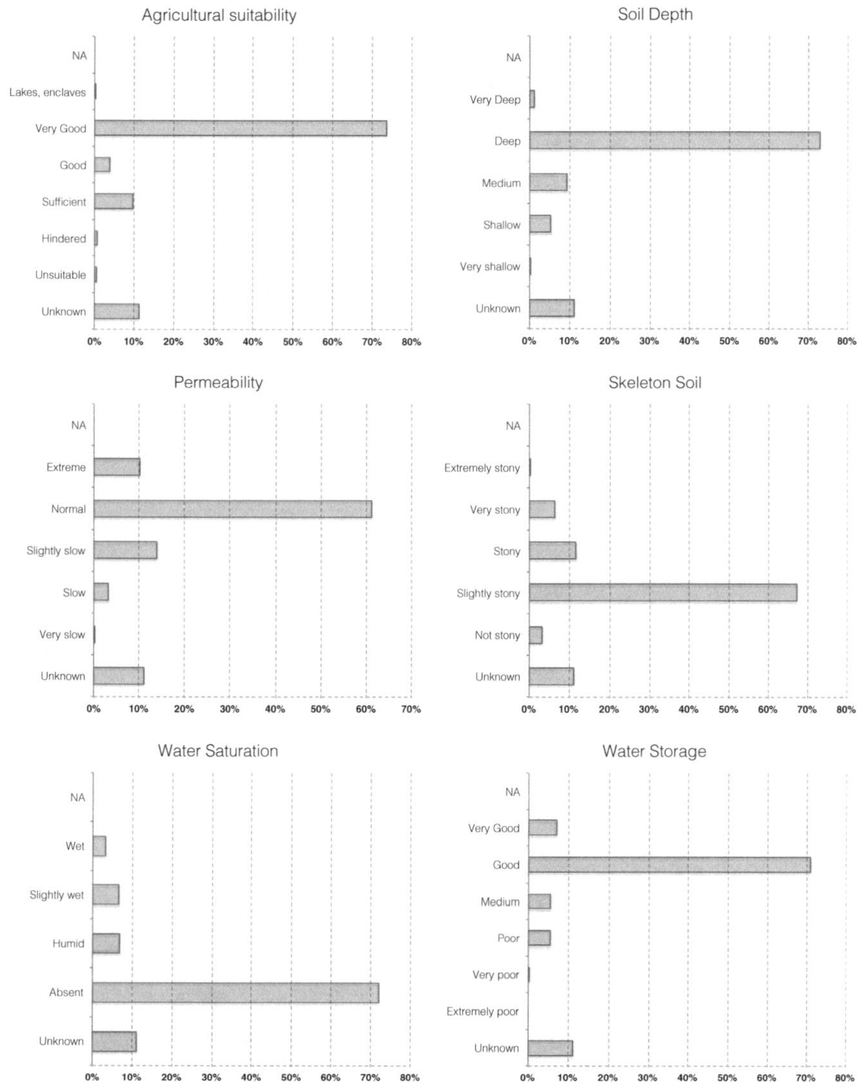

Fig. A.84 Comparative analysis of predicted high probability (>75%) areas using RF classification in Vaud (part 2)

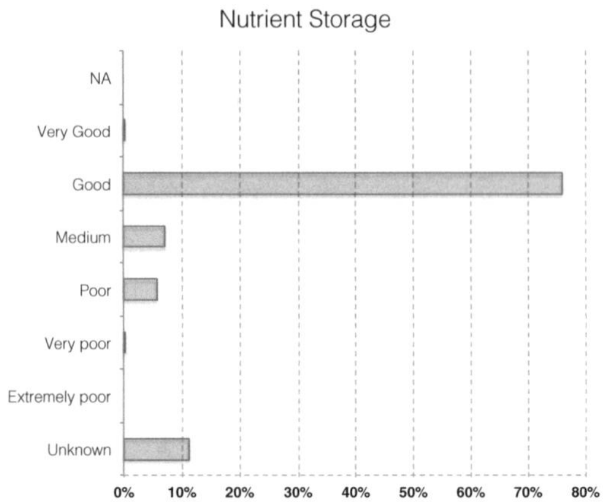

Fig. A.85 Comparative analysis of predicted high probability (>75%) areas using RF classification in Vaud (part 3)

Fig. A.86 Comparative analysis of predicted high probability (>75%) areas on geology classes using RF classification in Vaud

A.6 Canton of Zurich

A.6.1 Database Reclassification

See Table A.6.

Table A.6 Roman site typology classification for Zurich

Roman site typology database Zurich		
Class	English classification	Original classification
Fortifications	Fortification, bunker	Wehranlage/Bunker
Graves	Graves, burials	Grab/Bestattungen
Religious sites	Religious sites, holy sites	Kultplatz/Heiligtum
Roads	Roads	Strasse
	Traffic (roads)	Verkehr
Settlements	Industry, crafts	Industrie/Gewerbe/Handwerk
	Settlements	Siedlung
Single finds	Treasure, deposit (coins)	Hort/Depot
	Single finds	Einzelfund
Unknown	Undetermined	Unbestimmt
Water infrastructures	Logistics (water pipes, reservoirs)	Versorgung
	Other (water pipe)	Anderes

A.6.2 Environmental Variables

See Figs. A.87, A.88 and A.89.

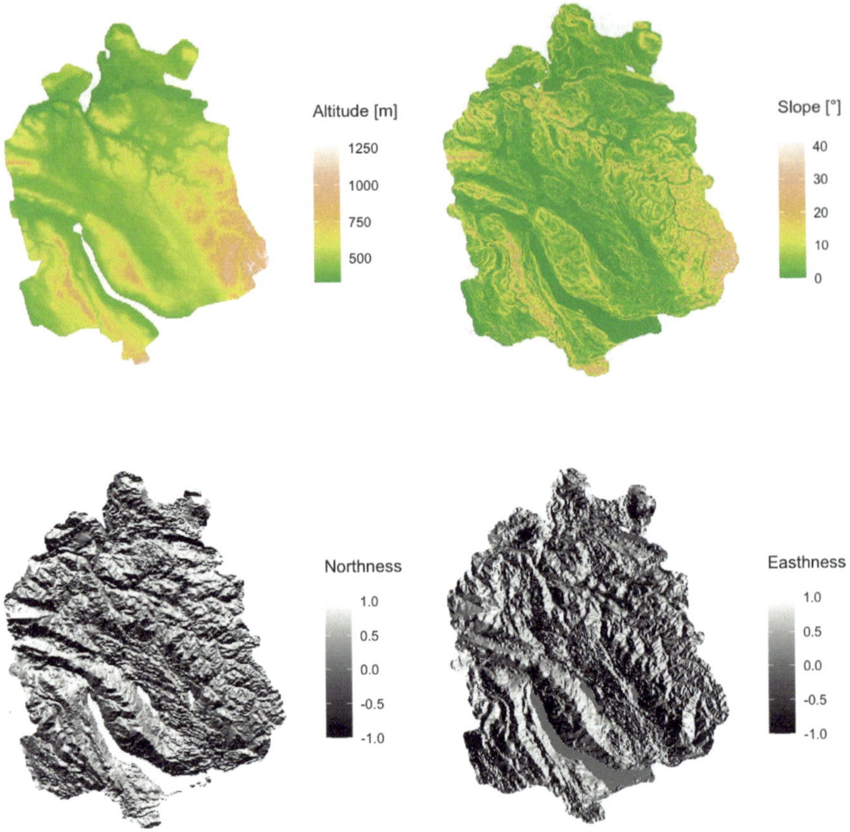

Fig. A.87 Environmental variables Zurich (part 1)

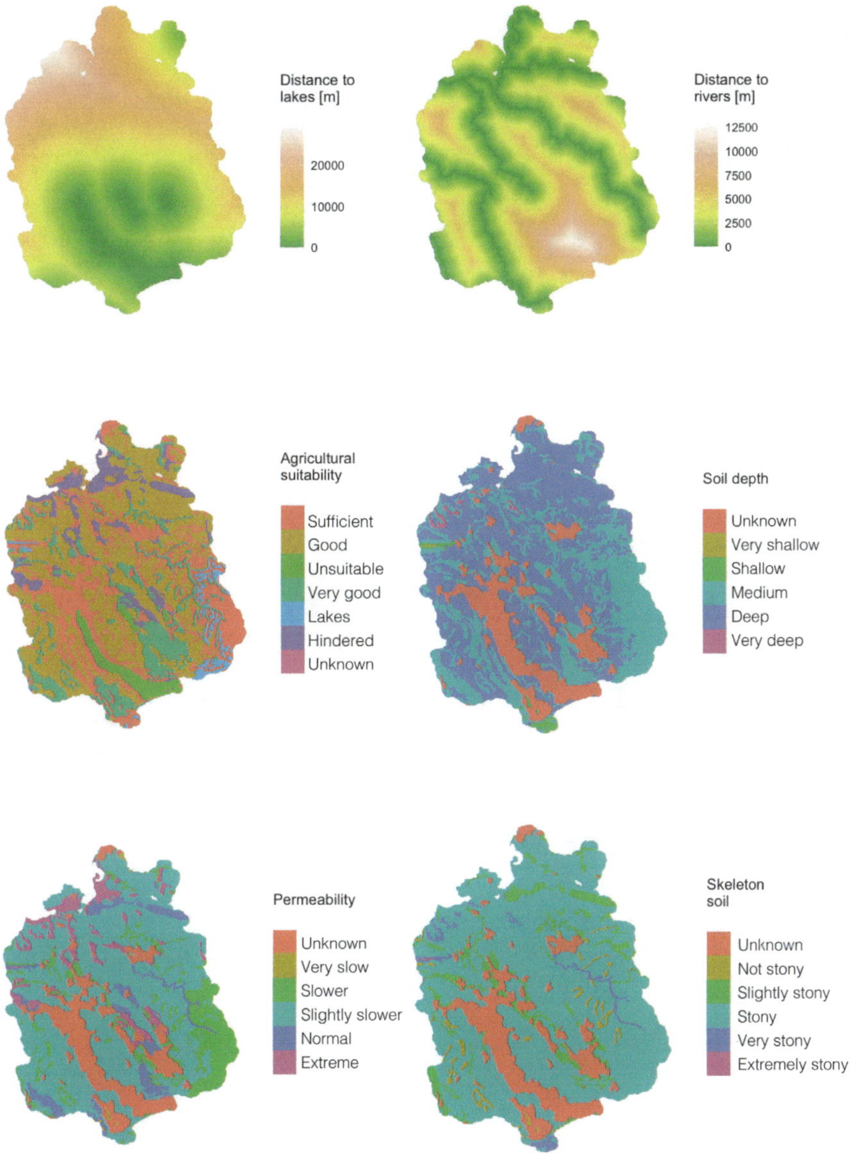

Fig. A.88 Environmental variable Zurich (part 2)

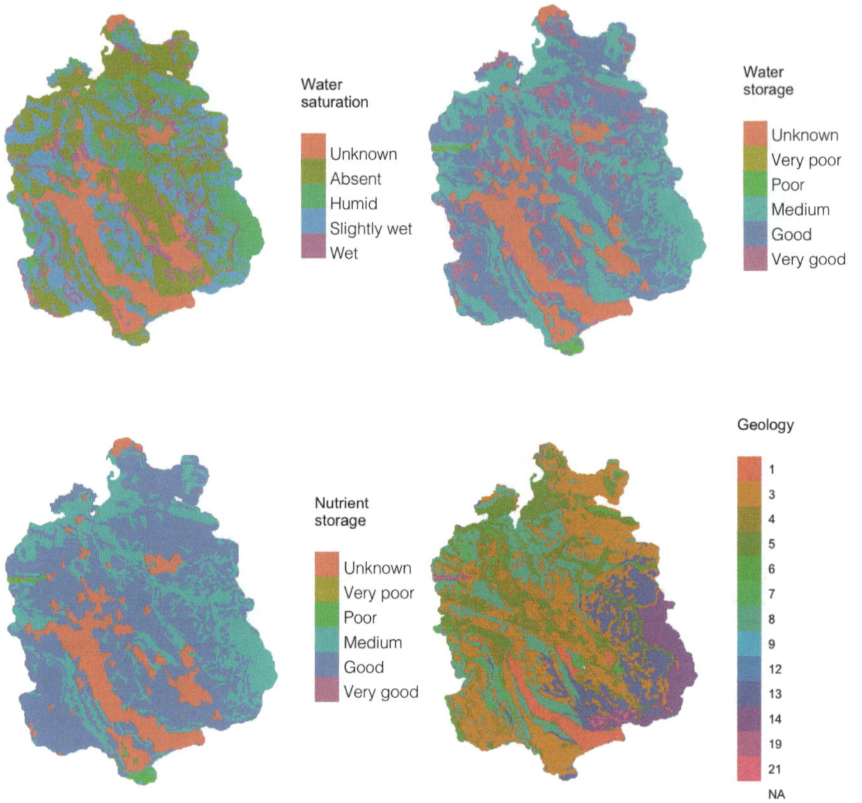

Fig. A.89 Environmental variables Zurich. For geology classes see Table 5.6 (part 3)

A.6.3 Locational Preference Analysis

See Figs. A.90, A.91, A.92 and A.93.

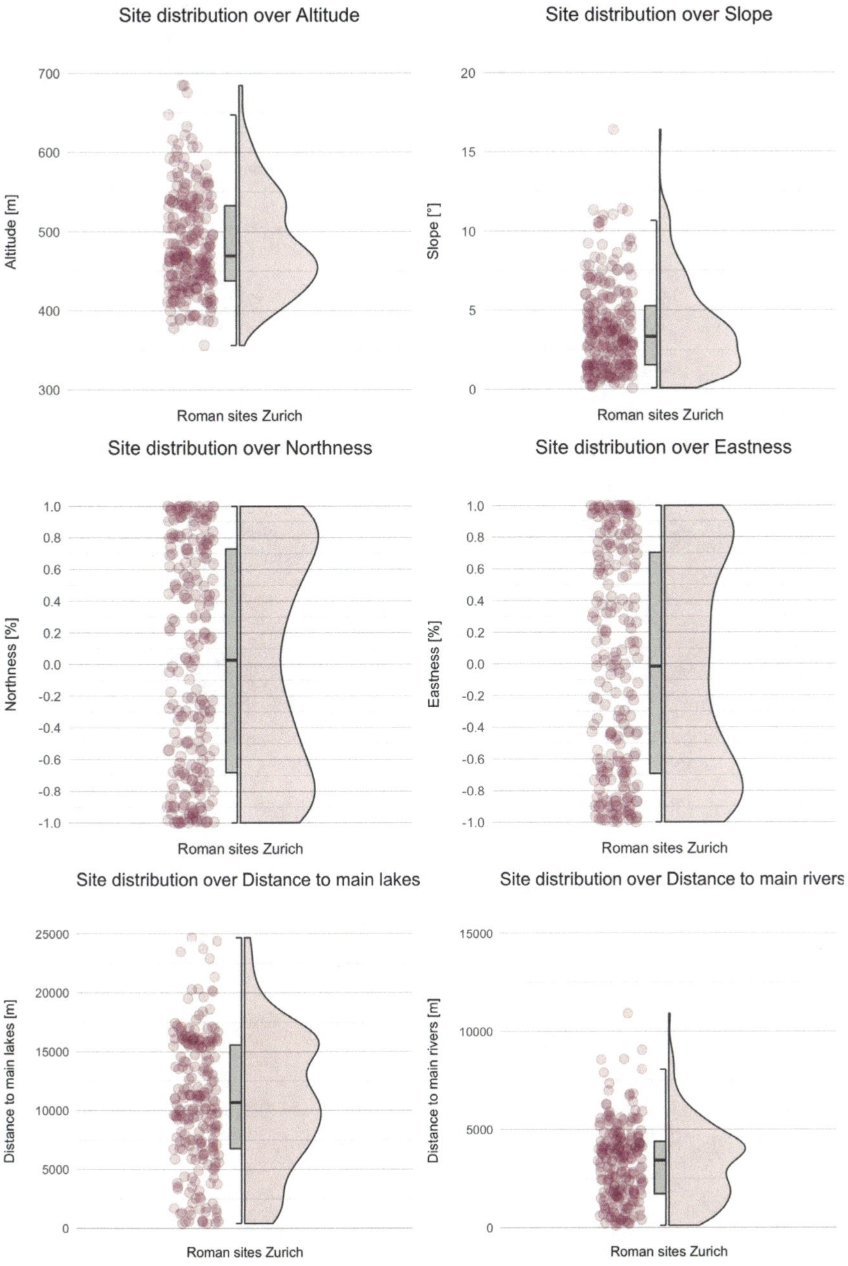

Fig. A.90 Locational preference analysis Zurich. The point jitters show all occurrences. The boxplot indicate the median value and the interquartile range IQR (25th (Q1) to 75th (Q3) percentile (half of all occurrences lie within the gray box), the whiskers show the "minimum" (Q1 − 1.5 * IQR) and "maximum" (Q3 + 1.5 * IQR) values. The colored area indicates the concentration of occurrences (Kernel probability density)

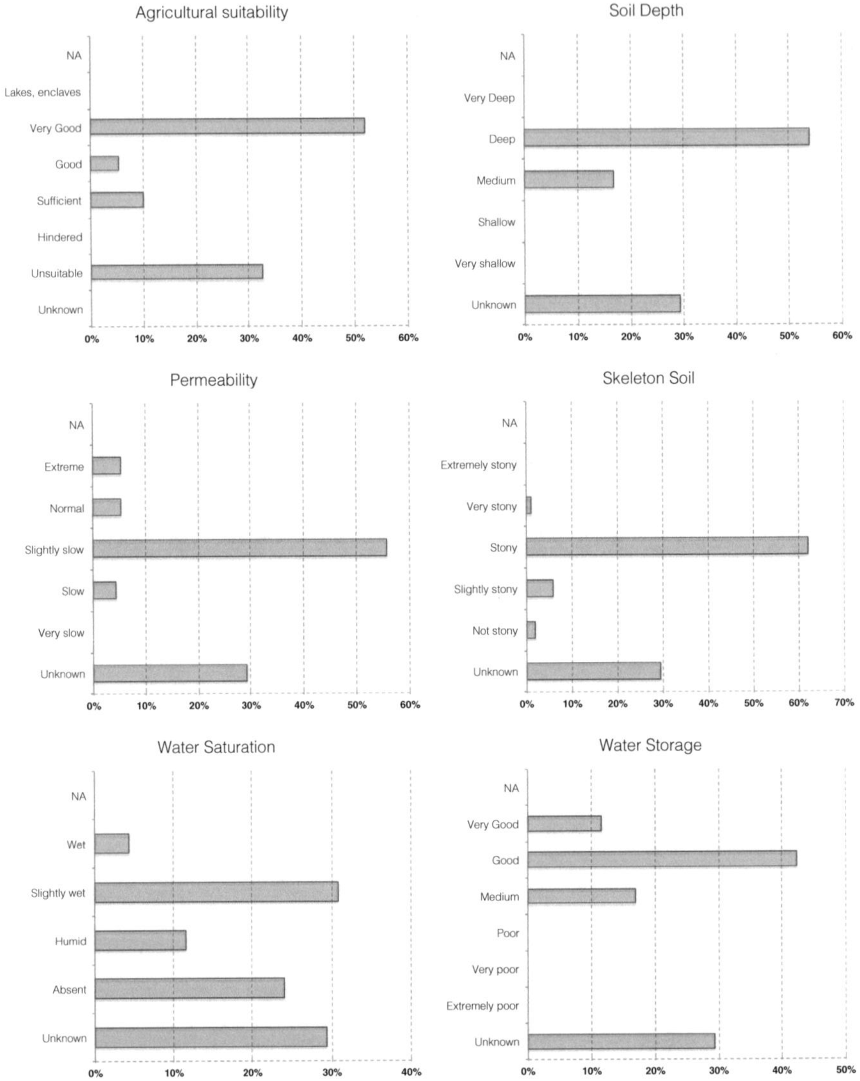

Fig. A.91 Locational preference analysis Zurich. Distribution of the Roman settlements over the agricultural suitability and the soil property classes (part 1)

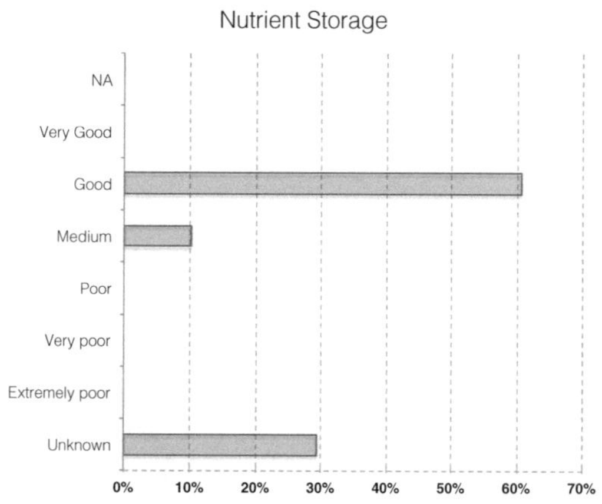

Fig. A.92 Locational preference analysis Zurich. Distribution of the Roman settlements over the agricultural suitability and the soil property classes (part 2)

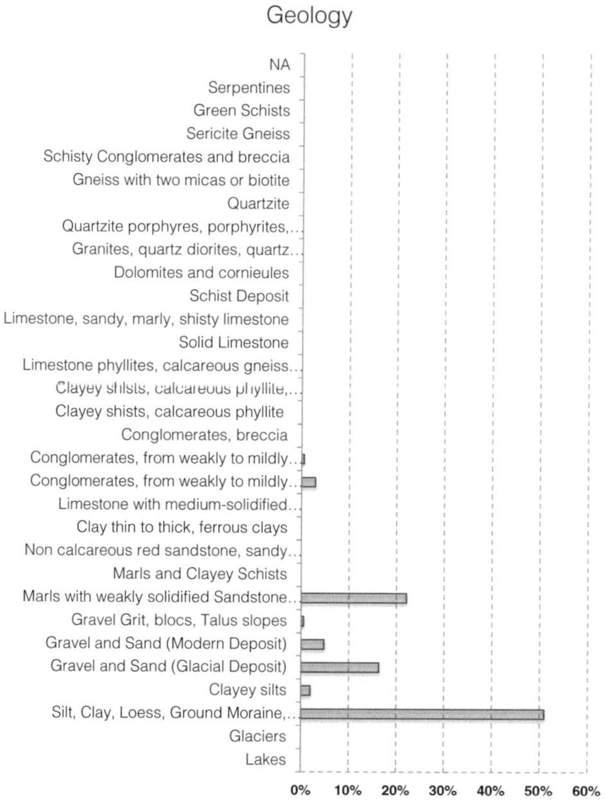

Fig. A.93 Locational preference analysis Zurich. Distribution of the Roman settlements over the geology classes

A.6.4 RF Classification Model Results

See Figs. A.94, A.95, A.96, A.97 and A.98.

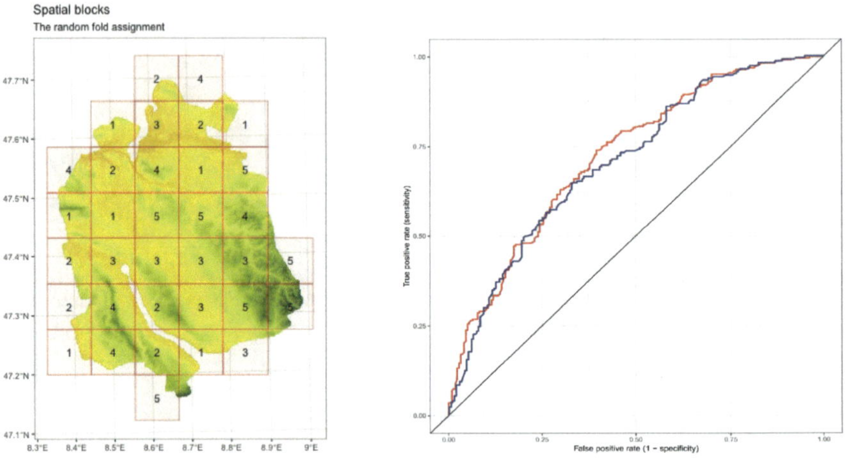

Fig. A.94 Left: Spatial blocks for *k-folds* cross validation Zurich. Number of folds: 5, size of blocks: 6000 m. Right: ROC curve and AUC for Zurich RF classification model. The red curve (AUC 0.72) resulted from the model using only the most important independent variables, the blue curve (AUC 0.70) resulted from the model using all the independent variables

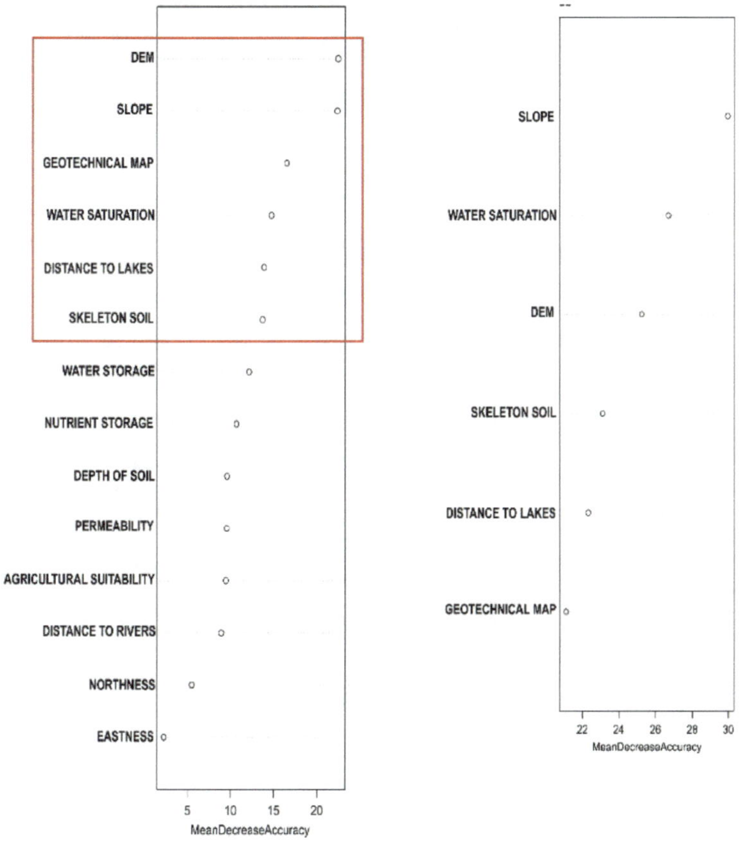

Fig. A.95 Left: Variable importance for the first RF classification model in Zurich using all variables and, to the right, using a selection (red box) of variables

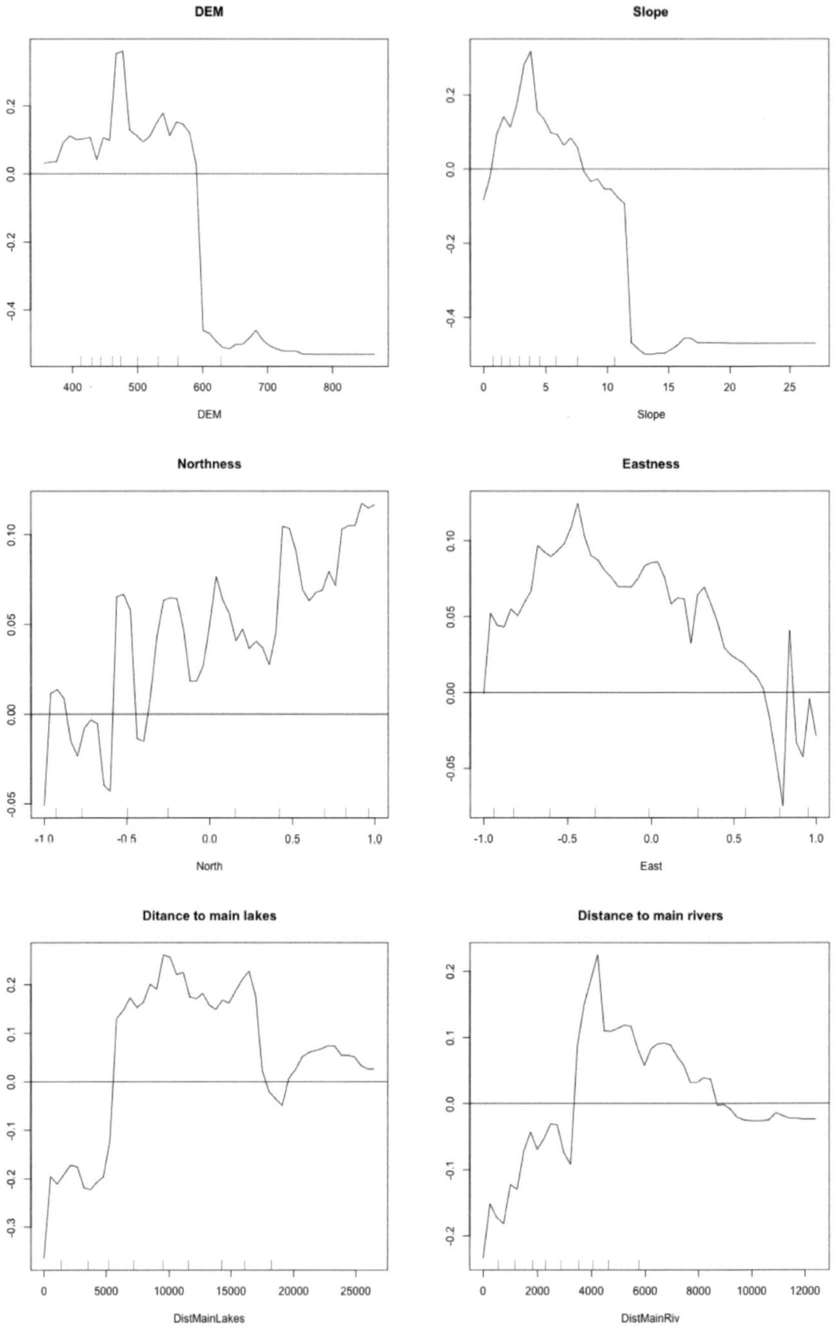

Fig. A.96 Partial dependence plots for RF classification model Zurich (part 1)

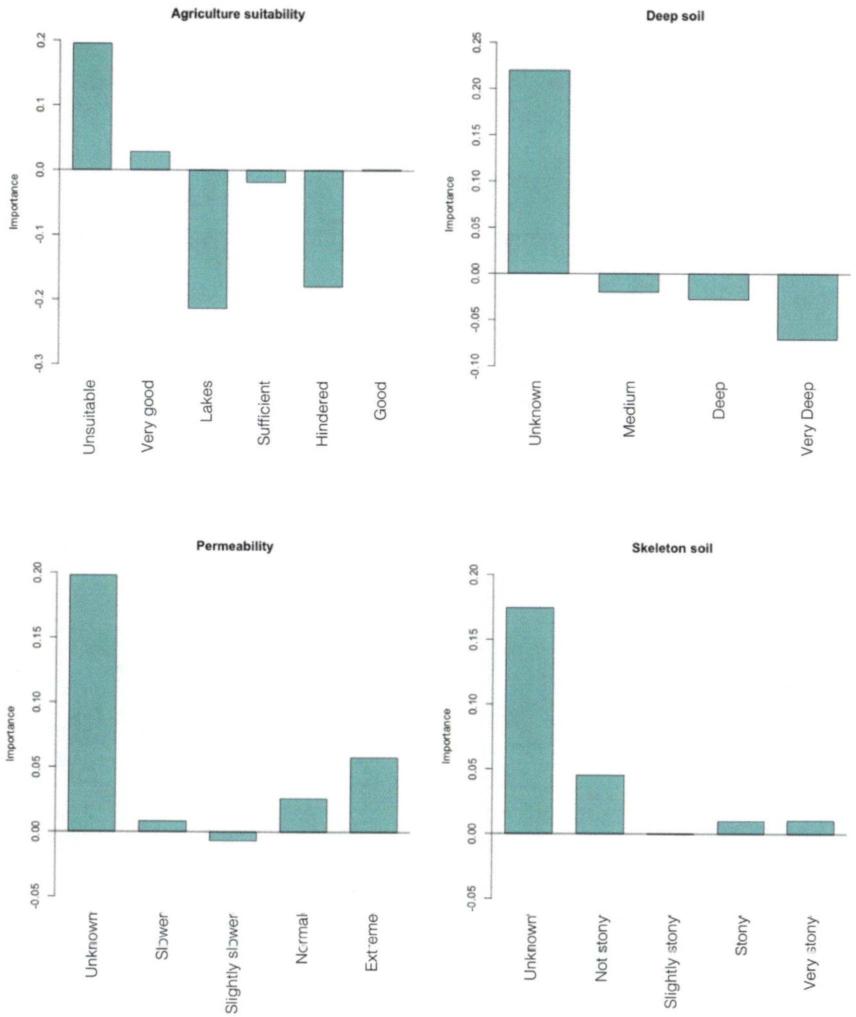

Fig. A.97 Partial dependence plots for RF classification model Zurich (part 2)

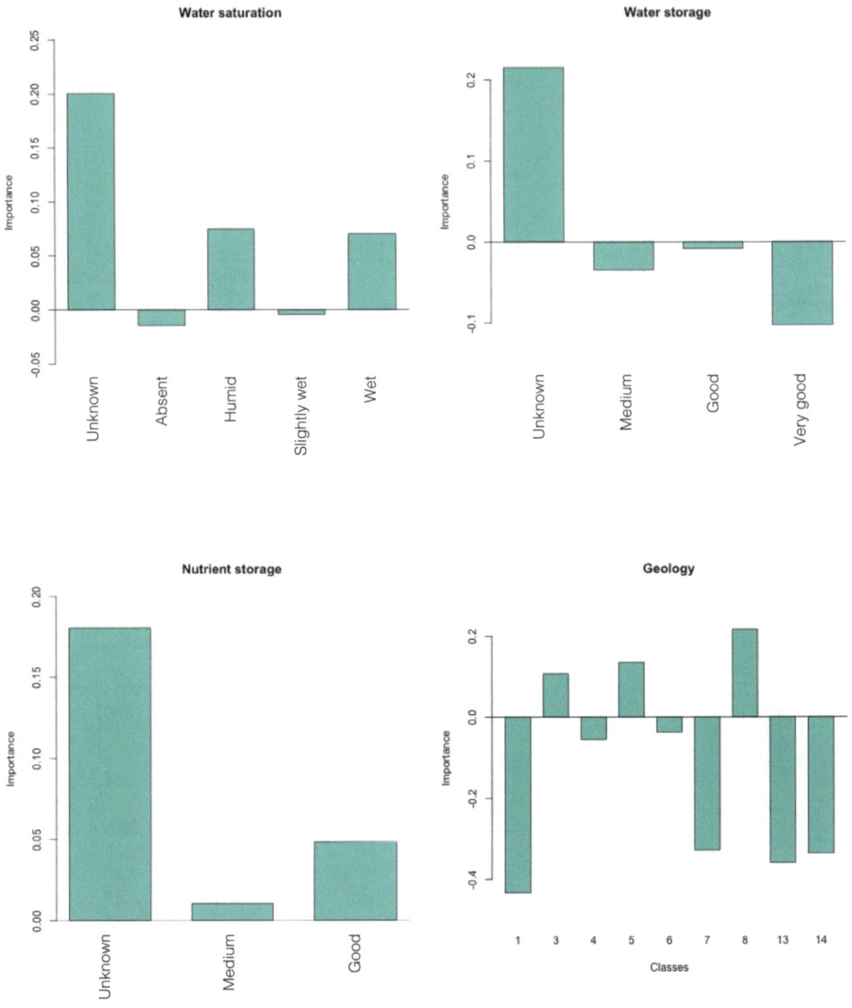

Fig. A.98 Partial dependence plots for RF classification model Zurich (part 3). For geology classes, see Table 5.6

A.6.5 Comparative Analysis of Predicted High Probability Areas—RF Classification

See Figs. A.99, A.100, A.101 and A.102.

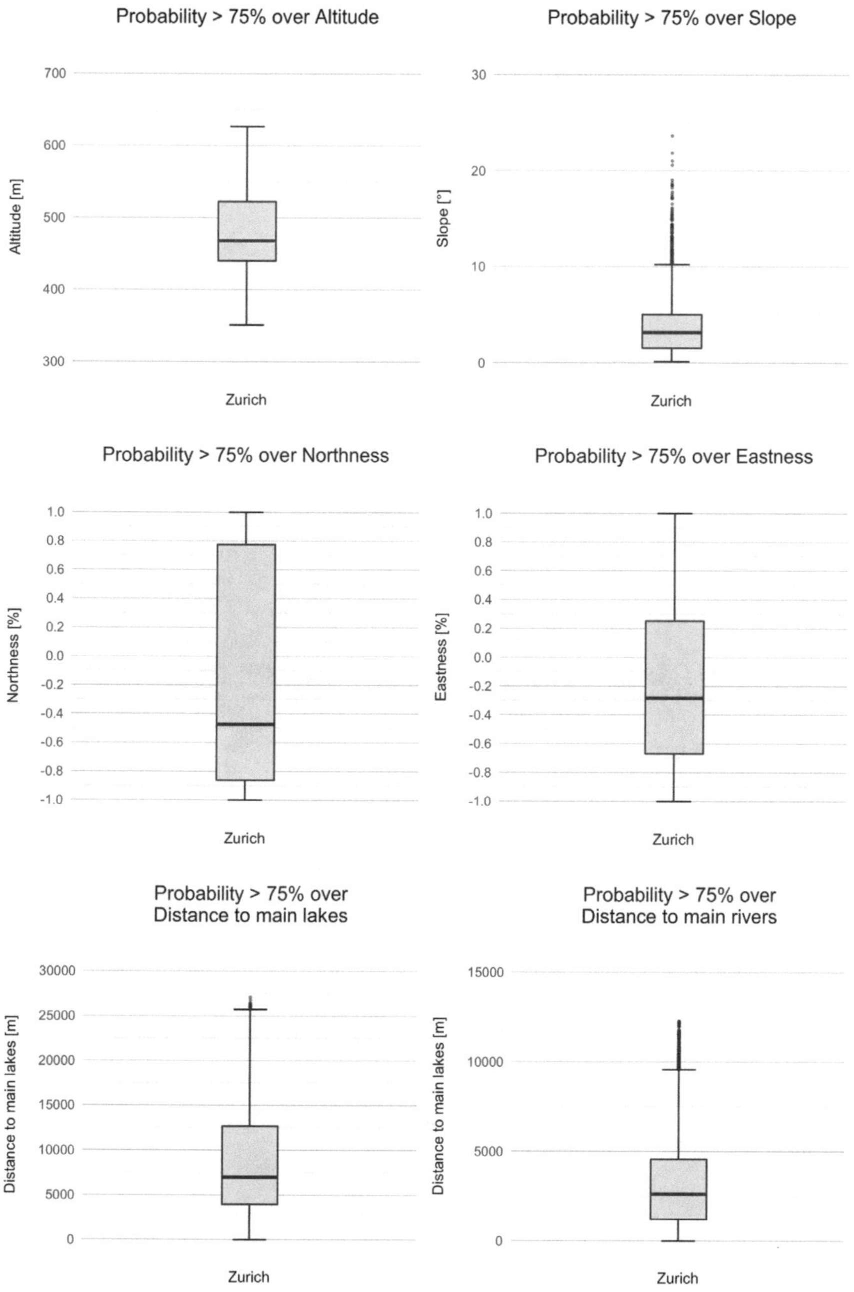

Fig. A.99 Comparative analysis of predicted high probability (>75%) areas using RF classification in Zurich (part 1). The boxplot indicate the median value and the interquartile range IQR (25th (Q1) to 75th (Q3) percentile (half of all occurrences lie within the box), the whiskers show the "minimum" (Q1 − 1.5 * IQR) and "maximum" (Q3 + 1.5 * IQR) values. Points show outliers

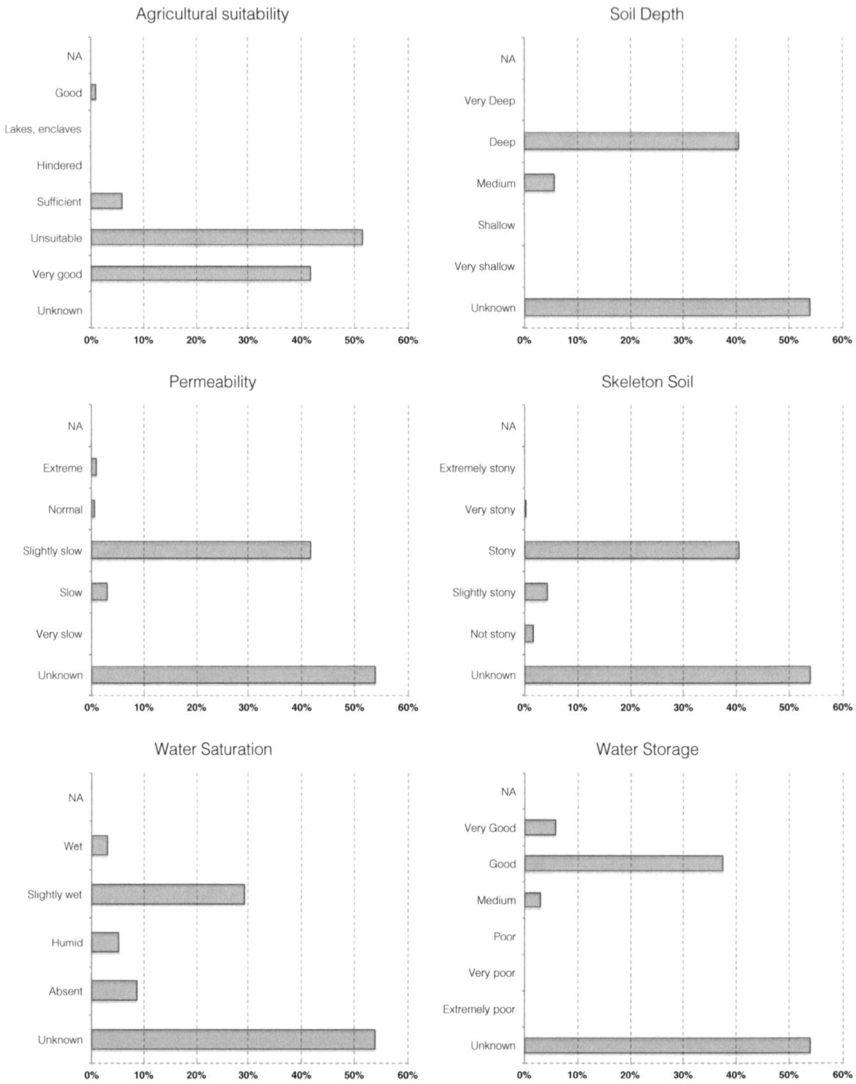

Fig. A.100 Comparative analysis of predicted high probability (>75%) areas using RF classification in Zurich (part 2)

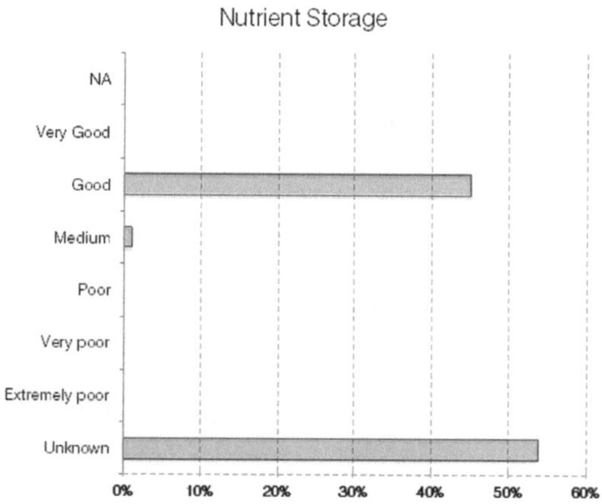

Fig. A.101 Comparative analysis of predicted high probability (>75%) areas using RF classification in Zurich (part 3)

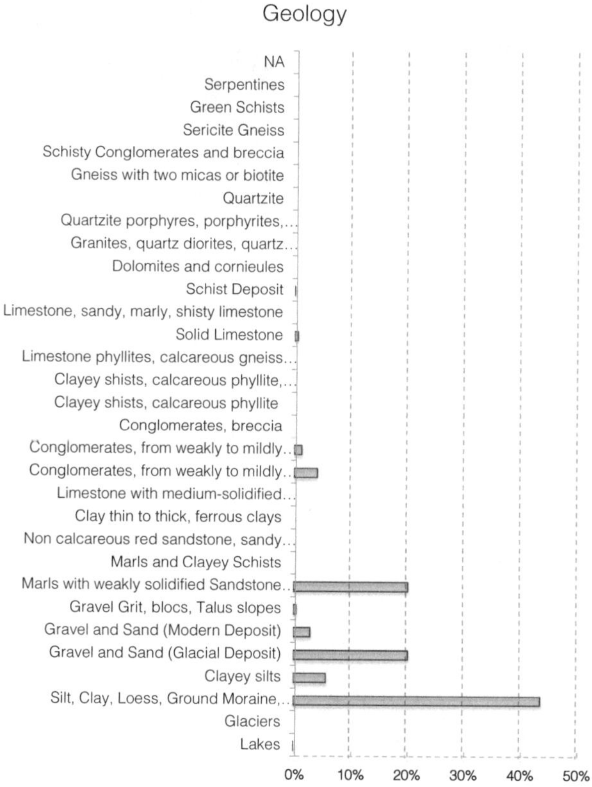

Fig. A.102 Comparative analysis of predicted high probability (>75%) areas on geology classes using RF classification in Zurich

A.7 Switzerland

A.7.1 Environmental Variables

See Figs. A.103, A.104 and A.105.

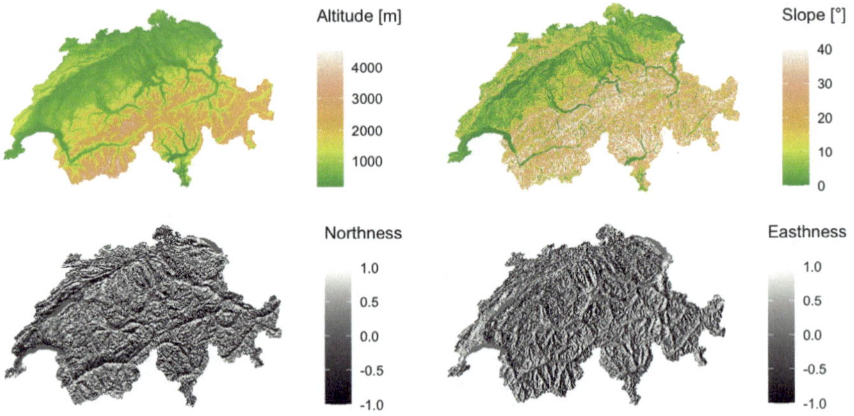

Fig. A.103 Environmental variables Switzerland (part 1)

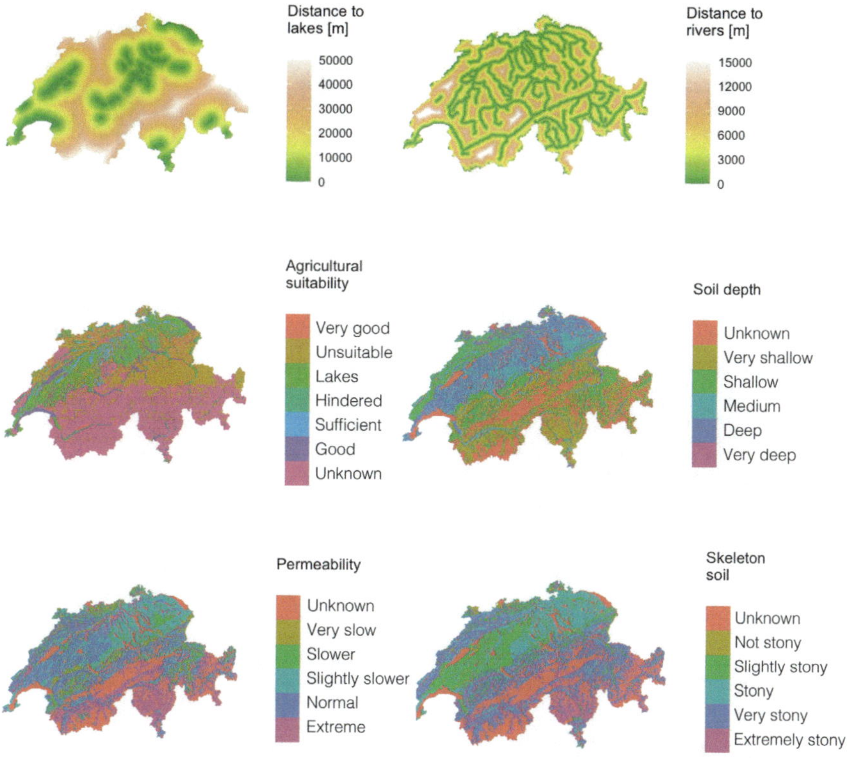

Fig. A.104 Environmental variables Switzerland (part 2)

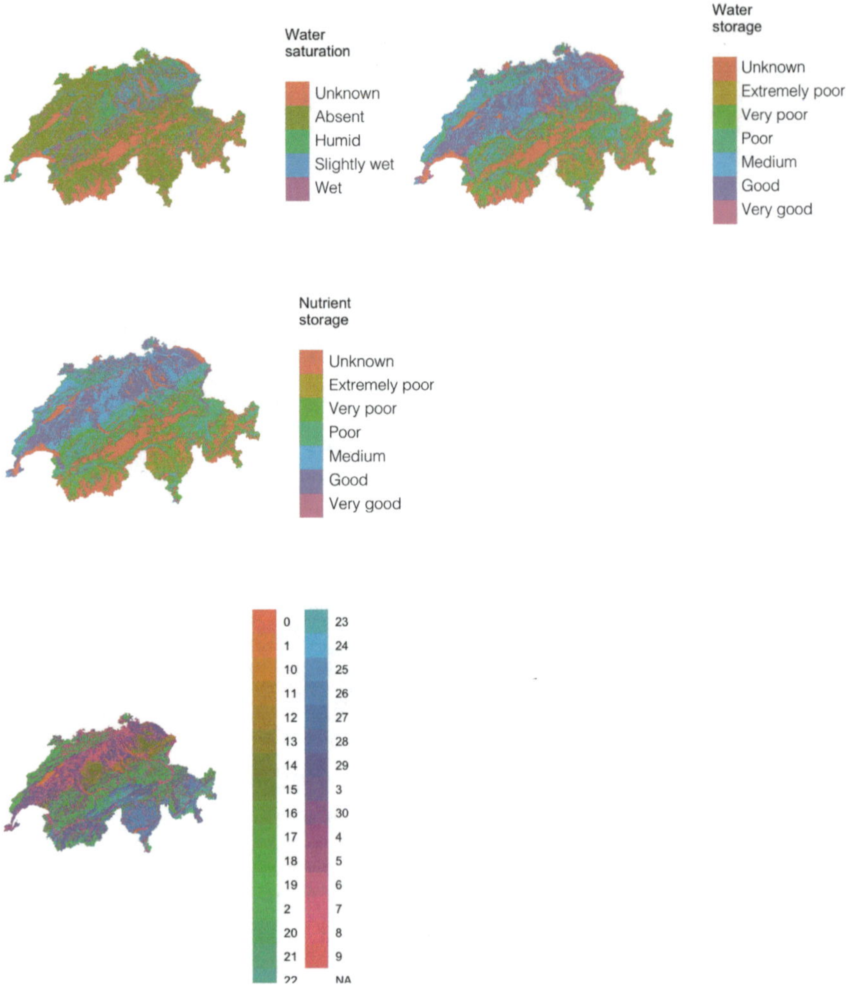

Fig. A.105 Environmental variables Switzerland. For geology classes see Table 5.6 (part 3)

A.7.2 Locational Preference Analysis

See Figs. A.106, A.107, A.108, A.109, A.110, A.111, A.112, A.113, A.114, A.115, A.116, A.117, A.118 and A.119.

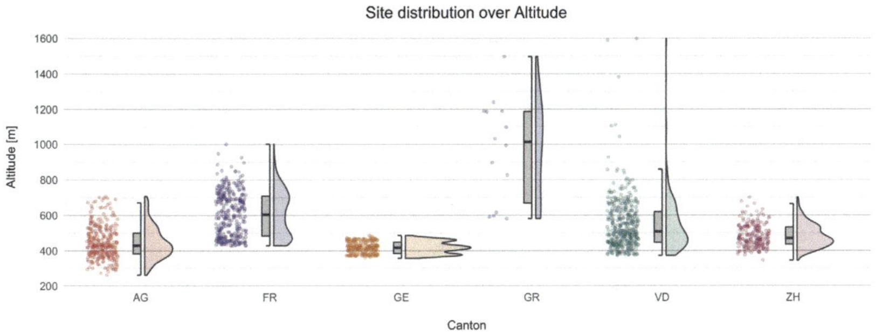

Fig. A.106 Distribution of all Roman settlements over Altitude per Canton. The point jitters show all occurrences. The boxplot indicate the median value and the interquartile range IQR (25th (Q1) to 75th (Q3) percentile (half of all occurrences lie within the gray box), the whiskers show the "minimum" (Q1 − 1.5 * IQR) and "maximum" (Q3 + 1.5 * IQR) values. The colored area indicates the concentration of occurrences (Kernel probability density)

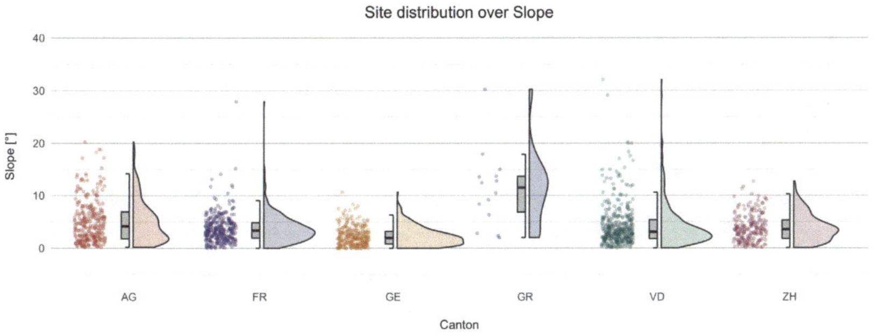

Fig. A.107 Distribution of all Roman settlements over Slope per Canton. The point jitters show all occurrences. The boxplot indicate the median value and the interquartile range IQR (25th (Q1) to 75th (Q3) percentile (half of all occurrences lie within the gray box), the whiskers show the "minimum" (Q1 − 1.5 * IQR) and "maximum" (Q3 + 1.5 * IQR) values. The colored area indicates the concentration of occurrences (Kernel probability density)

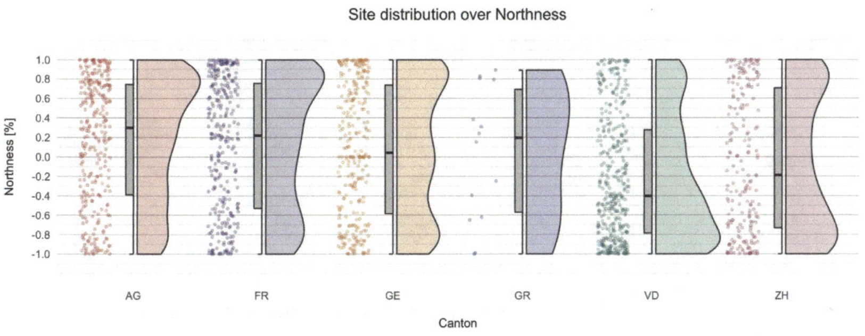

Fig. A.108 Distribution of all Roman settlements over Northness per Canton. The point jitters show all occurrences. The boxplot indicate the median value and the interquartile range IQR (25th (Q1) to 75th (Q3) percentile (half of all occurrences lie within the gray box), the whiskers show the "minimum" (Q1 − 1.5 * IQR) and "maximum" (Q3 + 1.5 * IQR) values. The colored area indicates the concentration of occurrences (Kernel probability density)

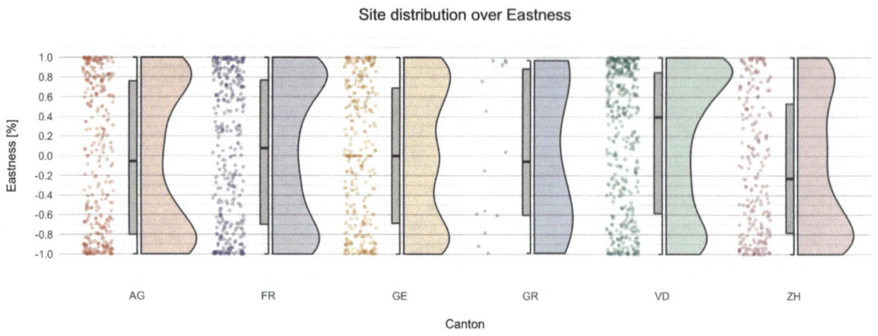

Fig. A.109 Distribution of all Roman settlements over Eastness per Canton. The point jitters show all occurrences. The boxplot indicate the median value and the interquartile range IQR (25th (Q1) to 75th (Q3) percentile (half of all occurrences lie within the gray box), the whiskers show the "minimum" (Q1 − 1.5 * IQR) and "maximum" (Q3 + 1.5 * IQR) values. The colored area indicates the concentration of occurrences (Kernel probability density)

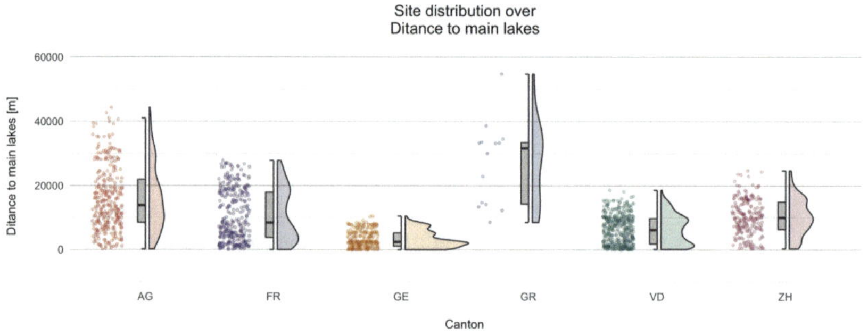

Fig. A.110 Distribution of all Roman settlements over the distance to main lakes per Canton. The point jitters show all occurrences. The boxplot indicate the median value and the interquartile range IQR (25th (Q1) to 75th (Q3) percentile (half of all occurrences lie within the gray box), the whiskers show the "minimum" (Q1 − 1.5 * IQR) and "maximum" (Q3 + 1.5 * IQR) values. The colored area indicates the concentration of occurrences (Kernel probability density)

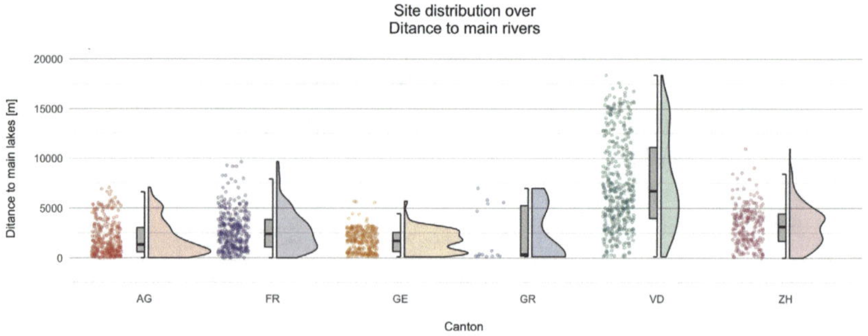

Fig. A.111 Distribution of all Roman settlements over the distance to main rivers per Canton. The point jitters show all occurrences. The boxplot indicate the median value and the interquartile range IQR (25th (Q1) to 75th (Q3) percentile (half of all occurrences lie within the gray box), the whiskers show the "minimum" (Q1 − 1.5 * IQR) and "maximum" (Q3 + 1.5 * IQR) values. The colored area indicates the concentration of occurrences (Kernel probability density)

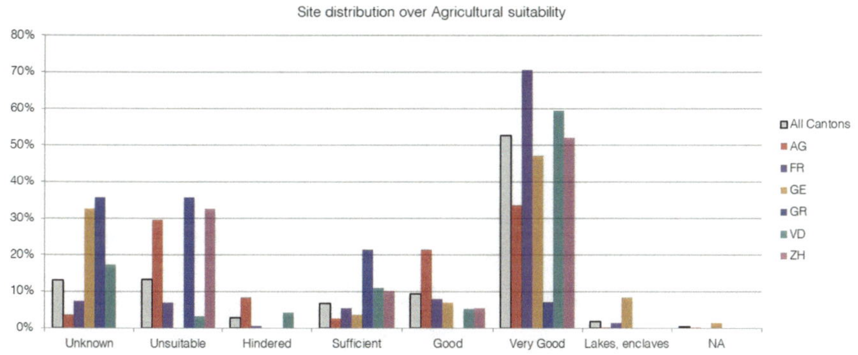

Fig. A.112 Distribution of all Roman settlements over the classes of agricultural suitability. The first column (all Cantons) shows the average values for the 6 computed Cantons

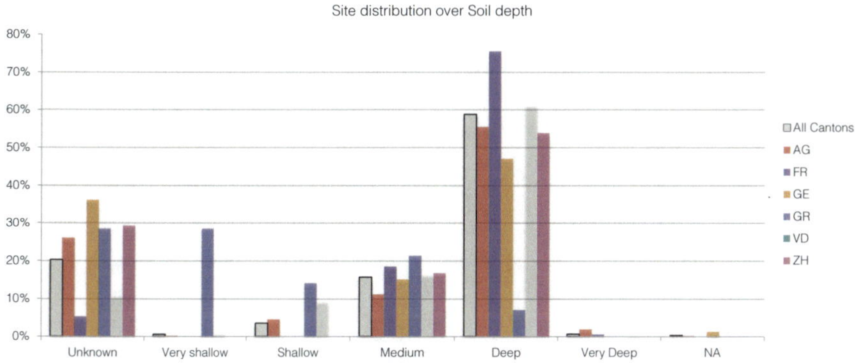

Fig. A.113 Distribution of all Roman settlements over the classes of soil depth. The first column (all Cantons) shows the average values for the 6 computed Cantons

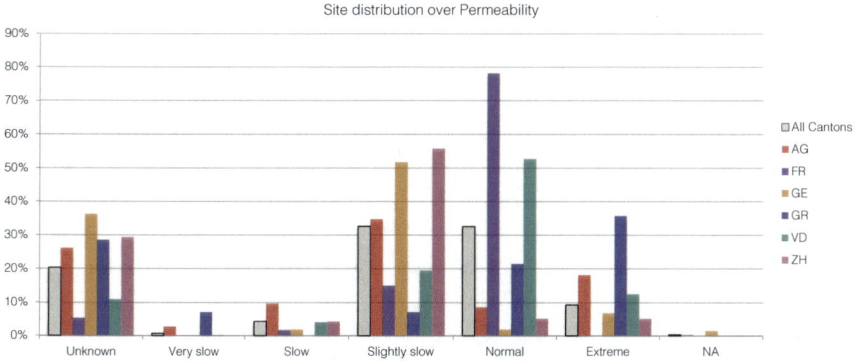

Fig. A.114 Distribution of all Roman settlements over the classes of permeability. The first column (all Cantons) shows the average values for the 6 computed Cantons

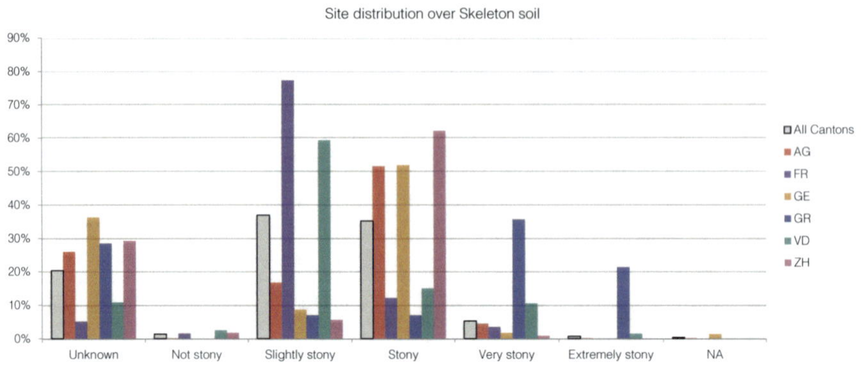

Fig. A.115 Distribution of all Roman settlements over the classes of soil skeleton. The first column (all Cantons) shows the average values for the 6 computed Cantons

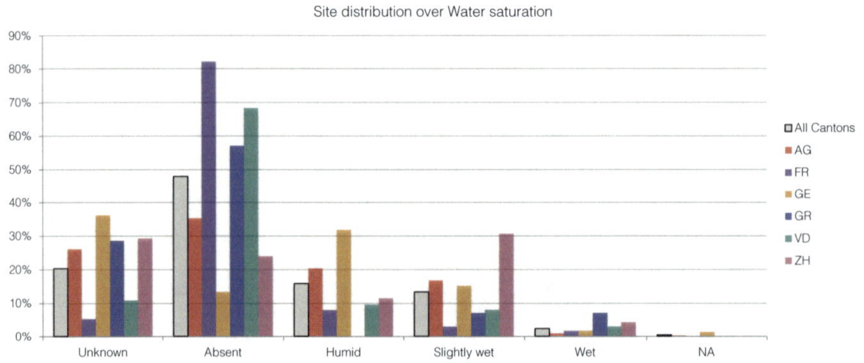

Fig. A.116 Distribution of all Roman settlements over the classes of water saturation. The first column (all Cantons) shows the average values for the 6 computed Cantons

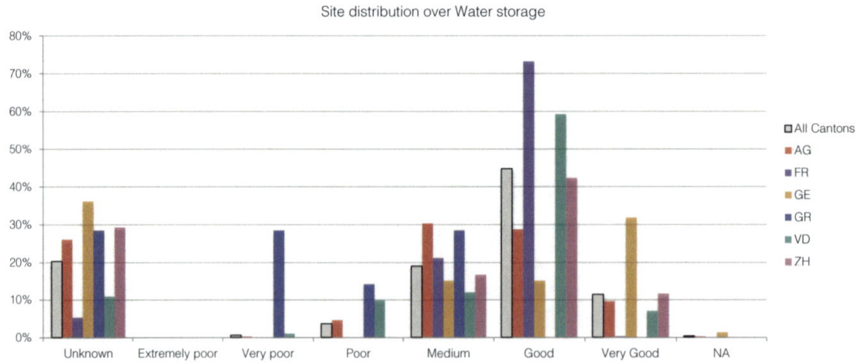

Fig. A.117 Distribution of all Roman settlements over the classes of water storage. The first column (All Cantons) shows the average values for the 6 computed Cantons

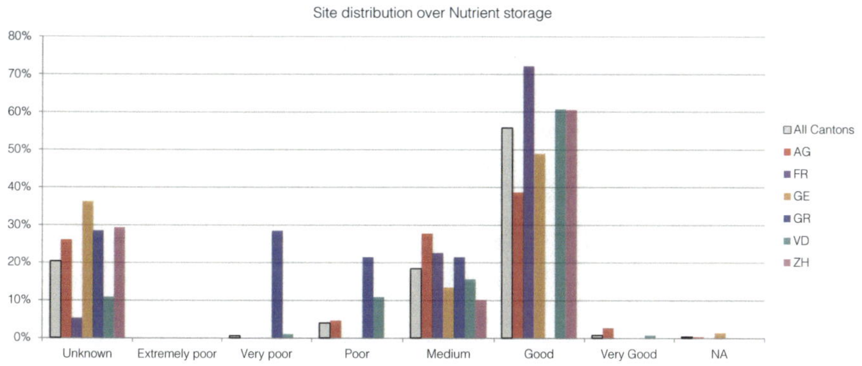

Fig. A.118 Distribution of all Roman settlements over the classes of nutrient storage. The first column (All Cantons) shows the average values for the 6 computed Cantons

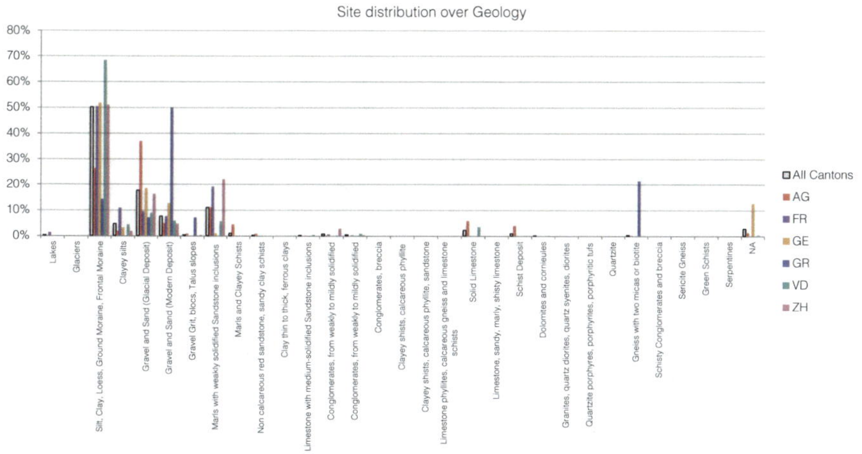

Fig. A.119 Distribution of all Roman settlements over the classes of geology. The first column (all Cantons) shows the average values for the 6 computed Cantons

A.7.3 RF Classification Model Results

See Figs. A.120, A.121, A.122, A.123 and A.124.

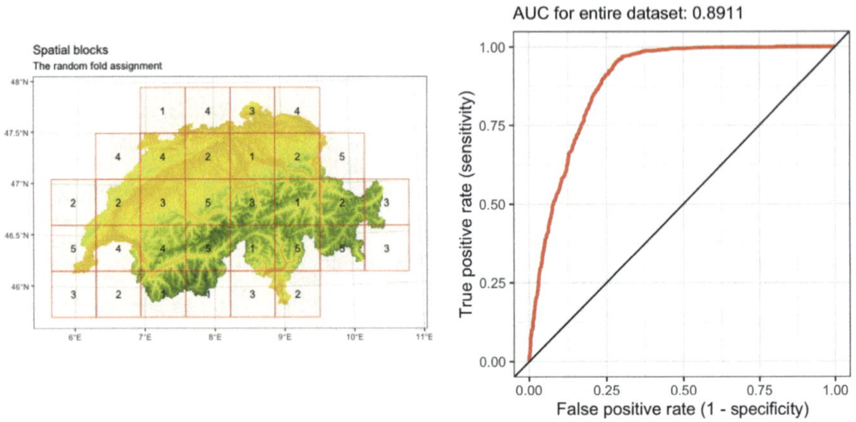

Fig. A.120 Left: Spatial blocks for *k-folds* cross validation Switzerland. Number of folds: 5, size of blocks: 50000 m. Right: ROC curve and AUC (0.89) for Switzerland RF classification model

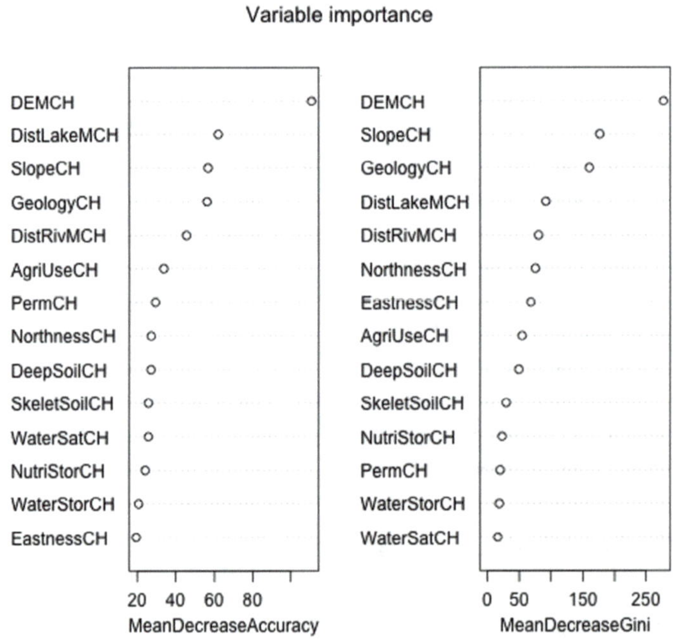

Fig. A.121 Variable importance for RF classification model Switzerland

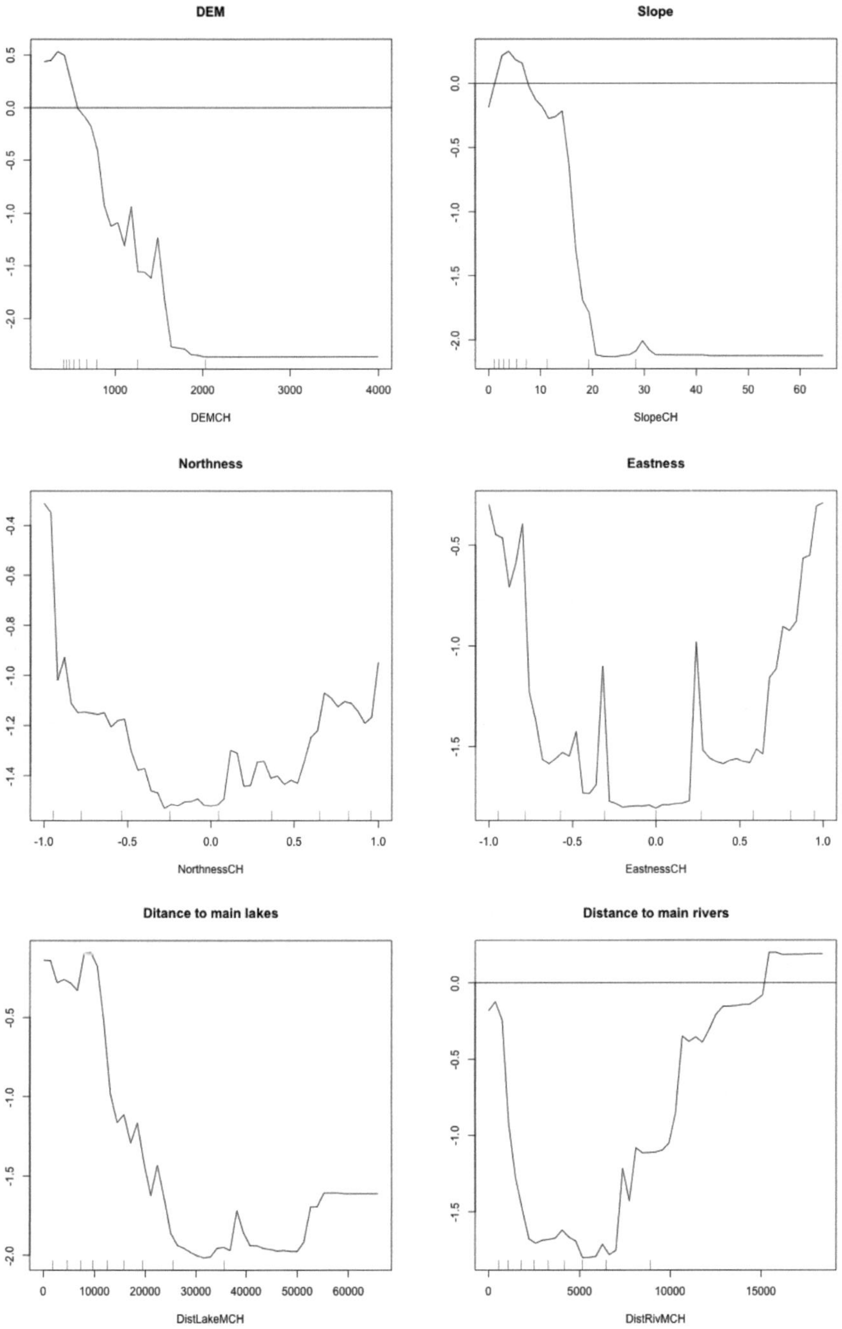

Fig. A.122 Partial dependence plots for RF classification model Switzerland (part 1)

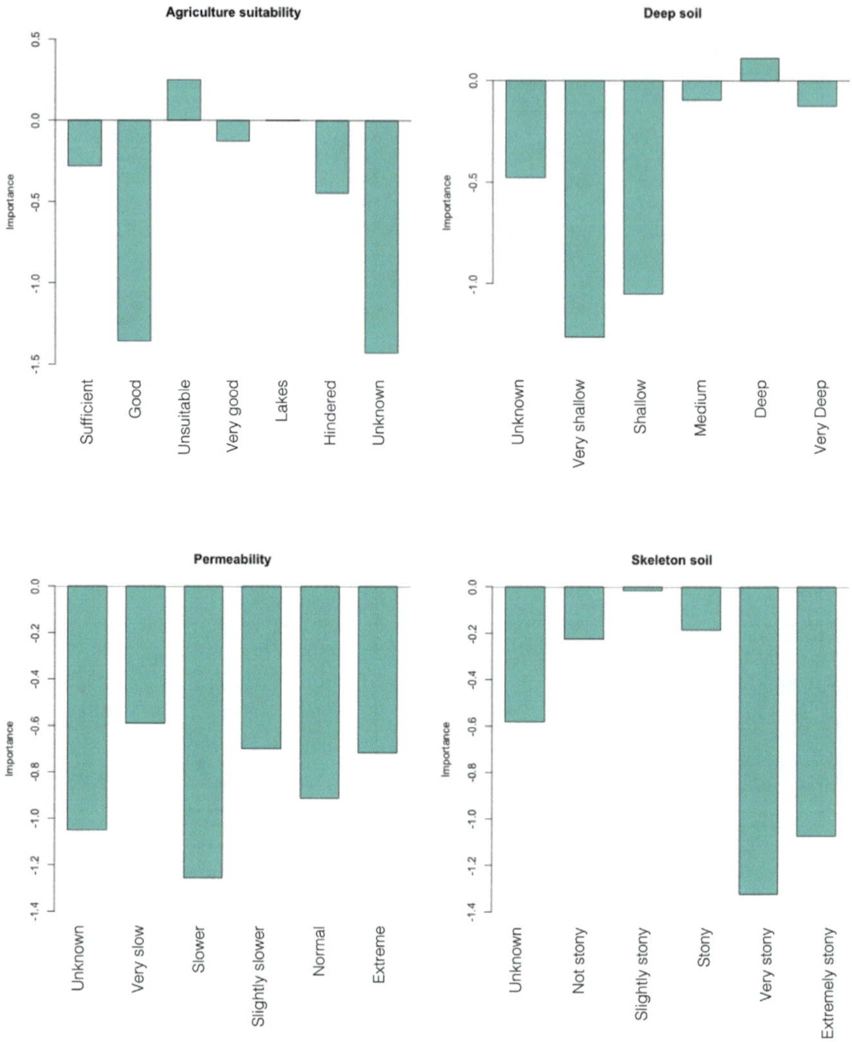

Fig. A.123 Partial dependence plots for RF classification model Switzerland (part 2)

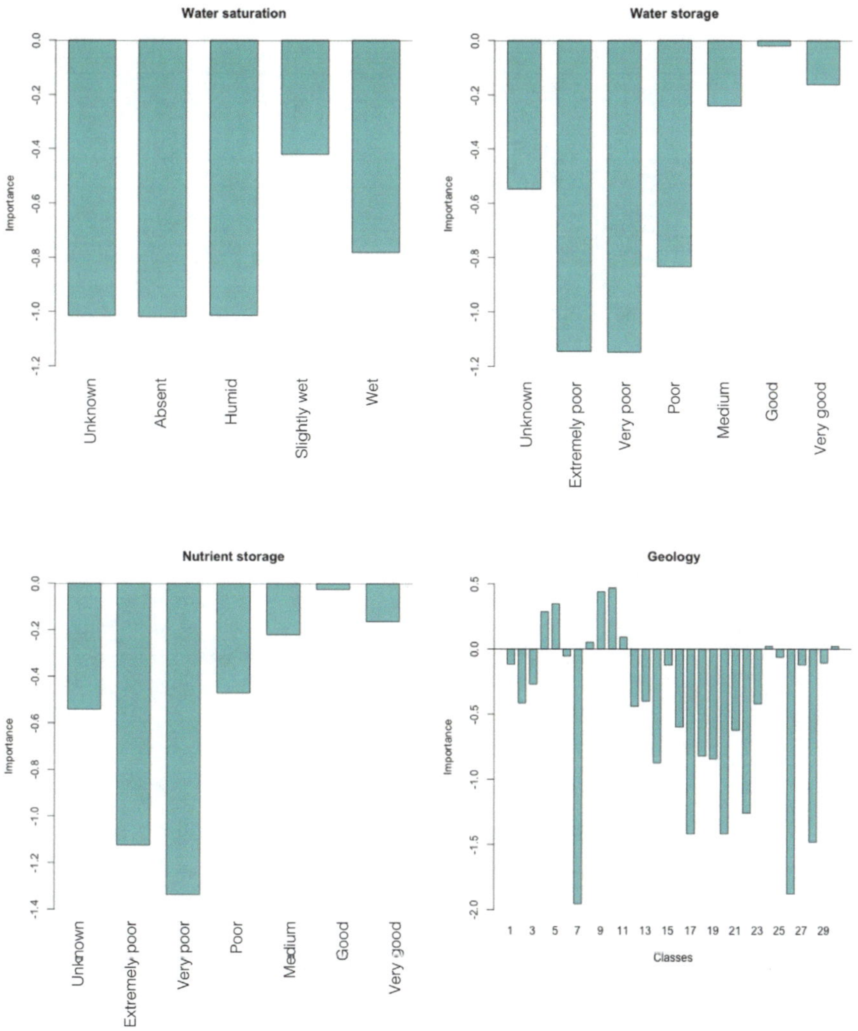

Fig. A.124 Partial dependence plots for RF classification model Switzerland (part 3). For geology classes, see Table 5.6

A.7.4 Comparative Analysis of Predicted High Probability Areas—RF Classification

See Figs. A.125, A.126, A.127, A.128 and A.129.

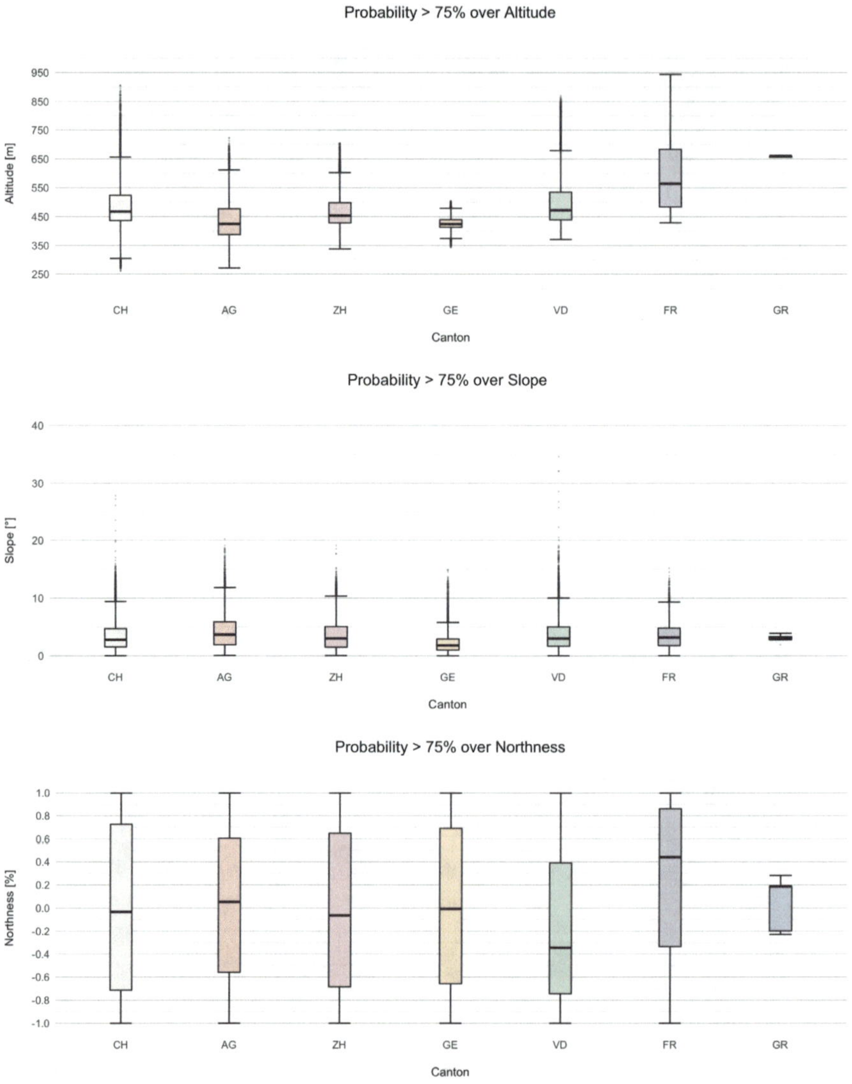

Fig. A.125 Comparative analysis of predicted high probability (>75%) areas using RF classification in Switzerland over altitude (top), slope (middle), northness (bottom). The boxplot indicate the median value and the interquartile range IQR (25th (Q1) to 75th (Q3) percentile (half of all occurrences lie within the box), the whiskers show the "minimum" (Q1 − 1.5 * IQR) and "maximum" (Q3 + 1.5 * IQR) values. Points show outliers. The first plot (CH) shows the distribution of high probability areas in the Cantons not considered individually in the analysis

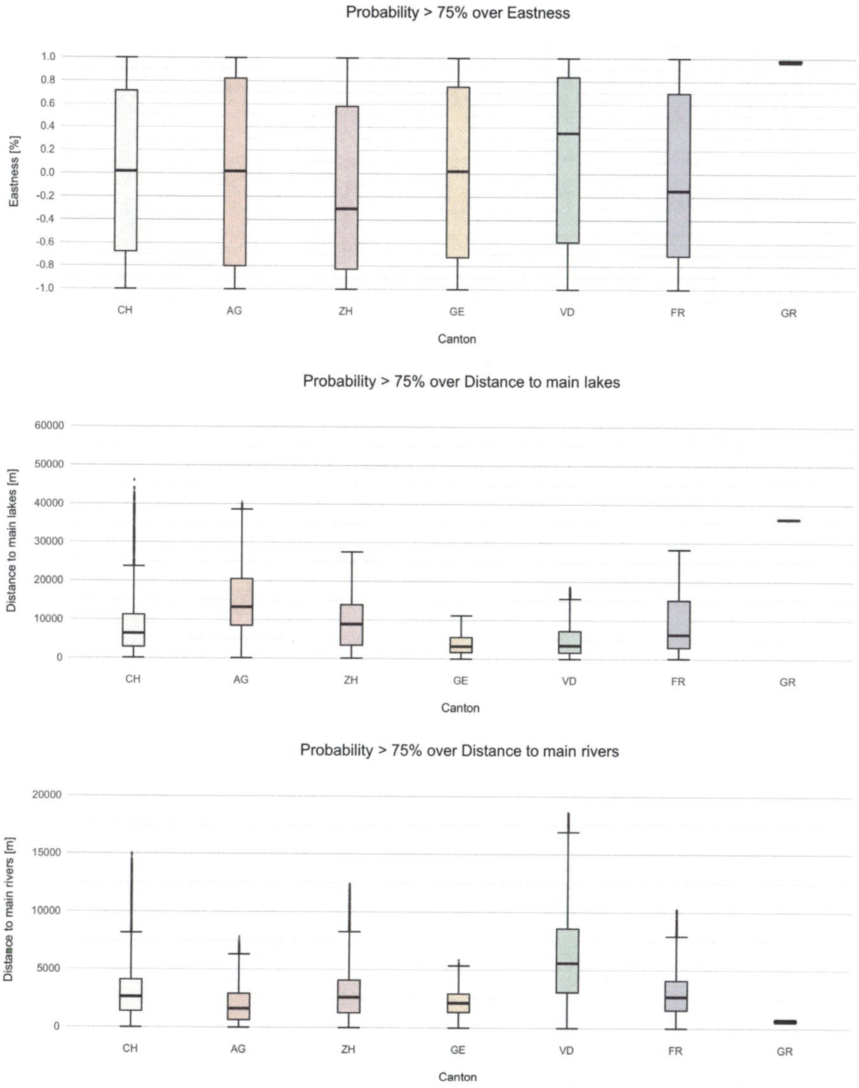

Fig. A.126 Comparative analysis of predicted high probability (>75%) areas using RF classification in Switzerland over eastness (top), distance to main lakes (middle), distance to main rivers (bottom). The boxplot indicate the median value and the interquartile range IQR (25th (Q1) to 75th (Q3) percentile (half of all occurrences lie within the box), the whiskers show the "minimum" (Q1 − 1.5 * IQR) and "maximum" (Q3 + 1.5 * IQR) values. Points show outliers. The first plot (CH) shows the distribution of high probability areas in the Cantons not considered individually in the analysis

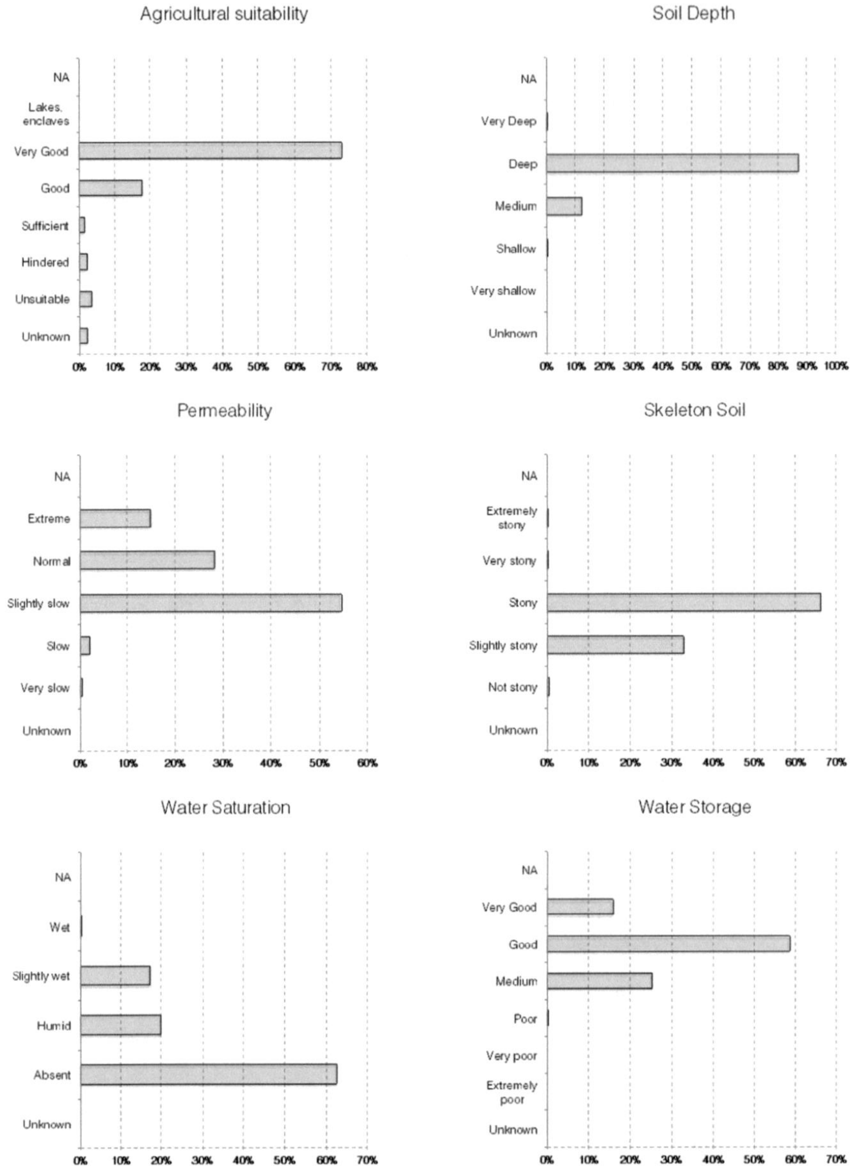

Fig. A.127 Comparative analysis of predicted high probability (>75%) areas using RF classification in Switzerland (part 2)

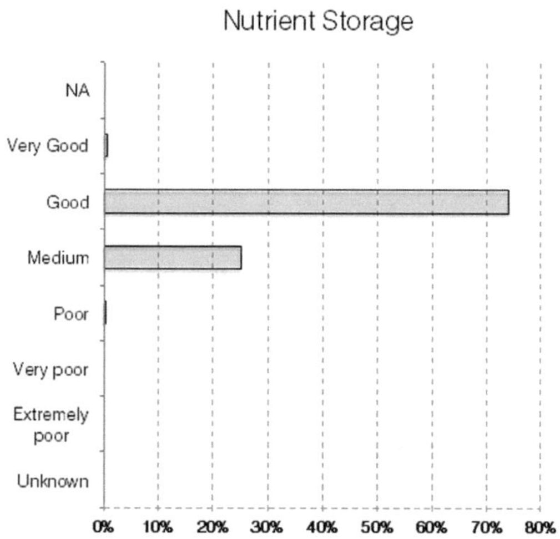

Fig. A.128 Comparative analysis of predicted high probability (>75%) areas using RF classification in Switzerland (part 3)

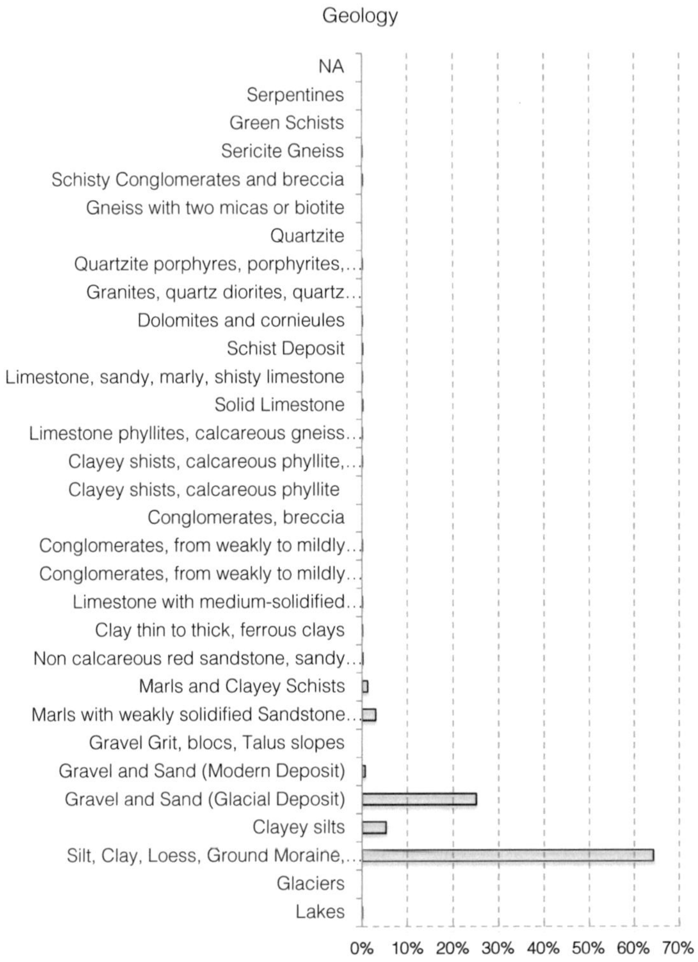

Fig. A.129 Comparative analysis of predicted high probability (>75%) areas on geology classes using RF classification in Switzerland

About the Author

Dr. Maria Elena Castiello Dr. Maria Elena Castiello is currently a Postdoctoral Research Fellow at the German Archaeological Institute (DAI) in Berlin, Germany, in Dr. Ferran Antolin's Natural Science team. She works at the edge of Natural Sciences and Humanities exploring the effectiveness of quantitative and computational methods in archaeological research.

Prior to her current role, Maria Elena has worked at the Archéorient Laboratory of the French National Center for Scientific Research (CNRS) and the Université Lumière Lyon 2, in Lyon, France, in a transdisciplinary research project focusing on Simulation modeling, Archaeology, Geography and Anthropology.

This book is the result of Maria Elena Castiello's Ph.D. Thesis in Archaeological Sciences and of several collaborations established with the Swiss Archaeological Cantonal Departments and the Department of Earth Surface Dynamics, at the Faculty of Geosciences, University of Lausanne, Switzerland. Maria Elena's degree was awarded in 2020 by the University of Bern, Switzerland, and was co-supervised by the Instituto de Ciencias del Patrimonio (Incipit-CSIC), Santiago de Compostela, Spain.

Before her research activities in Switzerland, France and Germany, Dr. Castiello obtained her Bachelor and Master degrees in archaeology at Sapienza Università di Roma in 2008 and 2012 respectively. She further spent a research stay at the University of Vienna, Austria, in 2011. She obtained a second Master with a focus on digital technologies applied to archaeology at the same university in Rome, in 2014. During the academic *cursus*, Maria Elena was granted several Swiss and international funding. In 2020, she received a grant by the European Commission (Horizon 2020 action) to extend her research on conceptual and computational modeling for archaeological and cultural heritage data at the Incipit CSIC, Spain.

M. E. Castiello, *Computational and Machine Learning Tools for Archaeological Site Modeling*, Springer Theses, https://doi.org/10.1007/978-3-030-88567-0

Dr. Castiello's current scientific interests include the application of advanced techniques of multi-agents simulation modeling, artificial intelligence and geospatial analysis, with the aim to shed light and to characterize the complex human–environment relationships, long-term climate change effects on population dynamics and site distribution, over different times and geographical areas.

Maria Elena's full list of publications is available at: https://dainst.academia.edu/MariaElenaCastiello.

References

1. Achino KF, Barceló JA (2019) Spatial prediction: reconstructing the "spatiality" of social activities at the intra-site scale. J Archaeol Method Theory 26(1):112–134. https://doi.org/10.1007/s10816-018-9367-1
2. Altschul J, Patterson T (2010) Trends in employment and training in American archaeology. In: Ashmore W, Lippert D, Mills B (eds) Voices in American Archaeology. SAA Press, Washington, DC, pp 291–316
3. Altschul JH, Kintigh KW, Klein TH, Doelle WH, Hays-Gilpin KA, Herr SA, Kohler TA, Mills BJ, Montgomery LM, Nelson MC
4. Amarosi T, Buckland P, Dugmore A, Ingimundarson J, Mcgovern T (1997) Raiding the landscape: human impact in the Scandinavian north Atlantic. Hum Ecol 25(3):491–518
5. Anderson D, Bissett T, Yerka S, Wells J, Kansa E, Kansa S, Myers K, Demuth R, White D (2017) Sea-level rise and archaeological site destruction: an example from the southeastern United States using DINAA. PLOS One 12(11):e0188142. https://doi.org/10.1371/journal.pone.0188142
6. Anichini F, Bini M, Fabiani F, Gattiglia G, Giacomelli S, Gualandi ML, Pappalardo M, Sarti G (2011) MAPPA project. Methodologies applied to archaeological potential predictivity. In: MapPapers 1en-I, pp 23–43
7. Arnoldus-Huyzendveld A, Citter C, Pizziolo G (2015) Predictivity—postdictivity: a theoretical framework. In: Campana S, Scopigno R (eds) Keep the revolution going, proceedings of 43rd computer applications and quantitative methods in archaeology. Atti del convegno internazionale, Siena, Oxford pp 593–598
8. Assael Y, Sommerschield T, Prag J (2019) Restoring ancient text using deep learning: a case study on Greek epigraphy. https://arxiv.org/abs/1910.06262
9. Attema PAJ, Burgers G-JLM, van Leusen PM (2011) regional pathways to complexity. Settlement and land-use dynamics in early Italy from the bronze age to the republican period. In: Derks T, Roymans N (eds) Villa landscapes in the Roman North. Amsterdam Archaeological Studies
10. Bachmann F, Michel R (1998) Quel avenir pour l'archéologie préventive en Suisse après les Grands Travaux? Les Nouvelles De L'archéologie 73:27–34
11. Baena J, Blasco C, Recuero V (1995) The spatial analysis of Bell Beaker sites in the Madrid region of Spain. In: Lock G, Stančič Z (eds), pp 101–116
12. Barceló JA (2008) Computational intelligence in archaeology. Comput Intell Archaeol. https://doi.org/10.4018/978-1-59904-489-7

M. E. Castiello, *Computational and Machine Learning Tools for Archaeological Site Modeling*, Springer Theses, https://doi.org/10.1007/978-3-030-88567-0

13. Barton CM, Ullah IIT, Bergin SM, Mitasova H, Sarjoughian H (2012) Looking for the future in the past: long-term change in socioecological systems. Ecol Model 241:42–53
14. Batáry P, Dicks LV, Kleijn D, Sutherland WJ (eds) (2015) The role of agri-environment schemes in conservation and environmental management. Conservat Biol 29:1006–1016
15. Bedon R, Chevallier R, Pinon P (1988) Architecture et urbanisme en Gaule romaine (52 av. J.-C. - 486 ap. J.-C.). Paris (2 vols)
16. Berg E (2007) Using a GIS-based database as a platform for cultural heritage management of sites and monuments in Norway. In: CAA2006. Digital discovery. Exploring new frontiers in human heritage. Computer applications and quantitative methods in archaeology. Proceedings of the 34th conference, Fargo, United States, April 2006, pp 345–351. http://proceedings.caaconference.org/files/2006/29_Berg_CAA2006.pdf
17. Bevan AH, Lake M (eds) (2013) Computational approaches to archaeological spaces. Left Coast Press, Walnut Creek, US
18. Binford LR (1964) A consideration of archaeological research design. Am Antiq 29(4):425–441
19. Boisaubert JL, Bugnon D, Mauvilly M (eds) (2008) Archéologie et Autoroute A1, destines croisés. Archéologie Fribourgeoise 22
20. Bonham-Carter GF (1994) Geographic information systems for geoscientists. Modelling with GIS. Pergamon Press, Oxford
21. Bonnet C (1986) Chronique des découvertes archéologiques dans le canton de Genève en 1984 et 1985. Genava n. 34:47–68
22. Bork H-R, Lang A (2003) Quantification of past soil erosion and land use. Land cover changes in Germany. In: Lang A, Hennrich K, Dikau R (eds) Long term hillslope and fluvial system modelling. Concepts and case studies from the Rhine river catchment. Berlin, Springer, 101, pp 231–239
23. Chen L, Priebe CE, Sussman DL, Comer DC, Megarry WP et al (2019) Enhanced archaeological predictive modelling in space archaeology. arXiv:1301.2738 [stat.AP]
24. Carranza EJM, van Ruitenbeek FJA, Hecker C, van der Meijde M, van der Meer FD (2008) Knowledge-guided data-driven evidential belief modeling of mineral prospectivity in Cabo de Gata, SE Spain. Int J Appl Earth Obs Geoinf 10:374–387
25. Castella D, Flutsch L (1991) Avenches, En Chaplix. Revue Historique vaudoise, p 34
26. Cecamore C, Castiello ME (2014) Un modello speditivo per la carta del Rischio Relativo nei Beni Culturali, in Atti della 15a Conferenza Italiana Utenti Esri. GEOmedia, [S.l.] 18(2), giugno 2014. ISSN 2283-5687 https://www.mediageo.it/ojs/index.php/GEOmedia/article/view/873/801
27. Caesar *de Bello Gallico* Book I. http://www.softwareparadiso.it/studio/letteratura/De_Bello_Gallico/libro_I.html
28. Cato the Elder, De Agricultura (On Agriculture, transl. H.B. Ash 1936) Loeb Classical Library. https://penelope.uchicago.edu/Thayer/E/Roman/Texts/Cato/De_Agricultura/J*.html
29. Chen L, Comer D, Priebe C, Sussman D, Tilton J (2013) Refinement of a method of identifying probably archaeological sites from remotely sensed data. In: Comer D, Harrower M (eds) Mapping archaeological landscapes from space. Springer, New York, NY, pp 251–258
30. Clevis Q, Tucker GE, Lock G, Lancaster ST, Gasparini N, Desitter A, Bras RL (2006) Geoarchaeological simulation of meandering river deposits and settlement distributions: a three-dimensional approach. Geoarchaeology 28(8):843–874
31. Comer D, Harrower M (eds) (2013) Mapping archaeological landscapes from space. Springer briefs in archaeology. Springer, New York, NY
32. Comer DC, Harrower MJ (2013) The History and future of geospatial and space technologies in archaeology. In: Comer D, Harrower M (eds) Mapping archaeological landscapes from space. Springer, New York, pp 1–8
33. Crumley C, Kolen J, Kleijn DEM, Van Manen N (2017) Studying long-term changes in cultural landscapes: outlines of a research framework and protocol. Landscape Res 42(8):880–890

34. Cushman D, Sebastian L (2008) Integrating archaeological models: management and compliance on military installations. Preservation research series 7. Rio Rancho (NM): SRI Foundation. Deal, k., 2017. Wildlife and natural resource management, 4th edn. Cengage Learning, United States
35. Dangermond J (1987) The maturating of GIS and the new age for geographic information modelling (GIMS). In: Aangeenbrug RT, Schiffman YM (eds) International geographic information system (IGIS): the research agenda, NASA symposium, Arlington, Virginia, vol 2, pp 55–66
36. Darvill T, Fulton A (1998) The monuments at risk survey of England 1995. School of Conservation Sciences, Bournemouth University, Bournemouth and English Heritage, London
37. David B, Thomas J (2008) Handbook of landscape archaeology. Left Coast Press, Walnut Creek, CA
38. Della Casa P (2007) Transalpine pass routes in the Swiss Central Alps and the strategic use of topographic resources. Preistoria Alpina 42:109–118
39. Della Casa P (2013) Switzerland and the Central Alps. In: Harding A, Fokkens H (eds) The Oxford handbook of the European Bronze Age. Oxford University Press, Oxford, pp 702–718
40. Della Casa P (2018) The Leventina prehistoric landscape (Alpine Ticino Valley, Switzerland). Chronos, Zürich
41. Demoule JP (2020) Aux origines, L'Archéologie. Une science au cœur des grands débats de notre temps (eds) La Découverte, Paris
42. De Vries P (2007) Archaeological predictive models for the Elbe Valley around Dresden, Saxony, Germany. In: Layers of perception. Proceedings of the 35th computer applications and quantitative methods in archaeology conference, Berlin, Germany, April 2–6, 2007, Bonn, pp 1–9.
43. Dingwall L, Exon S, Gaffney V, Laflin S, Van Leusen M (eds) (2010) Archaeology in the age of internet. In: CAA97. Computer applications and quantitative methods in archaeology. Proceedings of the 25th anniversary conference, Birmingham, April 1997. BAR International Series 750. Archeopress, Oxford, pp 35–52
44. Doran J (1970) Systems theory, computer simulations and archaeology. World Archaeol 1(3):289–298
45. Doran J, Hodson FR (1975) Mathematics and computers in archaeology. Edinburgh University Press, Edinburgh
46. Doran J (1990) Computer-based simulation and formal modelling in archaeology: a review. In: Voorrips A (ed), pp 93–114
47. Driessen P, Deckers J (2001) Lecture notes on the major soils of the world. FAO, Rome
48. Fish SK, Kowalewski SA (eds) (1990) The archaeology of regions: the case for full-coverage survey. Smithsonian Institution Press, Washington
49. Dunning PC (2010) L'archéologie en Suisse: une et unique? In: NIKE-Bulletin 6/2010, Patrimoine culturel sous pression, pp 38–45
50. Earley-Spadoni T (2017) Spatial history, deep mapping and digital storytelling: archaeology's future imagined through an engagement with the digital humanities. J Archaeol Sci 84:95–102
51. Erlandson JM, Rick TC (2010) Archaeology meets marine ecology: the antiquity of maritime cultures and human impacts on marine fisheries and ecosystems. Ann Rev Mar Sci 2:231–251
52. Fleming A (2006) Post-processual landscape archaeology: a critique. Cambridge Archaeol Rev 16(3):267–280
53. Flutsch L (2005) L'époque Romaine, Ou La Méditerranée Au Nord Des Alpes. Lausanne: Presses polytechniques et universitaires romandes (Le savoir suisse Histoire)
54. Fu T, Ma L, Li M, Johnson BA (2018) Using convolutional neural network to identify irregular segmentation objects from very high-resolution remote sensing imagery. J Appl Remote Sens 12(1):64
55. Gillings M (2012) Landscape phenomenology, GIS and the role of affordance. J Archaeol Method Theory 19:601–611

56. Graves McEwan D, Millican K (2012) In search of the middle ground: quantitative spatial techniques and experiential theory in archaeology. J Archaeol Method Theory 19(4):491–494
57. Green C (2011a) Winding Dali's clock: the construction of a fuzzy temporal-GIS for archaeology. BAR International Series 2234. BAR Publishing, Oxford
58. Green C (2011b) It's about time: temporality and intra-site GIS. In: Jerem E, Redö F, Szeverényi V (eds) CAA 2008: on the road to reconstructing the past. Archaeolingua, Budapest
59. Habermehl D (2011) Settling in a changing world. Villa development in the northern provinces of the Roman empire. Amsterdam PhD thesis VU University Amsterdam
60. Hafner A (1995) Die frühe Bronzezeit in der Westschweiz, Funde und Befunde aus Siedlungen, Gräbern und Horten der entwickelten Frühbronzezeit. Ufersiedlungen am Bielersee: vol 5. Staatlicher Lehrmittelverlag Bern, Bern
61. Haldimann M-A (1999) Genève. In: Tarpin M et al (eds) Le Bassin lémanique galloromain; Leveau, Ph. Ed, Dossier: Le Rhône romain. Dynamiques fluviales, dynamiques territoriales, Gallia 56, pp 33–44
62. Hambrecht G, Gibbons K et al (2018) Archaeological sites as distributed long-term observing networks of the past (DONOP). Quatern Int. https://doi.org/10.1016/j.quaint.2018.04.016
63. Hammer E, Ur JA (2019) Near Eastern landscapes and declassified U2 aerial imagery. Adv Archaeol Pract 1–20
64. Hargis C, Bissonette J, David J (1997) Understanding measures of landscape pattern. In: Bissonette J (ed) Wildlife and landscape ecology: effects of pattern and scale. Springer, New York, NY, pp 231–261
65. Harrower MJ (2013) Methods, concepts and challenges in archaeological site detection and modeling. In: Comer DC, Harrower MJ (eds) Mapping archaeological landscapes from space. Springer, New York, pp 213–218
66. Harrower MJ (2016) Water histories and spatial archaeology: ancient Yemen and the American West. Cambridge University Press, Cambridge
67. Hartley J, McWilliam K (2009) Computational power meets human contact. In: Hartley J, McWilliam K (eds) Story circle. Wiley-Blackwell, Malden, MA and Oxford, pp 3–15
68. Hatzinikolaou EG (2006) Quantitative methods in archaeological prediction: from binary to fuzzy logic. In: Mehrer MW, Wescott KL (eds) GIS and archaeological site location modelling. Taylor & Francis, New York, pp 437–446
69. Hatzinikolaou EG, Hatzichristos T, Siolas A, Mantzourani E (2003) Predicting archaeological site locations using GIS and fuzzy logic. In: Doerr M, Sarris A (eds) The digital heritage in archaeology: computer applications and quantitative methods in archaeology. Archive of Monuments and Publications, Hellenic Ministry of Culture, Heraklion, pp 169–178
70. Heilen M (2005) An archaeological theory of landscapes. PhD dissertation, University of Arizona. University Microfilms, Ann Arbor, MI
71. Heilen M, Altschul J (2014) Modeling sustainability and resilience through the investigation of critical habitats: a view from the US southwest. Paper presented at the 20th Annual Meeting of the European Association of Archaeologists, Istanbul, Turkey
72. Heilen M, Altschul J, Reddy S, Heckman R, Norris S (2016) Locational and significance modeling at San Clemente Island, California. Technical Report 15–58. Statistical Research, Redlands, CA
73. Hill AC, Limp F, Casana J, Laugier EJ, Williamson M (2019) A new era in spatial data recording: low-cost GNSS. Adv Archaeol Pract 7(2):169–177
74. Hodder I, Mol A (2016) Network analysis and entanglement. J Archaeol Method Theory 23(4):1066–1094
75. Horisberger B, Matter A (2004) Vom römischen Gutshof zur mittelalterlichen Siedlung. Zwei frühmittelalterliche Grubenhäuser und weitere mittelalterliche Befunde im römischen Gutshof Dällikon ZH. JbSGUF 87:141–162
76. Hu D (2012) Advancing theory? Landscape archaeology and geographical information systems. Pap Inst Archaeol 21:80–90. https://doi.org/10.5334/pia.381
77. Huggett J (2015) Challenging digital archaeology. Open Archaeol 1(1):79–85. https://doi.org/10.1515/opar-2015-0003

78. Jacomet S, Jacquat C, Maise C, Wick L et al (1999) Climat, environnement, économie agricole et alimentation. SPM IV 1999, pp 93–115
79. Kaenel G (2009) Helvètes des villes.../Towns of the Helvetii. In: L'âge du Fer dans la boucle de la Loire. Les Gaulois sont dans la ville. Actes du XXXIIe Colloque de l'Association française pour l'étude de l'âge du fer, Bourges, 1er-4 mai 2008. Tours: Fédération pour l'édition de la Revue archéologique du Centre de la France, pp 383–395 (Supplément à la Revue d'archéologique du centre de la France, 35)
80. Kamermans H, Wansleeben M (1999) Predictive modelling in Dutch archaeology, joining forces. In: Barceló JA, Briz I, Vila A (eds) New techniques for old times—CAA98. Computer applications and quantitative methods in archaeology. BAR International Series 757. Archaeopress, Oxford, pp 225–230
81. Kanevski M, Pozdnoukhov A, Timonin V (2009) Machine learning for spatial environmental data. In: Theory, applications and software. EPFL Press (eds) vol 05, p 368, EPFL Press.
82. Kanevski M (2013) A methodology for analysis and modelling of spatial environmental data. In: GEOProcessing 2013: the fifth international conference on advanced geographic information systems, applications, and services, Nice, France. Peer-reviewed, Think Mind, pp 105–107
83. Kansa EC, Kansa SW (2013) Open archaeology. We all know that a 14 is a sheep: data publication and professionalism in archaeological communication. J East Mediterranean Archaeol Herit Stud 1(1):88–97
84. Kintigh KW, Ammerman AJ (1982) Heuristic approaches to spatial analysis in archaeology. Am Antiq 47(1):31–63
85. Knapp AB, Ashmore W (1999) Archaeological landscapes: constructed, conceptualized, ideational. Archaeol Landscape Contem Perspect 1–30
86. Kouchoukos N, Wilkinson T (2007) Landscape archaeology in Mesopotamia: past, present, and future. In: Stone EC (ed) Settlement and society: essays dedicated to Robert McCormick Adams. Cotsen Institute of Archaeology, Los Angeles, pp 1–18
87. Krist FJ, Brown DG (1994) GIS modeling of Paleo-Indian period caribou migrations and viewsheds in Northeastern lower Michigan. Photogramm Eng Remote Sens 60(9):1129–1138
88. Landeschi G (2018) Rethinking GIS: three-dimensionality and space perception in archaeology. World Archaeol 1–16
89. Lane PJ (2015) Archaeology in the age of the Anthropocene: a critical assessment of its scope and societal contributions. J Field Archaeol 40(5):485–498
90. Lauricella A, Cannon J, Branting S, Hammer E (2017) Semi-automated detection of looting in Afghanistan using multispectral imagery and principal component analysis. Antiquity 91(359):1344–1355
91. Lewis K (2018) Finding archaeology in 2017: what is archaeology and why are we doing it? Why should we be doing it? Am Anthropol 120(2):291–314
92. Llobera M, Fabrega-Alvarez P, Parcero-Oubina C (2011) Order in movement: a GIS approach to accessibility. J Archaeol Sci 38(4):843–851
93. Lake MW, Woodman PE, Mithen SJ (1998) Tailoring GIS software for archaeological applications: an example concerning viewshed analysis. J Archaeol Sci 25(1):27–38
94. Lock GR, Harris T (1996) Danebury revisited: an English iron age hillfort in a digital landscape. In: Aldenderfer MS, Maschner HDG (eds) Anthropology, space, and geographic information systems. Oxford University Press, New York, pp 214–240
95. Lock GR (ed) (2000) Beyond the map: archaeology and spatial technologies. IOS Press, Amsterdam
96. Lock G, Kormann M, Pouncett J (2014) Visibility and movement: towards a gis-integrated approach. In: Polla S, Verhagen P (eds) Computational approaches to the study of movement in archaeology: theory, practice and interpretation of factors and effects of long term landscape formation and transformation. Walter de Gruyter GmbH & Co KG, Berlin, pp 23e42
97. Märker M, Angeli L, Bottai L, Costantini R, Ferrari R, Innocenti L, Siciliano G (2008) Assessment of land degradation susceptibility by scenario analysis: a case study in Southern Tuscany, Italy. Geomorphology 93:120–129

98. Märker M, Pelacani S, Schröder B (2011) A functional entity approach to predict soil erosion processes in a small Plio-Pleistocene Mediterranean catchment in Northern Chianti, Italy. Geomorphology 125:530–540
99. McAnany PA, Yoffee N (2009) Questioning collapse: human resilience, ecological vulnerability, and the aftermath of empire McAnany PA, Yoffee N (eds). Cambridge University Press, Cambridge
100. Mehrer M, Wescott K (eds) (2006) GIS and archaeological site location modeling. Taylor & Francis, Boca Raton, Florida
101. Meier H-R, Petzet M, Will T (2007) Heritage at Risk: special edition—cultural heritage and natural disasters risk preparedness and the limits of prevention. Paris
102. Menze BH, Ur JA (2012) Mapping patterns of long-term settlement in Northern Mesopotamia at a large scale. Proc Natl Acad Sci 109(14):E778–E787
103. Milo DS (1992) Le siècle: projet expérimental avorté, In: Périodes: la construction du temps historique, Actes du Ve colloque d'Histoire au présent, organisé par Dumoulin O. et Valéry, P. Paris, éditions de l'Ecole des Hautes Etudes en Sciences Sociales, pp 123–128
104. Motschi A, Wild W (2011) Städtische Siedlungen — Überblick zu Siedlungs - entwicklung und Siedlungs topografie: Zürich, Winterthur, Weesen In: Siedlungsbefunde und Fundkomplexe der Zeit zwischen 800 und 1350. Akten des Kolloquiums zur Mittelalterarchäologie in der Schweiz. Archäologie Schweiz AS, Schweizerische Arbeitsgemeinschaft für die Archäologie des Mittelalters und der Neuzeit SAM, Schweizerischer Burgenverein SBV (eds) Verlag Archäologie Schweiz, Basel, p 483
105. Mullins P (2019) The archaeology of nothing: grand challenges and everyday life. https://paulmullins.wordpress.com/2016/01/30/the-archaeology-of-nothing-grand-challenges-and-everyday-life/
106. Musa AB (2014) Logistic regression classification for uncertain data. J Math Stat Sci 2(2):2320–6047
107. Nicolucci F, Hermon S, Farinetti E (2004) Fuzzy logic application to survey data in a GIS environment. In: Beyond the artefact. Computer applications in archaeology. ArcheoLingua, Budapest (Hungary)
108. Niccolucci F, Hermon S (2010) A fuzzy logic approach to reliability in archaeological virtual reconstruction. In: Nicolucci F, Hermon S (eds) Beyond the artifact. Digital interpretation of the past. Proceedings of CAA2004, Prato 13–17 April 2004. Archaeolingua, Budapest, pp 28–35
109. Ortman SG, Parker JN, Peeples MA, Sabloff JA (2017) Opinion: fostering synthesis in archaeology to advance science and benefit society. Proc Nat Acad Sci 114(42):10999 LP–11002
110. Osborne JF (2014) Monuments and monumentality. In: Osbourne JF (ed) Approaching monumentality in archaeology. State University of New York Press, Albany, pp 1–22
111. Parcak SH (2019) Archaeology from space: how the future shapes our past. Macmillan, New York
112. Phillips P, Griffin JB, Williams S (2003) Archaeological survey in the Lower Mississippi Alluvial Valley 1940–1947. The University of Alabama Press
113. Pizziolo G, Sarti L (2015) Predicting prehistory: predictive models and field research methods for detecting prehistoric contexts. Museo e Istituto Fiorentino di Preistoria "Paolo Graziosi", Siena
114. Pliny (1938) Natural history, vol I: Books 1–2. Translated by H. Rackham. Loeb Classical Library 330. Harvard University Press, Cambridge, MA
115. Plog S, Plog F, Wait W (1978) Decision making in modern survey. In: Schiffer MB (ed) Advances in archaeological method and theory. Academic Press. New York, pp 383–420
116. Pohjankukka J, Pahikkala T, Nevalainen P, Heikkonen J (2017) Estimating the prediction performance of spatial models via spatial k-fold cross validation. Int J Geogr Inf Sci 31(10):2001–2019. https://doi.org/10.1080/13658816.2017.1346255
117. R Core Team (2014) R: a language and environment for statistical computing. R Foundation for Statistical Computing, Vienna, Austria. http://www.R-project.org/

118. Rankin C, Mog C, Jones S (2017) Parkaeology and climate change: Assessing the vulnerability of archaeological resources at Klondike Gold Rush National Historical Park, Alaska. Archaeol Rev Cambridge 32(2):56–77. https://doi.org/10.17863/CAM.23662
119. Ren S, He K, Girshick R, Sun J (2017) Faster R-CNN: towards real-time object detection with region proposal networks. IEEE Trans Pattern Anal Mach Intell 39:1137–1149
120. Richards-Rissetto H, Landau K (2014) Movement as a means of social (re)production: using GIS to measure social integration across urban landscapes. J Archaeol Sci 41:365–375
121. Ritter FL (1926–27) Les ponts romains de la Thielle et de la Broye, Annales Fribourgeoises, Fribourg
122. Ritter G (1981) Notice sur les vestiges des ponts romains a la Sauge. Bulletin de la Société de Science Naturelles de Neuchâtel, tome 19, 1889–1890, Neuchâtel
123. Roberts L (2016) Deep mapping and spatial anthropology. Humanities 5(1):5
124. Roberts P, Hunt C, Arroyo-Kalin M, Evans D, Boivin N (2017) The deep human prehistory of global tropical forests and its relevance for modern conservation. Nat Plants 3. https://doi.org/10.1038/nplants.2017.93
125. Rodriguez-Galiano VF, Chica-Olmo M, Chica-Rivas M (2014) Predictive modelling of gold potential with the integration of multisource information based on random forest: a case study on the Rodalquilar area, Southern Spain. Int J Geogr Inf Sci 28:1336–1354
126. Ryan M-L (2002) Beyond myth and metaphor: narrative in digital media. Poet Today 23:581–609
127. Sandweiss DH, Kelley AR (2012) Archaeological contributions to climate change research: the archaeological record as a paleoclimatic and paleoenvironmental archive. Annu Rev Anthropol 41:371–391
128. Scarlett SF, Lafreniere D, Trepal DJ, Arnold JDM, Pastel R (2018) Engaging community and spatial humanities for postindustrial heritage: the Keweenaw time traveler. Am Q 3:619–623
129. Schwab H (2002) Archéologie de la 2e Correction des Eaux du Jura, vol 3. Les artisans de l'Age du Bronze sur la Broye et la Thielle. Fribourg, p 272
130. Segard M (2009) Les Alpes Occidentales Romaines, Développement urbain et exploitation des ressources des régions de montagne (Gaule Narbonnaise, Italie, provinces alpines). Nouvelle édition [en ligne]. Publications du Centre Camille Jullian, Aix-en-Provence
131. Seong H, Son H, Kim C (2018) A comparative study of machine learning classification for color-based safety vest detection on construction-site images. KSCE J Civ Eng 22(2018):4254–4262
132. Shennan S (1997) Quantifying archaeology. Edimburgh University Press, Edimburgh
133. Simpson I, Adderley WP, Guðmundsson G, Hallsdóttir M, Sigurgeirsson M, Snæsdóttir M (2002) Soil limitations to Agrarian land production in premodern Iceland. Human Ecol 30:423–443. https://doi.org/10.1023/a:1021161006022
134. Smith ME (2017) Social science and archaeological enquiry. Antiquity 91(356):520–528
135. Soja EW (1989) Postmodern geographies: the reassertion of space in critical social theory. Verso, New York
136. Stover J (2017) There is no case for the humanities. Am Aff 1(4). https://americanaffairsjournal.org/2017/11/no-case-humanities/
137. Studer J, David-Elbiali M, Besse M (dir) (2011) Paysage… Landschaft… Paesaggio. L'Impact des activités humaines sur l'environnement du Paléolithique a la période romaine. Actes du colloque du Groupe de travail pour les recherches préhistoriques en Suisse (GPS/AGUS), Museum d'histoire naturelle, Genève, 15–16 mars 2007
138. Tarpin M (2002) L'héroïque et le quotidien : Hannibal et les autres dans les Alpes. Annales Valaisannes 7:7–19
139. Thérond D (2007) European preventive archaeology. Papers of the EPAC meeting 2004, Vilnius, Edited by Katalin Bozóki-Ernyey, p 9
140. Thétaz RG, Kellenberger M (2018) Tendances et défis. Faits et chiffres relatifs au Projet de territoire Suisse (Ufficio federale dello sviluppo territoriale (ARE), Ed)
141. Trepal D Scarlett SF, Lafreniere D (2019) Heritage making through community archaeology and the spatial humanities. J Commun Archaeol Heritage 1–19

142. Trifković V (2006) Persons and landscapes: shifting scales of landscape archaeology. In: Lock G, Molyneaux BL (eds) Confronting scale in archaeology. Issues of theory and practice. Springer, New York, pp 217–324
143. Tugby D (1965) Archaeological objectives and statistical methods: a frontier in archaeology. Am Antiqu 31(1)
144. Van der Leeuw S, Redman C (2002) Placing archaeology at the center of socio-natural studies. Am Antiq 67:597–605
145. Van Lanen RJ, Groenewoudt BJ, Spek T et al (2018) Route persistence. Modelling and quantifying historical route-network stability from the Roman period to early-modern times (AD 100–1600): a case study from the Netherlands. Archaeol Anthropol Sci 10:1037–1052. https://doi.org/10.1007/s12520-016-0431-z
146. Van Leusen PM (1995) GIS and archaeological resource management. A European agenda In: Lock G, Stancic Z (eds) Archaeology and geographical information systems. A European perspective. Taylor and Francis, London
147. VanPool TL, Leonard RD (2011) Quantitative analysis in archaeology. Wiley-Blackwell, Chichester
148. Varro. de Re Rustica (On Agriculture (transl HB Ash 1936)) Loeb Classical Library
149. Verhagen P, Gili S, Micó R, Risch R (1999) Modelling prehistoric land use distribution in the Rio Aguas Valley (SE Spain). In: Dingwall L et al (eds) Archaeology in the age of the internet. Proceedings of the CAA97 conference. BAR International Series 750
150. Verhagen P, Nuringer L, Tourneux F-P, Bertoncello F Jeneson K (2012) Introducing the human factor in predictive modelling. In: Proceedings of the 40th annual conference of computer applications and quantitative methods in archaeology. University of Southampton, UK, March 26–30, pp 379–388
151. Vink A (1975) Land use in advancing agriculture. Springer, Berlin
152. Yamin Y, Bescherer KM (eds) (1996) Landscape archaeology: reading and interpreting the American historical landscape. University of Tennessee Press
153. Watt A, Eng N (2014) Database design, 2nd edn. BC campus, Victoria, BC. https://opentextbc.ca/dbdesign01/
154. Whitley TG (2000) Dynamical systems modeling in archaeology: a GIS approach to site selection processes in the Greater Yellowstone Region. Unpublished Dissertation, Department of Anthropology, University of Pittsburgh, PA
155. Whitley TG (2002a) Modeling archaeological and historical cognitive landscapes in the Greater Yellowstone Region (Wyoming, Montana, and Idaho, USA) using geographic information systems. In: Burenhult G (ed) Archaeological informatics: pushing the envelope CAA2001. Computer applications and quantitative methods in archaeology. BAR International Series 1016, Oxford, pp 139–148
156. Whitley TG (2002b) Spatial variables as proxies for modeling cognition and decision-making in archaeological settings: a theoretical perspective. Paper presented at the 24th Annual Meeting of the Theoretical Archaeology Group, Manchester, United Kingdom, 21st–23rd December 2002. Wickham C (2011) Framing the Early Middle Ages: Europe and the Mediterranean, 400–800. Oxford University Press. https://doi.org/10.1093/acprof:oso/9780199264490.001.0001
157. Witcher RE, Kay SJ (2004) An application of predictive modelling in the Tiber Valley. In: Beyond the artifact digital interpretation of the past proceedings of CAA2004 Prato 13–17 April 2004
158. Yager RR, Filev DP (1994) Essentials of fuzzy modeling and control. Wiley, New York
159. Zeidler J (ed) (2001) Dynamic modeling of landscape evolution and archaeological site distributions: a three-dimensional approach. Center for Environmental Management of Military Lands, Fort Collins, CO

Made in the USA
Las Vegas, NV
03 November 2026

57424415R00175